AF690661

Ultra-Wideband Surveillance Radar

Related titles on radar:

Advances in Bistatic Radar Willis and Griffiths
Airborne Early Warning System Concepts, 3rd Edition Long
Bistatic Radar, 2nd Edition Willis
Design of Multi-Frequency CW Radars Jankiraman
Digital Techniques for Wideband Receivers, 2nd Edition Tsui
Electronic Warfare Pocket Guide Adamy
Foliage Penetration Radar: Detection and characterisation of objects under trees Davis
Fundamentals of Ground Radar for ATC Engineers and Technicians Bouwman
Fundamentals of Systems Engineering and Defense Systems Applications Jeffrey
Introduction to Electronic Warfare Modeling and Simulation Adamy
Introduction to Electronic Defense Systems Neri
Introduction to Sensors for Ranging and Imaging Brooker
Microwave Passive Direction Finding Lipsky
Microwave Receivers with Electronic Warfare Applications Tsui
Phased-Array Radar Design: Application of radar fundamentals Jeffrey
Pocket Radar Guide: Key facts, equations, and data Curry
Principles of Modern Radar, Volume 1: Basic principles Richards, Scheer and Holm
Principles of Modern Radar, Volume 2: Advanced techniques Melvin and Scheer
Principles of Modern Radar, Volume 3: Applications Scheer and Melvin
Principles of Waveform Diversity and Design Wicks *et al.*
Pulse Doppler Radar Alabaster
Radar Cross Section Measurements Knott
Radar Cross Section, 2nd Edition Knott *et al.*
Radar Design Principles: Signal processing and the environment, 2nd Edition Nathanson *et al.*
Radar Detection DiFranco and Rubin
Radar Essentials: A concise handbook for radar design and performance Curry
Radar Foundations for Imaging and Advanced Concepts Sullivan
Radar Principles for the Non-Specialist, 3rd Edition Toomay and Hannan
Test and Evaluation of Aircraft Avionics and Weapons Systems McShea
Understanding Radar Systems Kingsley and Quegan
Understanding Synthetic Aperture Radar Images Oliver and Quegan
Radar and Electronic Warfare Principles for the Non-specialist, 4th Edition Hannen
Inverse Synthetic Aperture Radar Imaging: Principles, algorithms and applications Chen and Marotella
Stimson's Introduction to Airborne Radar, 3rd Edition Baker, Griffiths and Adamy
Test and Evaluation of Avionics and Weapon Sytems, 2nd Edition McShea
Angle-of-Arrival Estimation Using Radar Interferometry: Methods and applications Holder
Biologically-Inspired Radar and Sonar: Lessons from Nature Balleri, Griffiths and Baker
The Impact of Cognition on Radar Technology Farina, De Maio and Haykin
Novel Radar Techniques and Applications, Volume 1: Real Aperture Array Radar, Imaging Radar, and Passive and Multistatic Radar Klemm, Nickel, Gierull, Lombardo, Griffiths and Koch
Novel Radar Techniques and Applications, Volume 2: Waveform Diversity and Cognitive Radar, and Target Tracking and Data Fusion Klemm, Nickel, Gierull, Lombardo, Griffiths and Koch
Radar and Communication Spectrum Sharing Blunt and Perrins
Systems Engineering for Ethical Autonomous Systems Gillespie
Shadowing Function from Randomly Rough Surfaces: Derivation and applications Bourlier and Li
Photo for Radar Networks and Electronic Warfare Systems Bogoni, Laghezza and Ghelfi
Multidimensional Radar Imaging Martorella
Radar Waveform Design Based on Optimization Theory Cui, De Maio, Farina and Li
Micro-Doppler Radar and its Applications Fioranelli, Griffiths, Ritchie and Balleri
Maritime Surveillance with Synthetic Aperture Radar Di Martino and Antonio Iodice
Electronic Scanned Array Design Williams

Ultra-Wideband Surveillance Radar

Mark E Davis

The Institution of Engineering and Technology

Published by SciTech Publishing, an imprint of The Institution of Engineering and Technology, London, United Kingdom

The Institution of Engineering and Technology is registered as a Charity in England & Wales (no. 211014) and Scotland (no. SC038698).

© The Institution of Engineering and Technology 2021

First published 2020

This publication is copyright under the Berne Convention and the Universal Copyright Convention. All rights reserved. Apart from any fair dealing for the purposes of research or private study, or criticism or review, as permitted under the Copyright, Designs and Patents Act 1988, this publication may be reproduced, stored or transmitted, in any form or by any means, only with the prior permission in writing of the publishers, or in the case of reprographic reproduction in accordance with the terms of licences issued by the Copyright Licensing Agency. Enquiries concerning reproduction outside those terms should be sent to the publisher at the undermentioned address:

The Institution of Engineering and Technology
Michael Faraday House
Six Hills Way, Stevenage
Herts, SG1 2AY, United Kingdom

www.theiet.org

While the author and publisher believe that the information and guidance given in this work are correct, all parties must rely upon their own skill and judgement when making use of them. Neither the author nor publisher assumes any liability to anyone for any loss or damage caused by any error or omission in the work, whether such an error or omission is the result of negligence or any other cause. Any and all such liability is disclaimed.

The moral rights of the author to be identified as author of this work have been asserted by him in accordance with the Copyright, Designs and Patents Act 1988.

British Library Cataloguing in Publication Data

A catalogue record for this product is available from the British Library

ISBN 978-1-83953-077-7 (hardback)
ISBN 978-1-83953-078-4 (PDF)

Typeset in India by Exeter Premedia Services

The material in this text has been Publically Released under DISTAR Case 31857 (4 November 2019), Case 32507 (12 November 2019), and Case 32524 (3 March 2020).

Contents

List of figures

List of tables

Preface

Battlefield radars became a major technology thrust in the late 1960s based on several advances in radar technology. The military needed airborne radar systems for all weather, day-night surveillance of an area to protect borders. These requirements have also been extended since 1980 for monitoring areas against trafficking of people and animals, along with assessment of the destruction of earth resources. This has been a very difficult radar system engineering process, due to the need for characterizing the objects and clutter for high confidence surveillance.

Two US radar research organizations have been leaders in new system designs and development of radar signal processing algorithms: MIT Lincoln Laboratory in Lexington, Massachusetts, has focused on the military applications; and NASA Jet Propulsion Laboratory in Pasadena, California, have developed the earth resources and natural disaster monitoring. A third source of radar system concepts and necessary program support has been the US Defense Advanced Research Projects Agency (DARPA) in Arlington, Virginia. Much of the content of this text has been drawn from the author's association with these leaders in the field.

The real impetus for affordability of these wide-area surveillance systems has been the rapid development of large-scale integrated circuits, high-performance computing, monolithic microwave integrated circuits and active element phased array radars. The rapid reduction in the size, weight and power of the components has enabled small systems that can be deployed on remotely piloted vehicles. Because of the leverage of monolithic radio frequency devices, all aspects of wide-area surveillance can operate with ultra-wideband characteristics in terms of spatial resolution, bandwidth characteristics and wide-angle surveillance operation. Most importantly, the economics of scale for commercial communications and direct broadcast radio frequency systems has made the new radar technology feasible for deployment in both military and commercial applications.

This text will provide a brief history of the devices, algorithms and radar systems as they apply to an air-to-ground surveillance radars over many decades of frequency. It will also outline the rapid development of algorithms for detecting, locating and characterizing both fixed and moving objects. As a result of the extraordinary improvements in real-time computation, these algorithms have become available for remote applications with dissemination of imagery to a distant user.

The author would like to acknowledge all of the pioneers in the development of wide-area radar surveillance systems and technology. The past 25 years have been a very enjoyable journey in the development of ultra-wideband radar design, testing and operation in a contested radio frequency environment. He would like to recognize the patience and support of his family, wife Diane and two sons Colin and Shelby. They have greatly supported and encouraged this international learning, collaboration and travel in the global radar community.

List of acronyms

Acronym	Meaning	Usage
ACR	Area coverage rate	Measure of radar operation for rate of collecting target returns over unit area
ADC	Analog-to-digital converter or conversion	Device or technique that converts an analog signal to digital words for processing
AESA	Active electronic scanned array	Antenna with active transmit and receive components at each radiating element for rapid scan and beam control
AF	Ambiguity function	A measure of match filter effectiveness with time delay and Doppler effects in the radar return
AFRL	Air Force Research Laboratory	USAF laboratory responsible for research and development of radar technology
AIRSAR	Airborne Synthetic Aperture Radar	NASA JPL airborne system
ALARM	Airborne Long-range Alerting Radar for MTI	Early battlefield surveillance radar for GMTI operation from a slow moving helicopter
AM	Amplitude modulation	Modulation technique on interfering signals using amplitude modulation
AMBR	Ambiguity ratio	Level of unwanted signal in detection process due to either antenna or waveform range and Doppler ambiguities
ASIC	Application specific integrated circuit	Custom integrated circuits for significant reduction of digital processing SWAP
ATD	Advanced Technology Demonstrator	Prototype system with form and function approaching operational unit
ATI	Along-track interferometry	Radar technique for discriminating moving targets in dual channel SAR images

Acronym	Meaning	Usage
BPA	Back projection algorithm	FOPEN SAR image formation algorithm
CEPT	Conference of European Postal and Telecommunications	European agency that regulates spectrum use by RF transmitters
CFAR	Constant false alarm rate	Receiver technique that sets a detection threshold to limit false alarm rate as a function of the SNR
CLS	Coherent least squares	Adaptive processing technique that evaluates least squares of signal error from ideal response
CNR	Clutter-to-noise ratio	Ratio of the clutter signal return to the noise in a resolution cell
COTS	Commercial off-the-shelf	Hardware or signal processing elements that can be adapted from commercially available products
DAC	Digital to analog converter	Component to convert digital data into an analog signal
DARPA	Defense Advanced Research Projects Agency	US research organization
DBS	Doppler beam sharpening	Method of improving cross-range resolution in ground mapping by using the Doppler surface clutter spread
DEM	Digital elevation map	Description of local terrain height
DMZ	Demilitarized zone	Geographic area with little or no passage for surface activity
DPCA	Displaced phase center array	Antenna designed for GMTI operation with multiple phase centers to cancel ground clutter
DTED	Digital terrain elevation data	Geospatial standard for representing terrain elevation on Earth's surface
EIRP	Equivalent isotropically radiated power	Power radiated by a device all directions, e.g., 0 dBi gain
ERIM	Environmental Research Institute of Michigan	Research organization in Ann Arbor, MI, with early SAR systems
ESA	Electronically scanned array	Antenna implementing electronic scan by individual control of elements or small array components

Acronym	Meaning	Usage
ESUS	Electronically-agile Solid-state Universal Surveillance	UN Navy concept AESA using space density taper on transmit
F/A	Frequency agile	Waveform where the frequency is changed between pulses or dwells for reducing scintillation effects
FCC	Federal Communications Commission	Governing body in United States for communications regulations
FDMA	Frequency division multiple access	Communications waveform that uses a sequence of frequencies to minimize RFI
FDR	Frequency-dependent rejection	NTIA technique for assessing total receiver rejection of UWB signal
FEBA	Forward edge of a battle area	Military term for line between armies
FFT	Fast Fourier transform	Technique to filter signals using a fast estimation of the Fourier coefficients
FIR	Finite impulse response	Digital signal processing technique for filter construction
FJB	Frequency jump burst	Technique for transmitting UWB signal by use of timed sequence of multiple pulses
FM	Frequency modulation	Modulation technique on interfering signals
FOA/FOI	Swedish Defence Research Agency	CARABAS Radar developer in Linköping, Sweden
FOPEN	Foliage penetration	Radar mode to detect objects under foliage
FORESTER	FOPEN Reconnaissance, Surveillance, Tracking, and Engagement Radar	FOPEN GMTI system on A-160 UAS
FPGA	Field Programmable Gate Array	Integrated circuit for high speed signal processing, but with rapid change of control of digital algorithms
GeoSAR	Geospatial Synthetic Aperture Radar	Dual frequency mapping Radar developed by NASA JPL
GFLOPS	One billion floating-point operations per second	Measure of throughput of a signal processor
GHz	Billion cycles per second	Measure of signal frequency or bandwidth

Acronym	Meaning	Usage
GMTI	Ground moving target indication	Technique for detecting targets moving in a ground environment
GNSS	Global Navigation Satellite System	Class of satellites in orbit around Earth for obtaining position, velocity and altitudes
GPS	Global positioning system	US developed GNSS system, available around the globe
GRCA	Ground reference coverage area	Region of surveillance by airborne radar which is referenced to a specific ground area
HF	High frequency	Radio frequencies between 3 and 30 MHz
HOWLS	Hostile Weapons Locator System	Early multimode battlefield radar system with ESA antenna
HPC	High-performance computer	Subsystem with multiple fast processing units, with parallel computations for high speed signal processing
ICM	Internal clutter motion	Modulation of clutter by environment to present small Doppler frequency effects
IEEE	Institute of Electrical and Electronics Engineers	International professional society that publishes technical papers and establishes electronic standards
IF	Intermediate frequency	Receiver downconversion technique with a frequency intermediate between transmit and baseband
IFFT	Inverse fast Fourier transform	Signal processing technique to estimate time samples from signal Fourier coefficients
IFSAR	Interferometric SAR	Radar mode utilizing two SAR spatial phase centers to measure phase change for moving target or geographic effect detections
IFSAR-E	Interferometric SAR elevation	Terrain height mapping Radar developed by ERIM
IMU	Inertial measurement unit	Subsystem that measures platform velocity, acceleration, and orientation
INS	Inertial navigation system	Subsystem that determines platform position, velocity, time, and orientation over flight path

Acronym	Meaning	Usage
InSAR	Interferometric Synthetic Aperture Radar	Radar mode using SAR returns from two antennas for terrain characterization
IPR	Impulse response	Metric of SAR image response to a waveform of specified bandwidth
I/Q	In-phase and quadrature	Measurement of the orthogonal components of a coherent signal
ISLR	Integrated sidelobe ratio	Ratio of waveform or antenna sidelobes energy to energy in mainlobe
JPL	Jet Propulsion Laboratory	NASA research organization in Pasadena, CA, developed AIRSAR and GeoSAR
JSTARS	Joint Surveillance and Target Attack Radar System	X-band GMTI Radar system developed by US Air Force
LFM	Linear frequency modulation	Radar waveform where frequency is modulated linearly in time
LMS	Least means square	Adaptive processing technique that minimizes the square of a relevant signal processing metric
LULC	Land use land cover	Data of types of clutter in a geographic map; used for characterizing surface regions
MASR	Multiple aperture surveillance radar	Early battlefield surveillance radar for GMTI with DPCA cancellation technology
MB/MSR	Multibeam, modular survivable radar	Concept radar developed to demonstrate a multiple beam ESA, that implemented multimode surveillance radar with emerging technologies
Mbps	Million bits per second	Signal conversion or transfer rate in terms of digital bits
MDV	Minimum discernable velocity	GMTI characteristic of slow target detection
ME	Minimum entropy	Method of assessing the statistics of clutter or radar return phase errors by determining the minimum entropy point

Acronym	Meaning	Usage
MF	Matched filter	Signal processing technique for correlating a radar return with its transmitted characteristics
MHz	One million cycles per second	Measure of signal frequency or bandwidth
MIC	Microwave integrated circuits	Integrated substrates combining microwave signal transmission between active and passive components to reduce radar SWAP
MIMIC	Millimeter and microwave integrated circuits	US program to develop RF integrated circuits for radar and communications miniaturization and improved performance
ML	Maximum likelihood	Signal condition measure using maximum SNR measure
MMIC	Monolithic microwave integrated circuits	Custom integrated circuits for significant reduction of microwave component SWAP
MNR	Multiplicative noise ratio	SAR image quality metric
MOCOMP	Motion compensation	SAR processing step to remove platform motion from the synthetic aperture collection
MOP	Measure of performance	Critical requirements for Radar system to meet operational requirements
MSE	Mean square error	Estimate of an average of the square of the signal error in a sequence of samples
MSR	Modular survivable radar	Early concept development radar to test multibeam, multimode radar prototype, employing MMIC and VHSIC technology
MTI	Moving target indication	Radar technique for discriminating moving targets in clutter
NASA	National Aeronautics and Space Administration	US Agency to develop space and aerospace systems
NES0	Noise Equivalent Sigma-0	Figure of measurement in SAR to characterize the ability to detect signals in noise at a 0 dB threshold

Acronym	Meaning	Usage
NTIA	National Telecommunications and Information Administration	US regulatory body for frequency use and licensing
OFDMA	Orthogonal frequency division multiple access	Communications waveform that uses an orthogonal sequence of frequencies to minimize cross channel interference
OFR	Off-frequency rejection	NTIA technique for assessing rejection of UWB signal outside receiver passband
OTR	On-tune rejection	NTIA technique for assessing rejection of UWB signal within receiver passband
PGA	Phase gradient autofocus	Technique for improving SAR image quality by assessing the phase variation due to MOCOMP errors
PRF	Pulse repetition frequency	Frequency of radar transmitted pulses
PRI	Pulse repetition interval	Time interval between radar transmit pulses
Radar	Radio detection and ranging	Technique of imaging and detecting objects through use of RF waveforms
REFORGER	Return of forces to Germany	NATO exercises in Germany
RF	Radio frequency	General class of emitters operating in bands set aside for radio operation
RFI	Radio frequency interference	Effect on radar image by other RF sources within line of sight
RLSTAP	Rome Laboratory Space Time Adaptive Processing	Modeling and simulation environment developed by AFRL and CAESoft to simulate multiple-channel Radars
RMA	Range migration algorithm	FOPEN SAR image formation algorithm
ROC	Receiver operating characteristics	Technique for measuring target detection and false alarms
ROI	Region of Interest	Area in SAR image derived from CFAR to enable further characterization of target versus clutter
RPV	Remotely piloted vehicle	Air vehicle that does not have a pilot on board. Also referred to as unmanned air vehicle (UAV)

Acronym	Meaning	Usage
SAR	Synthetic aperture radar	Radar image with improved cross-range resolution by flying a synthetic aperture
SCR	Signal-to-clutter ratio	Method of charactering a return signal from clutter energy in resolution cell
SDT	Space density taper	AESA concept by varying element density to achieve low sidelobe levels on transmit
SINR	Signal-to-interference-plus-noise ratio	Measure of radar signal to the sum of background noise and all interference
SIR	Signal-to-interference ratio	Ratio of the energy in the signal-to-total-noise-plus-interference
SLR	Sidelobe-minimizing receiver	Adaptive matched filter that reduces the temporal or Doppler sidelobes
SNR	Signal-to-noise ratio	Measure of difference between a signal return to the noise level in the resolution cell
SOTAS	Standoff Target Acquisition System	Early X-band GMTI system
STAP	Space-time adaptive processing	Method to adaptively cancel background interference by using both spatial and temporal channel degrees of freedom
SWAP	Size, weight and power	Measure of radar subsystems in terms of mass and energy required
TCR	Target-to-clutter ratio	Ratio of energy in target return to clutter in the same resolution cell
TOC	Tactical Operations Center	Ground processing and radar control system to support military operations
T/R	Transmit and receive	Element design that combines a transmit and receive function behind the radiating surface
TV	Television	Commercial transmitter creating RFI in many UWB radar systems
UAS	Unmanned air system	Airborne platform without a pilot on board
UHF	Ultra-high frequency	300–1,000 MHz frequency band
ULA	Uniform linear array	Model of ESA with a uniformly spaced elements in one dimension

Acronym	Meaning	Usage
USGS	United States Geographical Service	Organization in US with responsibility for mapping and characterization of Earth's surface
UWB	Ultra-wideband	Radar operation with greater than 50% bandwidth
VHF	Very high frequency	30–300 MHz frequency band
VHSIC	Very high speed integrated circuit	US Defense Department program to develop custom digital integrated circuits to reduce SWAP of avionics subassemblies
VLSIC	Very large scale integrated circuit	Very high integration of digital processing on a single chip for control and computation of sensors
V/STOL	Vertical/short take-off and landing	Aircraft capable of flying from and to a small area without forward motion
WiFi	Wireless fidelity	Local UWB communications techniques

About the author

Mark E. Davis is a Life Fellow of the IEEE, winner of the 2008 Warren D. White Award and the 2018 IEEE Denis Picard Medal. He retired from the GE Aerospace and Air Force Research Laboratory. He was Program Manager at DARPA on Foliage Penetration Radar, and the author of Foliage Penetration Radar – Detection and Characterization of Objects Under Trees, published by SciTech in 2011.

Chapter 1
History of airborne surveillance radars

This book concentrates on the development and applications of ultra-wideband (UWB) surveillance radars, specifically a radar that surveys an area of the Earth's surface from an airborne platform to detect and characterize cultural features or man-made objects. Furthermore, an UWB radar will operate with an antenna and waveforms that support very-high-resolution spatial and Doppler measurements. This category of radar had a rather late start compared to other surveillance radar applications, predominantly due to the immaturity of both radar components and signal-processing algorithms to meet these radar requirements. The early airborne surveillance radar efforts were large systems deployed for either ground moving target indication (GMTI) or synthetic aperture radar (SAR). These systems were large because of the technology of the time, and their system operation was predominantly experimental and on aircraft flying in controlled airspace.

There are several classes of surveillance radars [1]. Three classes exist based on their location with respect to the Earth's surface: ground-based, where the radar is located on the surface (ground or sea) and has little or no motion of the radar platform; airborne, where the radar is on a platform that flies a significant elevation over the surface with modest to high platform velocity; and space-based, where the radar that obeys Kepler's law to stay in an orbit around the Earth at several kilometers per second.

These three radar platforms are similarly characterized by the types of targets the radar attempts to detect, track and characterize. Stationary targets including buildings, roads and land characteristics are typically detected using SAR. Moving targets include ground vehicles, maritime vessels and airborne targets. There are distinct radar modes that utilize Doppler waveforms to detect moving targets and distinguish the returns from fixed targets or clutter returns. And finally, there is an emerging surveillance radar that can attempt to detect objects underground or hidden by foliage.

Battlefield surveillance was a major focus in the radar community to provide all weather, day–night detection and tracking of military activity [2]. As these systems became more effective, the adversaries were developing tactics to defeat the microwave GMTI and SAR systems. This was particularly true in the potential conflicts in central Europe (Fulda Gap) and in actual conflicts in South east Asia with its dense tropical forests. It became common that the adversary would hide under trees or apply camouflage nets to avoid detection by the microwave radars. Each of these counters to the surveillance systems has spurred new technologies to recover the radar operational needs.

As a result of the last issue, several systems were built and rapidly sent to the Viet Nam – both ground-based and airborne [3]. Chapter 3 will illustrate these systems and the innovations that were made to provide modest success at detecting targets in dense foliage. The moving targets were detected by ground-based systems at either very high frequency (VHF) or ultra-high frequency (UHF). Eliminating platform motion effects on moving target detection from airborne platforms was significantly beyond the state of the art at UHF frequencies.

Early surveillance SAR experiments were designed to exploit the coherent return of man-made objects when compared with specular return from natural clutter. The Doppler returns from man and animals in the forest provide a cue that objects were moving in the dense clutter. Depending on the geometry and types of foliage, these motions could be observed by a trained operator. Digital signal processing was just becoming a reality, and fixed and moving target detection results were encouraging. The mobility of airborne platforms coupled with the early development of remotely piloted vehicles (RPVs) gave the radar community the challenge of developing several tactical systems.

The history of battlefield surveillance has spanned over more than two centuries [4]. In their need for seeing enemy positions farther from the battle lines, tactics were developed around observer positions on high hills or eventually on lighter than air balloons. This was chronicled in Harper's Weekly in 1862 where balloons carried forward observers at altitude to survey the enemy position. However, there were several instances where the winds brought the forward observers over enemy lines, putting the observers in harm's way. Even today, tethered balloons are being used in battlefields to give persistent surveillance around a forward base.

During the South east Asia conflict, a GMTI radar system was developed for installation on a high hill with the antenna mounted on a very tall mast. This was developed at VHF to provide reasonable propagation through the dense foliage. Coherent radar operation was essential to differentiate the Doppler effect of enemy troops walking through the forest and suppression of the foliage clutter. However, these radar had limited range, and the observation of their deployment caused the enemy to "avoid" their detection. There was a distinct need for a more mobile approach to this technology.

The first mobile battlefield surveillance radar was developed by the US Army on the Standoff Target Acquisition System (SOTAS). Its early X-band prototype radar was integrated on a UH-1 helicopter, with low velocity to limit the ground clutter spread. The tests were conducted for several years starting in early 1974 in the Return of Forces to Germany exercise. A ground station had a display with GMTI detections presented on a ground map. The individual detections were referenced to that map, and they could be played back in the form of a "history display." These moving dots immediately gave the ground control cell the current position, direction and speed of the targets. By using these new surveillance radars in military exercises, the US Army obtained excellent lessons in military doctrine. An airborne radar could see moving brigades approaching a forward edge of a battle area (FEBA), along with a staging area where the vehicle motion stops. However, the enemy soon developed tactics of hiding these targets in foliage or camouflage, which made it difficult to see the enemy's stationary positions.

The first airborne radar designed to be carried on an RPV was the Ku-band Hostile Weapons Locator System (HOWLS), which was developed by MIT Lincoln Laboratory [3]. This system was utilized in the US Army netted radar program and was developed to detect and register artillery weapons at standoff ranges. The use of the future installation on an RPV was to provide threats of enemy action against piloted aircraft.

These early surveillance radar systems were all medium bandwidth (approximately 10%) due to the available radio frequency (RF) antenna and waveform technology at the time. As a result, the radar systems had modest resolution to detect objects in background and volumetric clutter. However, as more accurate characterization of the objects was needed, development of wider-bandwidth components was provided by the industry. In addition, there was a parallel development of digital signal processing for image and communication applications. These tools were rapidly recognized and employed in the radar fields. This increase in the bandwidth for radar required more spectrum allocation. These requirements for more bandwidth soon came to be known as UWB, and it had unintended consequences in the parallel development of communications capabilities.

UWB RF systems are defined by the IEEE Standard 1675 to be greater than 25% or over 500 MHz of instantaneous bandwidth. With these characteristics, it soon became a serious issue with the frequency allocation regulations in almost all developed countries. Within the same technical environment, UWB radar was shown to have many applications in both the military and civilian communities. Significant improvements in range and cross-range resolution for discrimination and characterization of objects from clutter were recognized for foliage and ground penetration. The foliage penetration enabled the users to detect and locate buildings, communications lines, paths and streams, and even persons under this volumetric cover. These capabilities have direct applications for military and border security objectives. They can also be used for detecting changes in the forests due to pollution, erosion and illegal harvesting of forests. The ground penetration applications provide the ability to detect buried mines, underground tunnels and safety measurements in mining. The development of wide-area surveillance radars with bandwidth and polarimetric characteristics have also been used for land use and land characterization for Earth observation.

An early concept for battlefield surveillance was a small aircraft or RPV flying over at tactical area, as depicted in Figure 1.1. The radar would fly at a reasonable standoff distance and be able to illuminate a swath of the area to detect buildings, vehicles and lines of communications. It was important that the tactical objects be referenced to the cultural regions with roads, rivers and natural objects. This gave the user the ability to evaluate the detections with respect to areas of military significance. A radar would have sufficient power aperture to ensure detection probability on small targets. And the on-board receiver would have the ability to digitize the radar returns and send the data to a Tactical Operations Center (TOC).

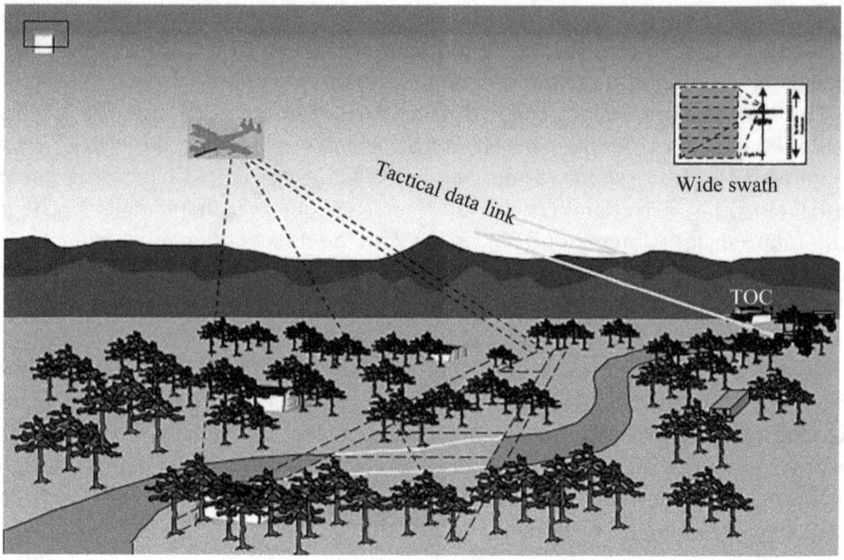

Figure 1.1 Early concept of battlefield surveillance radar [4]

These early systems required a direct line of sight from the radar to the TOC for real-time evaluation of the radar returns. The system engineering task was not just for UWB surveillance radar design, but also for the concept of operations. Questions were raised on how much of the system could be installed on either a small manned or unmanned vehicle, and whether it could be processed onboard in real time. The systems also needed accurate motion measurement and compensation of collection geometries to maintain coherence of the radar signals and provide geolocation accuracy. Needless to say, global positioning systems (GPS) were not available during the time of the system development. So, inertial measurement and guidance system were stressed for the long data runs to obtain georeferenced images and geolocation of both fixed and moving targets. Tactical data links did not have sufficient bandwidth to send down all of the data for image processing on the ground. And satellite communications were not available until after 1990.

There have been several key technology advances that have facilitated the development of UWB airborne surveillance radar. A short list of the advances and their applications to radar are given in Section 1.3. These advances have been instrumental in making these airborne systems more autonomous and affordable to the civilian and commercial markets.

GPS provide a worldwide knowledge of the position of the radar platform. Early GPS capability had a selective availability interface, where the positional accuracy was degraded by intentional degradation of the signals from the satellites. Positional accuracy for civilian use was limited to 100 m. In 2000, President Clinton issued an executive order to make GPS more attractive to civil and commercial users worldwide. Now GPS has accuracy within approximately 1 m worldwide. When combined with an inertial navigation system (INS), the position, velocity, orientation and time of the platform can be known accurately and with a high update rate. This technology has enabled airborne radars to be independent of expensive and bulky navigation systems that were originally required on airborne surveillance radar.

Coherent transmitter amplifiers were provided by tube assemblies, which had requirements for high-voltage power supplies, along with heavy and complicated cooling of the tubes and power supplies. These amplifiers certainly provided wideband, linear amplification of the radar signals. The early airborne surveillance radars all had employed this tube amplifier technology. However, they were expensive and heavy, limiting their applicability to small unmanned or civilian radars, and they often had a low mean time to failure. With the development of solid-state transmitters for active electronic scanned antennas (AESAs), the radar became significantly smaller and more reliable than previous tube amplifiers. The millimeter and microwave integrated circuits (MiMICs) program greatly accelerated the technology in terms of a couple of orders of magnitude reduction in size, weight and power [5]. A single integrated chip could provide the low noise amplification, digital control of amplitude and phase, and integration into the face of the array. This enables active element phased arrays and greatly improves the reliability and availability of the radar systems. The one drawback of solid-state transmit amplifiers is the need for operating in signal saturation for maximum power added efficiencies. New techniques were needed to provide the array amplitude control for low sidelobes.

It should be noted that the greatest advances in solid-state transmitter components for radar have come from the communications world. The volumes of devices for cell phones, cellular communications and wideband signal modulation for spectrum conservation have reduced the production cost of the solid-state AESAs.

The development of high-speed and bandwidth integrated circuits was important to transition from analog-to-digital signal processing. High dynamic range digitization of the RF signals provides the interface to digital signal processing. Early analog-to-digital converters (ADCs) provided modest bandwidth and dynamic range for radar signal processing. These ADCs were used on all coherent radars, but due to their size and complexity, only a small number of channels could be designed into the surveillance radar systems. The commercial needs for wideband communications in cellular radios have greatly improved the ADC bandwidth and dynamic range. More importantly, the small size and low power of the devices are such that they can be directly integrated into the active electronically scanned antenna (ESA) front end.

Along with the ADC, there have been advances in the digital-to-analog converters. These components have been used in many emerging radars for direct synthesis of waveforms, channel spectrum compensation to satisfy the RF environment need for interoperability, and preselection and channel match for real-time adaptive array operation.

One of the most pervasive technology advances has been with the signal-processing architectures of more intensive algorithms. The Very High-Speed Integrated Circuit program started the focus on very fast signal processing for real-time operation. The first applications were for personal computers and massively parallel signal processing for many applications from weather prediction, aircraft design for more efficient flight envelopes, design of adaptive radar and optics for deep space exploration. These high-performance computation tools enabled radar on-board processing for real-time detection and characterization of the scene and objects within the images.

But the first technical objective is to gain a significant assessment of the target and clutter characteristics – scattering and losses at the design frequency for each system. The frequencies considered for surveillance radars varied from VHF and UHF for foliage penetration to X-band and Ku-band for open-area SAR and GMTI wide-area surveillance. Several collection campaigns were undertaken to characterize the ground clutter statistics along with returns from military and civilian structures and vehicles, as covered in Chapter 2. A major issue in these collections was the spatial resolution to discriminate the man-made objects from clutter. All of these issues formed the basis for surveillance radar component and signal-processing techniques, which would enable sizing the radar systems and computers that could be build. It is important to remember that GPS and Moore's Law were coming into reality.

These technical advances have provided a continuous development of UWB surveillance radar, along with digital mobile communications, for many applications. These two applications are often in conflict in a dense RF environment. The following sections will provide the background on historic surveillance systems and the impact of technology advances over the past several decades.

1.1 Early airborne surveillance radar systems

The surveillance radar research community started to prototype new radar systems with the emerging technologies. Both National Aeronautics and Space Administration (NASA) and The Defense Advanced Research Projects Agency (DARPA) initiated several radar systems to determine the utility of new wide-area imaging capabilities. The main focus was for day–night, all-weather surveillance of the Earth's surface to provide an assessment of man-made and cultural objects. There was a distinct objective of testing the concept of operations of these systems before the military committed to the large expenses for deployment and operation.

The primary radar technical questions that were being asked in the 1970s were:

- Can multiple RF channel systems be used to provide the clutter cancellation and detection of very slow-moving targets?
- Can an ESA be flown on a small aircraft and be remotely controlled to detect and track artillery and moving targets?
- Can an operationally capable radar be flown and provide the tactical battlefield operational picture to disperse units?
- Can the combination of two RF radar systems on a single aircraft provide the detection of objects above and below the foliage canopy and provide interferometric measurement of the terrain height for detecting man-made objects?

Several prototypes were developed to answer the questions on how coherent, integrated multimode radars could be best developed and used. Some principal radars to be investigated include:

- SOTAS – Development of the first GMTI wide-area surveillance system to detect and characterize enemy moving targets in the FEBA [2].
- Multiple Aperture Surveillance Radar (MASR) – Development of multiple channel detection of slow-moving targets, by using highly correlated returns of targets buried in clutter [2].
- HOWLS – Development of ESA suitable for small RPVs to detect and characterize fixed and moving targets in an active fire scenario [3].
- Geographic synthetic aperture radar (GeoSAR) – Demonstrated a multiple, frequency interferometric radar for terrain elevation maps in the clear and under foliage, with a second objective of measuring the characteristics of land-use measurements of cultural clutter [6].

1.1.1 Stand-off target acquisition system

The first GMTI system for battlefield surveillance was developed by the US Army, originally named the Airborne Long-range Alerting Radar for MTI (ALARM). The US Defense Department was slowly becoming aware of the benefit of detecting large masses of slowly moving targets in eastern European geography. However, the history of "moving blobs" on early MTI radars needed significant testing for acceptance by the military users. The military commanders in both Europe and South Korea needed a continual MTI coverage that would be interpretable by the tactical users on the ground. This knowledge of enemy force dispositions and actions could be used by the commander to maneuver his own forces. And the need for accurate detection and tracking of these enemy maneuvers for artillery fire, directing attack helicopters and close air support would be beneficial to military tactics.

An ALARM experimental system was initiated in 1970, which would take advantage of the emerging coherent processing radar technologies and provide a data link of the radar detection returns to a ground processing center. The name was soon changed to SOTAS to stress the mobility and survivability of the system [2]. This radar was implemented using an X-band APS-94 side looking airborne radar by adding a moving target mode. The antenna could rotate 360° under the helicopter to provide battlefield surveillance with a high revisit rate. A helicopter airborne platform was necessary to minimize the platform motion and provide sufficiently low minimum detectable velocity. And since it operated from the UH-1 helicopter shown in Figure 1.2, it provided persistent surveillance of a tactical area. Because the helicopter operated with a slow radar platform velocity, it provided detection of targets with slow range rate (or Doppler) with respect to the radar. Based on many experimental tests, it was reported that the radar could "see" a small group of people at distances of 15 km.

The radar system along with the ground control center were taken to several exercises in North Korea and West Germany. The officers observing the exercises

Figure 1.2 Original SOTAS radar system on UH-1 helicopter [2]

could observe when forces moved off of the roads: when artillery peeled off to setup positions and pulled up to move to new positions, when air defense elements converged on moving columns to provide protection, and when the reinforcements arrived. These observations provided significant advantage to the surveillance radar operators and the command and control team.

SOTAS was an immediate success in the "war games." Figure 1.3 shows a ground center scope photo of the situation near the demilitarized zone (DMZ) in Korea. The software placed the detection of moving targets on to the situation map,

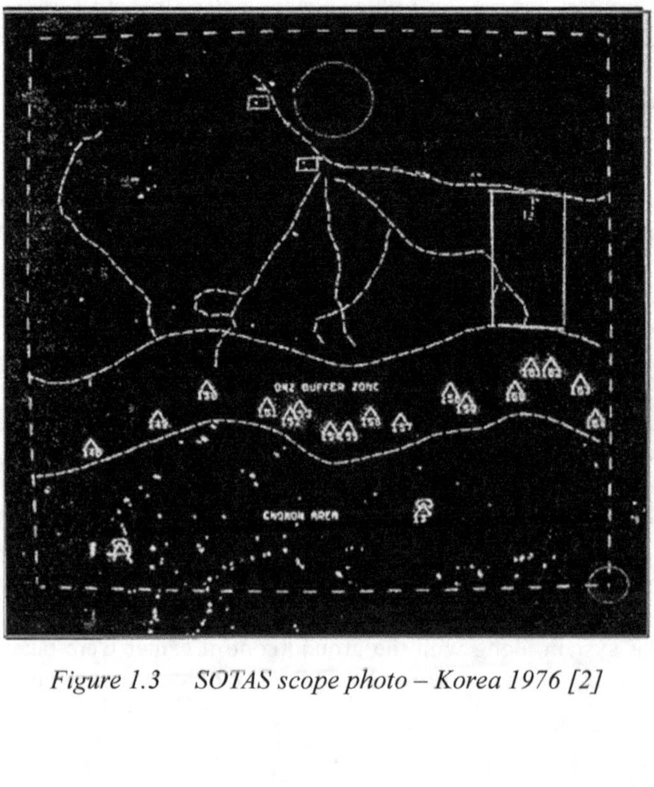

Figure 1.3 SOTAS scope photo – Korea 1976 [2]

and the operators could annotate the display with known areas and roads. Moving targets could be seen on both sides of the DMZ. And as the SOTAS Program Manager observed, "since there were no unannounced movers north of the DMZ, the false alarm performance could be measured." These capabilities were found to be breakthrough for support of military operations [2].

The SOTAS display was determined to be quite interpretable with its selectable map feature and the "history replay," where the last several minutes of data were presented to the operator. This called attention to the current position, direction and speed. Based on these quasi-operational tests, the Army started a full scale engineering development program in 1979. A new fully coherent radar to be mounted on the UH-64 Blackhawk helicopter was contracted. However, in 1980, due to program cost, schedule risks and a new administration, the program was terminated. SOTAS was a great success in terms of the new capabilities. But the hardware technology readiness was not sufficient for the military to obtain further funding.

1.1.2 Multiple aperture surveillance radar

In the early 1970s, the US Air Force was looking for a tactical radar that could detect and track slow-moving ground targets from a small aircraft. The Displace Phased Center Antenna (DPCA) technique had been analyzed to enable detection of moving targets by cancelling the stationary ground clutter. By shifting the effective radiation center in the antenna backward between pulses, moving targets could be detected. Based on these innovations, a concept was developed to build a precision, low-sidelobe ESA that could survey a region as the fixed wing aircraft flow parallel to a region of interest. The MASR was funded by the Air Force and developed at the MIT Lincoln Laboratory [3].

MASR was an experimental L-band GMTI system for wide-area scan of a future battlefield surveillance region and it was the first evidence of an airborne, developed DPCA techniques for very strong cancellation of stationary clutter and detection of ground moving targets. This surveillance radar prototype shown in Figure 1.4 was successful in demonstrating a phased array antenna with inter-channel match far superior to earlier applications. By switching the individual elements between pulses, short-term correlation of the clutter up to 46 dB could be cancelled, and slow-moving targets could be discriminated.

Prior to the MASR concept, GMTI systems used noncoherent processing with a clutter reference technique. Subsequent pulses were cancelled using an analog delay line by subtraction from one PRI to another. There was sufficient correlation between the pulses that the target Doppler would sufficient to leave the moving target residual after the cancellation. Furthermore, noncoherent integration could provide an increase in the signal to noise. However, the sensitivity of these noncoherent techniques was limited in sub-clutter visibility.

When the MASR was started, there were medium-scale integrated circuits that would provide improvement in the integration processing. This form of moving target detection was instrumental in providing a capability for detecting and locating moving

Figure 1.4 MASR prototype [3]

targets from an airborne air to ground surveillance radar system. The details of the MASR DPCA antenna and signal processing capabilities are given in Section 3.1.1.

1.1.3 Hostile weapons location system

The joint DARPA-US Army HOWLS began in 1974 and focused on the development of technique to locate and classify stationary indirect fire weapons in the battlefield clutter [3]. It was a novel concept for the Army to provide real-time assessment of movement of military vehicles in a battle area and provide real-time detection and characterization of artillery fire information to a small unit. The original use was to employ the technology applicable to a small RPV, such as the Amber prototype. This program was funded out of DARPA Tactical Technology Office, system integration was provided by MIT Lincoln Laboratory, and the radar was developed and operated by GE Aerospace Electronics Systems Department.

The Ku-band ESA shown in Figure 1.5 was deployed under a small twin-engine aircraft for long-range sector surveillance. And the prototype radar, installed within the aircraft cabin, provided a significant increase in search and track of targets in the area. The persistent surveillance of a wide area was maintained by rapid mode switching between fixed and moving target detection. The objective was to segregate the returns of small vehicles and artillery rounds from the clutter and to use Doppler processing to locate and characterize their location. This task was accomplished since the coherent radar could switch modes in 4 ms and maintain up to four interleaved modes, under control from the aircraft or remotely via a C-band data link. All data collected in the air were transferred in real time, displayed on a tactical control unit, and recorded in a ground control shelter, as shown in Figure 1.6.

Figure 1.5 HOWLS prototype radar [3]

The tactical display was located in the ground shelter for both algorithm verification and demonstrating the test objectives to users. The processing algorithms were prototyped offline based on the recorded data, and they were subsequently integrated into the ground station for display to the users. The military observers were particularly interested in the radar ground map that displayed the detection of targets, with updates to verify the movement of targets or firing of artillery. Georeferenced

Figure 1.6 HOWLS ground control and processing station and data link receiver [3]

images were provided to verify the location of both the firing units and the artillery impacts on the ground.

Several emerging air-to-ground modes were prototyped and tested for their applicability to this new radar concept. These modes provided the long coherent dwell with digital signal processing to detect and characterize many targets in the battlefield area. The first technique to be demonstrated was a frequency agile, noncoherent ground imaging mode for despeckling the clutter. By changing the frequency on a pulse-by-pulse or between coherent processing intervals (CPIs), the fading of targets was greatly reduced. HOWLS also had a suite of coherent Doppler processing modes to measure the separation of target returns and Doppler artifacts from background clutter. Furthermore, a Doppler beam sharpening mode increased the ground map resolution in short collection intervals. This improvement in the ground map provides for correlation of fixed surface features to moving targets.

The details of the HOWLS ESA are given in Chapter 4, and the description of the signal processing algorithms developed for fixed and moving target detection is given in Chapter 5.

1.1.4 Geographic synthetic aperture radar

The GeoSAR program was started in 1995 to develop a dual-frequency, airborne interferometric radar to address the needs of a variety of geographic mapping users in government and private industries [7]. It was funded by DARPA and developed by NASA Jet Propulsion Laboratory, with the quick objective of demonstrating rapid characterization of land-use characteristics in the open and under dense foliage. The two frequencies were X-band to map the tops of the trees and ground clutter (and buildings), and P-band (UHF) for penetration of dense foliage as shown for the aircraft in Figure 1.7. Both frequencies had interferometric cross-track processing to measure the digital elevation terrain data (DTED), both in the open and under the forest. The primary rationale for this development was to estimate terrain slopes and measure foliage characteristics in areas that microwave radar alone could not

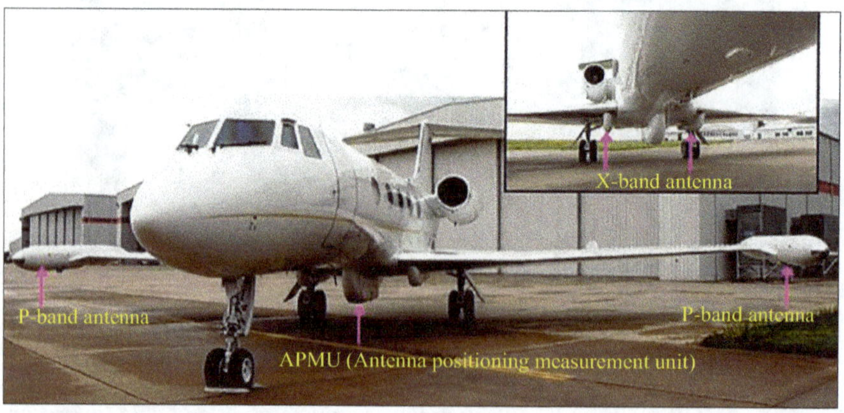

Figure 1.7 GeoSAR mapping system with dual-frequency IFSAR antennas [7]

measure. The combination of both frequencies has been instrumental in many geophysical experiments such as earthquake remediation, detection of downed aircraft in forested regions, and assessment of distress on forests and crop lands by drought, floods and illegal forestry cutting.

The P-band antenna system is mounted on two wing pods with a long interferometric SAR (IFSAR) baseline separation of 20 m. A complementary X-band antenna pair is mounted across the centerline of the aircraft with a 2.5-m baseline separation. The position and orientation of the four antennas is monitored by a laser baseline measurement system to provide the necessary IFSAR location, orientation and motion compensation. When combined with the GPS and INS subsystems, the GeoSAR mapping has survey grade accuracy of the channel data. All data are recorded onboard the aircraft for postflight processing [6].

The four IFSAR maps are collected simultaneously to obtain coincident images and interferometric phase data, as shown in Figure 1.8. The swath width of both images are 20 km, with simultaneous collection on both sides of the aircraft. This provided high area coverage rate of the area for efficient observation of the terrain below. The objective was a commercial application of terrain DTED collection from an airborne platform, but not requiring real-time processing. These data are then used for further exploitation for geoscience applications. In addition to the IFSAR data, fully polarimetric SAR data can be collected at P-band for characterization of the biomass. Since there are only four high-speed recording channels, the dual frequency and polarimetric IFAR collections are separately obtained.

The digital elevation map characteristics have been used worldwide for planning roads and flood plain engineering. The P-band radar is fully polarimetric to enable efficient characterization of objects under the foliage as well as measuring the health of the biomass. The planimetric accuracy is sufficient for topological mapping and planning. However, since the radar was UWB operation, significant development

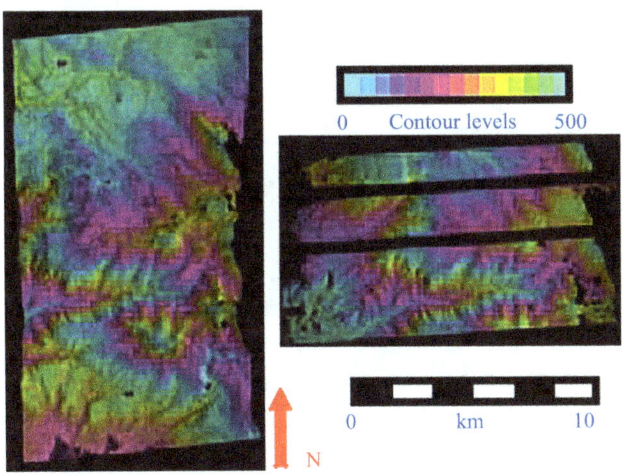

Figure 1.8 GeoSAR P-band (left) and X-band (right) data [7]

of spectrum compliance was required from the US National Telecommunications and Information Administration (NTIA) to obtain frequency allocation in populated areas [8].

This system has been in continuous operation by a commercial company between 2003 and 2018. Their international mapping has been well documented, along with improvements in the IFSAR processing for improvement in terrain characterization [7]. Details of the IFSAR processing are given in Chapter 6, and the spectrum compliance issues are presented in Chapter 7.

1.2 UWB definition

The definition of UWB is explicitly covered by the IEEE Standard 1675. The radar frequency application is defined as an operation of an RF system with greater than 500 MHz of bandwidth, or a percentage bandwidth over 25% [8]. This percentage bandwidth is measured by the ratio of the high and low frequencies to the center frequency of operation:

$$\text{Percent BW} = \frac{2\left(f_{\mathrm{H}} - f_{\mathrm{L}}\right)}{f_{\mathrm{H}} + f_{\mathrm{L}}} \tag{1.1}$$

This definition has been adopted by spectrum regulation organizations around the world. In the United States, it is the NTIA, and in Europe it is Comité Européen des Postes et Telecommunications. Any system that is characterized by UWB must comply with their regulations in terms of the design and employment of its operation. It can be seen in Figure 1.9 that frequencies below 2 GHz can readily be characterized by both the UWB criteria: percentage bandwidth and 500 MHz absolute

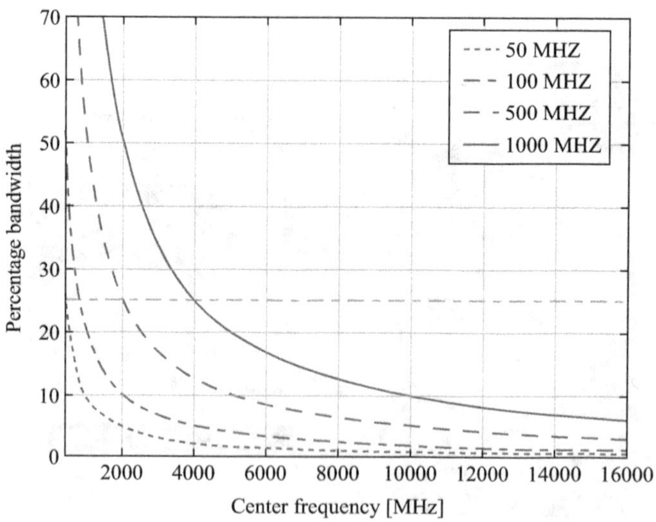

Figure 1.9 UWB radar definition versus frequency

bandwidth. Above 2 GHz, the limitation on percentage bandwidth is not as much of a problem even up to 1 GHz bandwidth. However, obtaining frequency authorization for over 500 MHz of radar bandwidth has been a real challenge due to the high density of other RF users.

Surveillance radar spectrum will often exceed both criteria, especially with the need for finer spatial resolution even at Ku-band. This factor in the UWB surveillance radar has become a major social politic issue, with the advent of digital and personal communications. Since 2008, the spectrum normally set aside for a radar has been sold off to the telecommunications industry. Chapter 7 will explore this issue in more depth, along with potential solutions for easing the cost of entry into many radar system applications.

The most important impact on UWB radar development and operation is the commercialization of digital/personal communications. Cellular radios and mobile phones have provided the world an improved capability to obtain and use information with immediate access. As a result, economics of the World Wide Web is much greater than the limited manufacture and deployment of UWB radars. This imbalance in the scale of small mobile radios to distributed radar systems has significantly complicated frequency use licensing. Regulatory organizations have all placed stringent requirements on spectrum and spectral purity.

UWB radar has the potential for improving many applications over the earlier battlefield surveillance radar: improved range and cross-range resolution, penetration of foliage and camouflage, detection of buried objects and unexploded ordinance, and characterizing land for Earth observation. These UWB radar systems were developed based on a rapid advancement in the key RF and computational technologies: GPS, solid-state transmitters, high dynamic range ADCs, and the implementation of high-performance computers. Moore's Law[1] has been credited with the continuous advance in all of these technologies over the years beyond 1975. However, as the same RF technology enabled wideband digital and personal communications, the ability to deploy UWB radar systems has been significantly curtailed. There are stringent requirements on the use of the RF spectrum and demands on improving the spectral purity of the systems.

Figure 1.10 illustrates several advanced development radars being tested from 1970 to 2000. It can be seen that more of these systems have become UWB. The early battlefield surveillance radars shown are not UWB. Because of the maturity of the technology, or risk in implementation, the spectral bandwidth was significantly lower than 500 MHz. However, it can be seen that the impetus for UWB was recognized in 1990. And more examples of UWB applications have continued for the past 25 years. These radar systems will be explored in the subsequent chapters, in terms of their technologies and technical requirements achieved.

[1]Moore's Law is the observation in 1965 by Gordon Moore that the number of transistors per square inch on integrated circuits has doubled every year since the integrated circuit was invented. Moore predicted that the trend would continue for the foreseeable future.

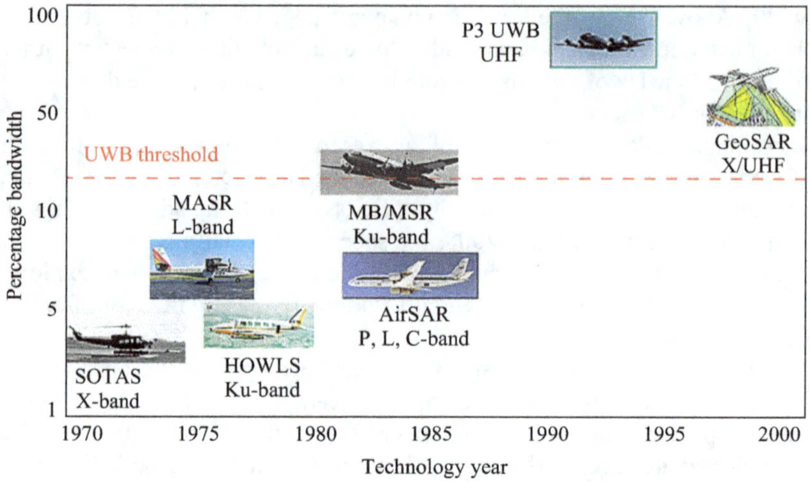

Figure 1.10 History of UWB radar for ground surveillance applications, 1970–2000

1.3 Technology enablers for UAS surveillance radars

Section 1 outlined three major technical enablers for the development of UWB surveillance radars. They are active ESA (AESA), monolithic microwave integrated circuits (MMICs) and high-performance computing. This section will provide some examples where these technologies have provided the component and subsystem capabilities for a long history of developing UWB surveillance radars. It can be said that Moore's Law has been extended to the RF community, in addition to the digital processing needs. There are significant commercial off-the-shelf (COTS) devices available, which reduce the cost and volumetric requirements for these new radar systems.

1.3.1 Active electronic scanned arrays

Advances in RF microwave integrated circuits (MICs) have significantly reduced the cost, size, weight and power of electrically scanned antennas [5]. A comparison of two ESA techniques is shown in Figure 1.11. The first application is a passive ESA, employing analog phase shifters in the feed system [3]. The second is an active ESA where the transmitter and receiver are located behind the radiation element [9].

The antenna on the left was developed for the HOWLS program in 1975, and it was constructed using a combination of waveguide and PIN diode phase shifters to provide the electronic beam steering antenna control [3]. The HOWLS transmitter was a high-efficiency traveling wave tube amplifier. However, all of the microwave control of the signal to transmit and receive passed through lossy circuits after the power amplifier and before the low noise receiver. There was no additional amplification, so the RF signal level was attenuated in both the transmit and receive paths.

HOWLS passive array
ca. 1977

Passive array

+ High-efficiency UWB transmitter (tube)
+ Reciprocal Tx and Rx signal paths
– High loss after Tx and before Rx NF
– Tx taper through feed structure
– Very heavy ESA assembly

Thales active array
ca. 2007

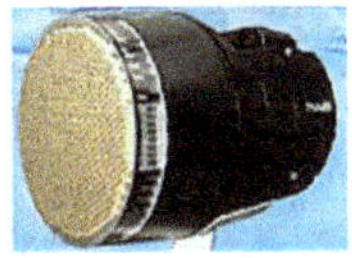

Active array

+ Small lightweight TR modules
+ Low loss after Tx and before Rx NF
+ Moderate efficiency solid-state Tx
– Difficult Tx taper for high efficiency
– Tx and Rx spectral mismatch

Figure 1.11 Phased array technology comparisons [3, 9]

As a result, the radar range equation had a two-way loss of signal in the beam forming process, and there is a significant reduction in the noise figure on receive. This made for a very inefficient radar operation. The main advantage of the passive ESA was that the aperture on transmit could be weighted to provide low sidelobes.

The antenna on the right is representative of an active element phased array, designed by Thales for the Rafale fighter [9]. The aperture is an active ESA, and each element had a small transmit receive module (TR module) immediately behind the radiating element. As a result, both the transmission loss and the noise figure were improved by the different in loss between the active ESA and passive ESA design. Furthermore, both the transmit and receive amplification and signal control can be integrated into the AESA assembly. The solid-state TR modules had moderate efficiency with the gallium arsenide (GaAs) devices available in 2000. The main disadvantage of this approach compared to a passive ESA is that the GaAs MIMIC devices need to be operated in saturation (Class C operation). As a result, the ability to do amplitude taper on transmit was limited. The second technical evaluation was the spectral match across the band, and scan angle will cause issues with the correlation of the signal for clutter mitigation or target direction. These effects are minor, since it is possible to do pre-emphasis on transmit and adaptive equalization on receive.

The advent of MMIC designs was significant in providing lightweight and small form factor for small manned and unmanned air vehicle operation. The benefits for power aperture efficiency for active versus passive ESA are detailed in Section 4.4.

1.3.2 Monolithic microwave integrated circuits

MICs were readily available for active ESA design and implementation. The design tools and implementation for well-matched circuits had been developed for airborne avionics in the 1970s. At lower frequencies (UHF to S-band), single-element TR modules could be implemented within the element spacing. Figure 1.12 shows the comparison between MIC and MMIC designs for active phased array design. The

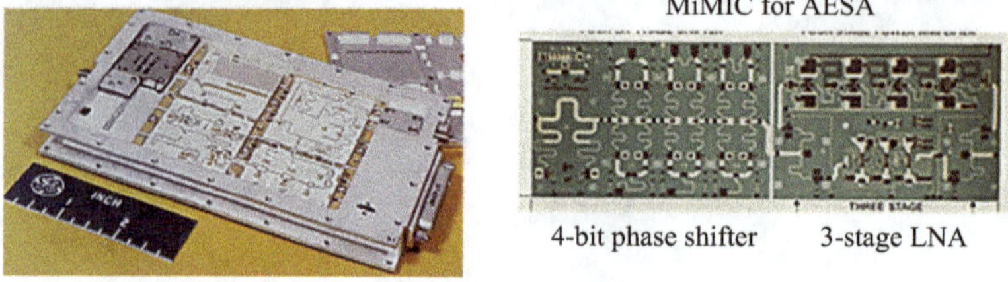

Figure 1.12 MICs for surveillance radar applications [5, 10]

assembly on the left was a 200-w peak power TR module with amplitude and phase control behind an L-band phased array antenna [9]. These assemblies were feasible with 40-dB element match over moderate bandwidths. This was sufficient for many airborne applications for reasonable aircraft. However, at the microwave frequencies, the need for thousands of TR modules at microwave frequencies made the cost of the antenna prohibitive.

Under the DARPA MIMIC program, the size of the transmitter low-power chain and the whole receive chain could be implemented in a small number of monolithic MIC chips [5]. With the reduction of the assembly costs and the reduced size, weight and power, AESA designs were developed for many airborne applications. Moreover, since the chips were significantly smaller than the MIC designs, the depth of the AESA was reduced to a few 10 s of centimeters. The TR modules were now feasible within a half-wavelength, square area behind the radiating face. The radar transmit and receive chips for an airborne fighter radar are shown on the right in Figure 1.12 [9]. Hence, AESA designs were considered for implementation in many new radar applications.

The further development of solid-state materials such as silicon germanium (SiGe) and gallium nitride (GaN) have improved the efficiency of the TR modules, along with wide instantaneous bandwidth for UWB applications. These new microwave device and integrated front ends continue to provide wider bandwidth and more sensitivity to emerging surveillance radars.

As the RF designs of AESA systems matured, the requirements for signal processing has greatly increased. In the front end of the radar, the signal generation, preselection of spectral purity and equalization for multichannel correlation require significant signal processing in the small form factor.

1.3.3 High-performance computers

The major advance in onboard processing came with the development of the very large-scale integrated (VLSI) circuits. This level of integration placed the computation of radar control and signal processing near to the front of the radar system. And their speed has significantly increased based on the large investment by the commercial sector. Embedded processing allowed the tight coupling of the radar hardware and the software, as illustrated in Figure 1.13. The radar architectures were able to increase the bandwidth of data coming into the front end of the radar, but to control

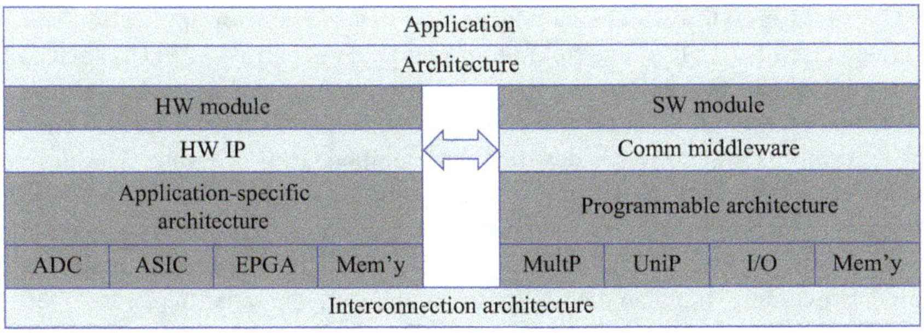

Figure 1.13 Embedded computing architecture [11]

the processing response to the changing environment. The wideband transfer of data from the hardware to the software computing has enabled more capability for radar applications [11].

The key components for efficient embedded processors are the development of high bandwidth, high dynamic range of the ADCs. Figure 1.14 shows the history of ADC development over two decades during the development of VLSI components. In 1990, ADCs with dynamic range of a modest 48 dB could be obtained only at 50 MHz sampling speed. This represents an improvement of 12 dB in dynamic range and 10 times improvement in sampling rate in 10 years. Furthermore, the rapid advance in digital signal processing hardware based on Moore's Law continues to improve the processor throughput by a factor of 2 every 18 months or, equivalently, by two orders of magnitude per decade [8]. These subsystem advances will be instrumental in the development of high area coverage rate, real-time UWB SAR sensors for the objective of small, unmanned platform operation.

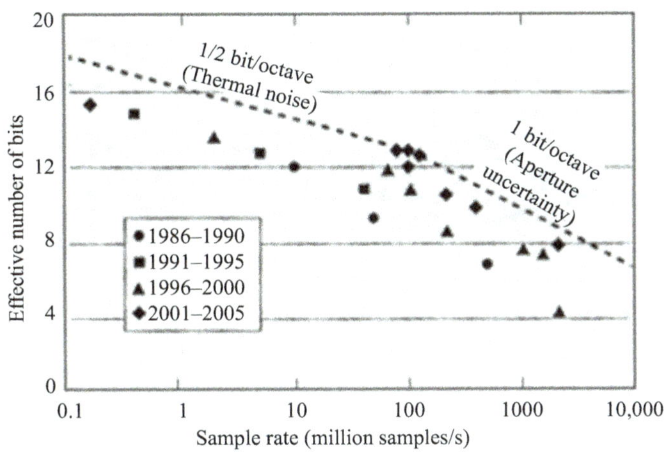

Figure 1.14 History ADC speed and effective number of bits (1986–2005) [11]

The rapid growth of memory density and new devices of application-specific integrated circuits (ASICs) and field programmable gate arrays (FPGAs) placed the digital control and processing in higher density integration. Figure 1.13 shows the historic bandwidth and dynamic range of components which are coupled with efficient algortithms and software developed for embedded processing. The development of extremely high-speed processing coupled with dense memory has led to significant reduction in both the volume and cost of onboard processing. In addition, new digital processing algorithms have enabled throughput for not only the classic Doppler processing, but also the ability to compensate the radar hardware over significant bandwidth. This coupling of hardware and software will be explored in the following chapters.

The advances in VLSI components and integrated communications within a system came with reduction of the volume of the on-board computers and a significant reduction of the power required in airborne radars. There are three basic computational technologies that have matured over the past four decades, as depicted in Figure 1.15. The two axes show variation in the density of computational tasks (billion operations per second GFLOPS) to input power and volume of the processing units. The computational processing unit (CPU) is the most flexible element of an onboard processor, with direct high-order software computing the tasking detection and characterization of the radar modes. This reprogrammable capability comes with a higher overhead in the interface between the processing elements and the onboard memory. The most capable processing technology is the ASIC that has high throughput processing designed for a (fixed) signal processing task. However, this speed comes with a lower flexibility in terms of tasks that can be changed between radar modes. Furthermore, the power requirements to obtain this processing speed has been slower in development. The middle technology for embedded processing

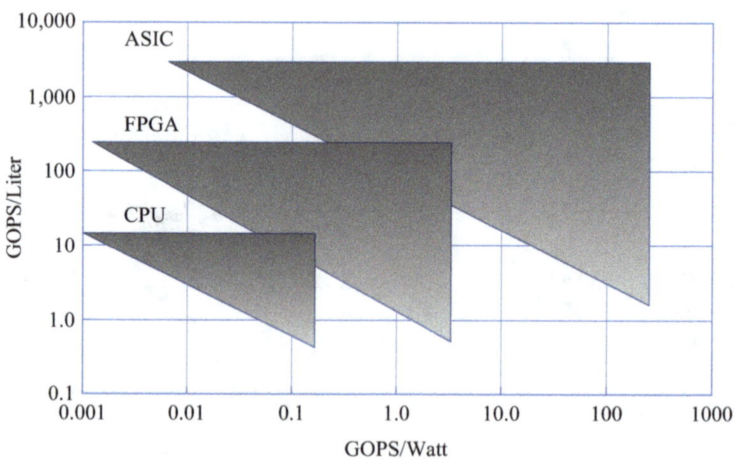

Figure 1.15 Alternative designs of processing technology for lower volume and power [11]

has been the FPGA, where the processing algorithms can be rapidly switched with firmware to enable multimode operation. Thanks to Moore's Law, this heuristic graph shown in Figure 1.15 of both GFLOPS/w and GFLOPS/l have been changing each year and a half.

For on-board signal processing of SAR and GMTI signals, the number of real-time, floating point operations per second (FLOPS) has continued to increase. Two examples of COTS processing are shown in Figure 1.16. In 2001, the FOPEN Advanced Technology Demonstration system had a 128-node PowerPC computer for UWB image formation. That system was over 2 cubic feet in size and cost approximately \$1M. Its capacity was 400 billion FLOPS (GFLOPS). In 2016, a single-board FPGA processor could be purchased at \$3K and process 480,000 logic operations and 2,800 digital signal processing cells at 200 million operations a second. This represents a 150:1 improvement in volume or GFLOPS, at a reduction of 300:1 in cost.

Radars require complex signal characteristics for all coherent applications. As a result, a complex addition requires two scalar additions for each step of the algorithm. And complex multiplication requires four scalar multiplications and two scalar additions. So, the estimation of signal processing requirements will require scaling of these mathematical operations to estimate the digital throughput.

One of the early examples of airborne radar digital signal processing came from the HOWLS program. Although this processing was implemented on the ground, it demonstrated the utility of digital processing for detecting moving targets. The simplistic radar has a CPI of N_p pulses and range bins of N_R for each received pulses, as illustrated in Figure 1.17. Each pulse is compressed with a time bandwidth product of N_{BT}, using the double convolution algorithm. It is assumed that the signal bandwidth is between 10 and 60 MHz, representative of the HOWLS architecture. The objective now is to understand the number of digital computations needed to form the range-Doppler representation of the returns.

The fast Fourier transform (FFT) is often used because of its reduction in the number of complex adds and multiplies versus a direct Fourier transform (DFT). The DFT has N^2 multiplies and adds, while the FFT has $N/2 \log_2 (N)$ multiplies and adds. The only real limitation of the FFT is that it has to be implemented in powers of 2 [12].

The expected computational count in signal processing is estimated by the number of complex processing operations need to be completed. For real-time operations, a

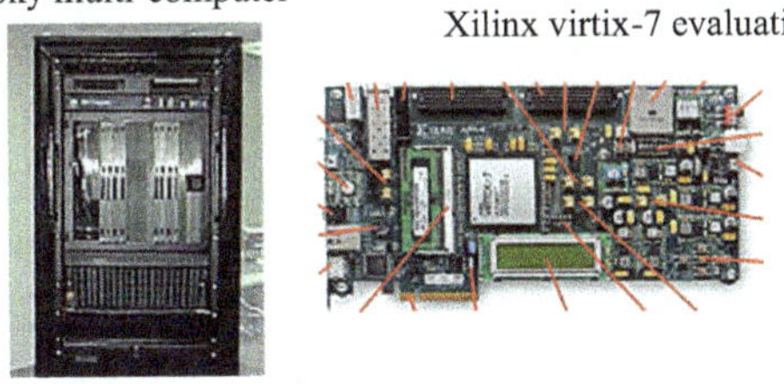

Figure 1.16 High-performance computer technology comparisons

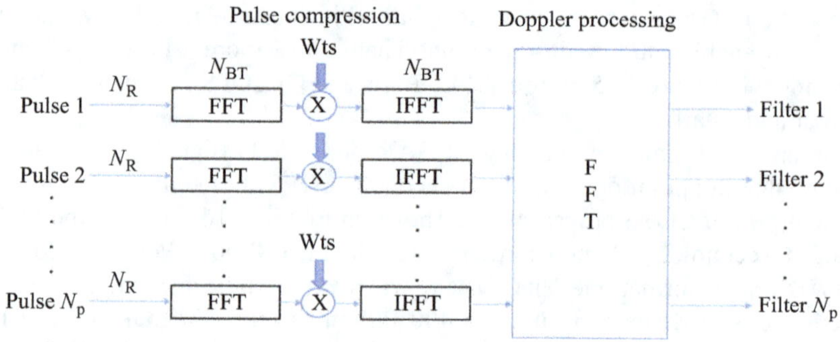

Figure 1.17 GMTI digital processing with FFT [12]

number of computations need to be accomplished within a CPI. An example of the input parameters for the processing in Figure 1.17 is shown in Table 1.1, where the number of pulses for GMTI is $N_p = 64$, and the BT of the pulse compression is given by a size varied from 64 to 256. The range on the ground varies from 1 to 10 km. So, the number of range samples are given by the range resolution δ_R:

$$\delta_R = \frac{c}{2B} \tag{1.2}$$

where c is the speed of light and B is the signal bandwidth.

The processing operations can be considered in the following stages. First, for the pulse compression of N_p pulses, the processing is given as [12]

$$\text{Proc\_pc} = N_p\,[(N_{BT}/2)\log_2(N_{BT}) + 4 + (N_{BT}/2)\log_2(N_{BT})] = N_p\,[N_{BT}\log_2(N_{BT}) + 4] \tag{1.3}$$

where N_{BT} is taken as [64 128 192 256]. Next, the number of range increments are given by

$$N_R = [1{,}000{:}1{,}000{:}10{,}000]/(c/2\ \text{BW}) \tag{1.4}$$

Finally, the N_p FFT for Doppler filtering is given by

$$\text{Proc\_dop} = N_p/2\log_2(N_p) \tag{1.5}$$

Given these factors, the total processing operations can be calculated for the range extent N_R and the bandwidth BW, as shown in Figure 1.18a. When the signal bandwidth is 64 MHz, the range resolution is 2.34 m, so that the number of range samples

Table 1.1 Parameters for GMTI operations count example

Parameter	Value	Parameter	Units
Pulse width	4 μs	Bandwidth	[16 32 48 64] MHz
Number pulses	64	BT values	[64 128 192 256]
Range distance	1,000:10,000 m	PRF	1,000

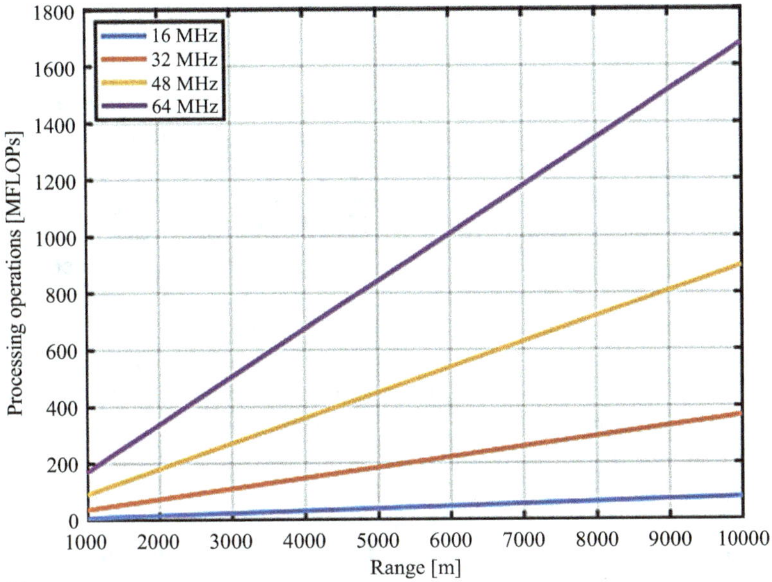

(a)

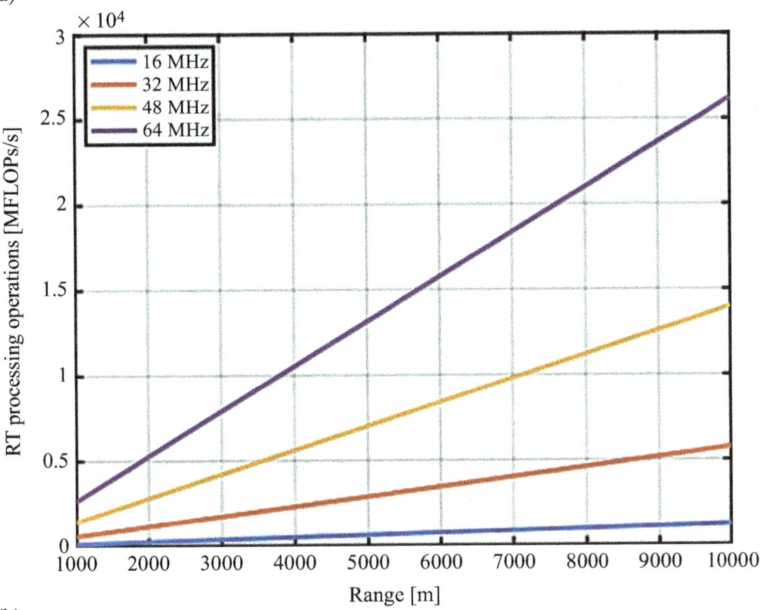

(b)

Figure 1.18 GMTI complex operations versus signal bandwidth. (a) Total operations count and (b) real-time operations throughput is 4,327. At this point, the processing count is 1.68 GFLOPs. For real-time processing, the CPI is 64 ms, and the real-time processing operations are 26 GFLOPs/s, as shown in Figure 1.18b.

During the HOWLS development time, there were no real-time processors that could keep up with this computational speed. However, the current state of the art in multicore, single-board processors is significantly higher than this metric. The signal bandwidth was limited to 30 MHz, and the signal processing was carried out offline on the ground. This would correspond to 200 KFLOPs/s. The UWB surveillance applications covered in this book require six orders of magnitude computation in a real-time onboard processor.

In today's COTS processors, significant SAR and GMTI processing can be obtained in small volume and weight form factor consistent with Class 3 UAV systems. Each of the radar modes will be examined in the following chapters, and the radar architectures and processing algorithms will be examined in terms of the GFLOP requirements.

1.4 Summary of airborne surveillance requirements

As a summary for this introduction to UWB surveillance radars, there has been significant advances in the past 40 years to enable this important capability for both military and geophysical applications. However, the short-range and low altitude of piloted airborne systems made practical battlefield radar systems unattractive. RPVs (or UAVs in today's vocabulary) were starting to be developed. More importantly, the availability of data links would enable significant processing and image interpretation on the ground. Data links provided line-of-sight connection to command and control centers at a safer distance to the airborne platform.

Both the military and scientific communities learned from the experiments in the late 1960s to mid-1980s. However, two system realities limited the growth of the technology – foliage attenuation limited the ranges for surveillance of forested areas from short to medium range, and manned aircraft could not be adequately protected at these ranges.

By the late 1980s, the radar community had determined that SAR would provide acceptable detection and characterization of ground targets and terrain features in both open and forested regions. The small antennas and modest power required at *P* through Ku-band was acceptable for experiments and might be possible for installation on RPVs. At the same time, new technologies for UWB signal generation, reception and processing required the need for emerging digital technology to provide UWB capabilities.

For GMTI, a very large of the antenna is required at microwave frequencies for both detection and localization of moving targets. The size of the antenna will be explored in Chapter 3, where it is shown that for slow-moving targets at microwave frequencies requires a few meters in length and over 10 m for UHF. The SOTAS development at X-band verified the benefit of stationary rotary wing operation for GMTI. Airborne Ground Moving Target FOPEN systems were another story. The size of the antenna at UHF and the lack of unmanned helicopters did not give rise to FOPEN MTI for more than two decades.

However, we will need to be advancing the technologies, since the operation of UWB radar systems are being regulated out of the spectrum needed for operation. Cognitive radio has been transitioned to operational capability by continuously sensing the RF environment and using the unused capabilities for short-term operation. The radar community needs to adapt these same engineering and regulation challenges to maintain its position in the radar capabilities. The rest of this document will show the history of advanced UWB radar, and some emerging algorithms and system implementation for being more compliant with the spectrum regulations.

The same technology that is being used in commercial and personal communications smartphones is applicable to today's radar needs. However, the regulations and frequency allocation process have changed significantly in the past 15 years. The radar community needs to develop new technologies to satisfy the regulators and also needs to enlighten the regulators to these new capabilities to "buy back" the spectrum. These issues will be covered in Chapter 7.

Commercial and personal communications are ubiquitous, with eCommerce as the major source of many business funding. Moreover, digital communications are important for security and life safety in populated regions. The issue facing UWB radar is recognizing that RF spectrum compliance standards are conservative and inflexible. As a result, the international radar community needs to adapt and develop new technologies – analogous to cognitive radio [13].

References

[1] Skolnik M.I. *Introduction to Radar Systems*. New York: McGraw Hill; 1962. p. 14.

[2] Fowler C.A. 'The standoff observation of enemy ground forces from project peek to JointSTARS-A prolusion'. *IEEE Aerosp. Electron. Syst. Mag.*. 1997;**12**(6):3–17.

[3] Bryant T.G., Morse G.B., Novak L.M., Henry J.C. 'Tactical radars for ground surveillance'. *MIT Lincoln Lab.* 2000;**12**(2):341.

[4] Davis M.E. *Foliage Penetration Radar – Detection and Characterization of Objects Under Trees*. Raleigh, NC: Scitech Publishing; 2011.

[5] DARPA MIMIC Program. MIMIC. Available from https://www.darpa.mil/about-us/timeline/microwave-and-millimeter-wave-integrated-circuit [Accessed 13 September 2020].

[6] Kramer H. GeoSAR (Geographic Synthetic Aperture Radar). 2002. Available from https://directory.eoportal.org/web/eoportal/airborne-sensors/geosar [Accessed 13 September 2020].

[7] Hensley S., Chapin E., Freedman A., Le C., Madsen S., Michel T. *First P-band results using the GeoSAR mapping system, Proceedings of the 2001 IEEE Radar Conference*. Atlanta, GA; 2001. pp. 126–8.

[8] Federal Communications Commission. *Understanding the fcc regulations for low power, non-licensed transmitters*. OET Bulletin No. 63, FCC Office of Telecommunications Technology, Columbia, MD, October; 1993.

[9] Mancuso Y. *Thales components and technologies for T/R modules, Proceedings of the third European Microwave Integrated Circuits Conference*; Amsterdam, the Netherlands; 2008. pp. 270–3.

[10] Davis M.E., Smith J.K., Grove C.E. 'L- band T/R module for airborne phase array'. *Microwave Journal*. 1977;**20**:54–60.

[11] Martinez D.R., Bond R.A., Vai M.M. *High Performance Embedded Computing Handbook – A Systems Perspective*. Boca Raton, FL: CRC Press; 2008.

[12] Stearns S.D. *Digital Signal Processing with Examples in Matlab*. Boca Raton, FL: CRC Press; 2003. pp. 315–22.

[13] McHenry M., Livsics E., Nguyen T., Majunmdar N. *XG dynamic spectrum sharing field test results, Proceedings of the 2nd IEEE International Symposium on New Frontiers in Dynamic Spectrum Access Networks*. Dublin, Ireland; 2007.

Chapter 2
Surveillance radar detection

Surveillance radars have been used for detecting fixed and moving objects within a ground area or region for several decades. The basic process is to survey the surface area and detect potential targets for subsequent characterization and investigation for image interpretation objectives. These applications can be either military or civilian in nature. In this chapter, the focus of the airborne radar environment will be on ground surveillance. This objective is in contrast to air surveillance of objects in the air or space, which are not covered in this text.

The platform for the surveillance radar can be either a piloted aircraft or an unpiloted air vehicle. Historically, the surveillance radar platform was an aircraft with sufficient power and volume to be controlled on-board by radar operators. However, with the advent of high-speed digital computers and digital data links, the transition to unmanned vehicles has been rapidly growing. This chapter presents the basic radar operation and the impact of clutter and propagation effects on the detection of several classes of objects. These basic radar system tools will be used in subsequent chapters as the ultra-wideband (UWB) radar analysis and algorithms are developed.

2.1 Radar waveform

A surveillance radar will transmit a signal to detect objects in the radar area of regard (AOR), and receive the radar return to detect and discriminate an object from the background clutter. The radar waveform is transmitted to provide the sensor with received signal resolution characteristics necessary for target detection and characterization. This waveform has several resolution attributes for discriminating the objects from noise or background clutter, e.g. range, cross-range and Doppler of the target return. Besides the radar waveform design, there are many system objectives to trade-off in designing the radar. Those include the desired range for target detection, the discrimination of the target from background clutter, the waveform signal processing to mitigate the effects of clutter and interference in the environment, and the desire for a high area coverage rate (ACR) for revisiting the surveillance volume while the radar is examining the AOR. The following sections will set up the formalism in the radar system design from a historical viewpoint and provide the emerging radar system capabilities based on recent and future technology advances.

2.1.1 Radar range and Doppler

It is important to understand the fundamental characteristics of pulsed radar waveforms. A pulsed radar is traditional for wide-area surveillance (WAS). The transmit pulse characteristics provide the ability to determine the range to a scattering object. Figure 2.1 illustrates the principle time measurements for a radar pulse to a target (tgt), along with the ranging parameters. The two-way propagation time ΔT of the signal from the radar to a point in the surveillance volume, as shown in Figure 2.1, determines the range R to the target:

$$\Delta T = \frac{2R}{c} \tag{2.1}$$

It is also important that the radar has an accurate clock to measure this propagation time for accurate ranging to any object in the scene. The width of the pulse T_P determines the minimum range R_{min} to the target, and the bandwidth of the signal B provides a range resolution capability.

$$R_{min} = \frac{cT_P}{2} \tag{2.2}$$

$$B = \frac{1}{T_P} \tag{2.3}$$

The minimum range is important in monostatic radars; the receiver needs to be blanked, so that the transmitted signal does not blind the radar receiver. And historically, the radar transmitted a pulse of sufficient duration to provide energy to each pulse, which affects the fine range resolution. This was a tradeoff of energy versus resolution. However, by modulating the pulse in either frequency or amplitude, both fine range resolution and lower sidelobe can be obtained.

In order to increase the total energy in the transmission, multiple pulses are integrated for improved radar operation. The separation between transmitted pulses is called the pulse repetition interval T. This interval T determines the waveform pulse repetition frequency PRF, which in turn sets the time interval of target detection for measuring the range rate of the target with respect to the radar. It also sets the unambiguous range R_{UA} that can be measured to the target.

$$\text{PRF} = \frac{1}{T} \tag{2.4}$$

$$R_{UA} = \frac{cT}{2} = \frac{c}{2\text{PRF}} \tag{2.5}$$

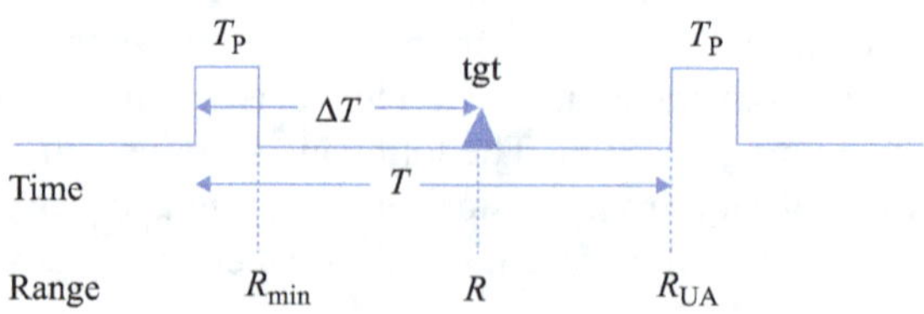

Figure 2.1 Radar pulse principle time measurements

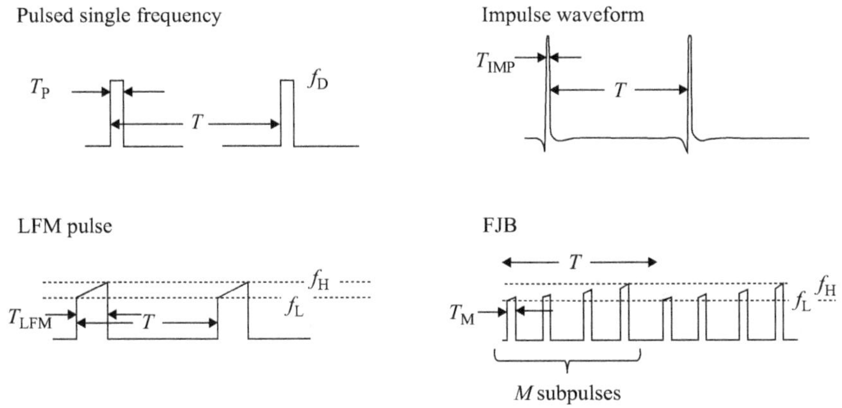

Figure 2.2 Candidate waveforms for bandwidth resolution synthesis

Finally, the ratio of pulse width T_p to the *P*PRF determines the duty factor D_f of the waveform, a critical factor in both the target detection probability and the surveillance ACR.

$$D_f = \frac{T_P}{T} \tag{2.6}$$

Many improvements in waveform design have occurred in the early days of radar development, as shown in Figure 2.2. The linear frequency modulation (LFM) waveform provides a much wider bandwidth than just a short pulse. By sweeping the transmit frequency during the pulse, the effective bandwidth is increased. With higher bandwidth, the range resolution is improved within a given pulse width, and without shortening the pulse width or reducing the duty factor. This radar pulse modulation technique enabled medium-range resolution for location of the target and discrimination of the object from the surrounding clutter.

Coherent processing was a major enabler for surveillance radar, in that it allowed the radar to measure the phase on a target return with respect to one's own platform. The phase manifests itself from the range rate to the radar based on the Doppler effect. Doppler is a change in frequency f when the signal return to the radar platform. A simple expression for Doppler f_d is

$$f_d = \frac{2v}{c} f = \frac{2v}{\lambda} \tag{2.7}$$

For a radar on a moving platform, the Doppler shift is proportional to the relative velocity between the two platforms. If the angle between the two platform velocity vectors is φ, then the Doppler shift f_d due to the range rate becomes

$$f_d = \frac{2v}{\lambda} \cos \varphi \tag{2.8}$$

The magnitude of the Doppler is positive when the target is moving toward the radar (closing range) and negative when it moves away (opening range). It should be noted that not only the target creates Doppler, but the ground under the aircraft does as well.

As with ambiguous range measurements, there is a corresponding Doppler ambiguity. The maximum Doppler that can be unambiguously measured as determined by the radar PRF is

$$f_{\mathrm{d,max}} = \frac{\pm \mathrm{PRF}}{2} \tag{2.9}$$

This is the condition for the measurement of Doppler ambiguity. If the target velocity v_{max} is moving faster than the ambiguous Doppler, its return will fold back into the Doppler space. As a result, for detecting targets with large radial velocities, there is a minimum PRF_{min} for detecting the target unambiguously.

$$\mathrm{PRF}_{\mathrm{min}} = 2f_{\mathrm{d,max}} = \frac{4v_{\mathrm{max}}}{\lambda} \tag{2.10}$$

The pulse characteristics discussed above were applicable to early surveillance radars, where the range resolution was considered to be in the form of moderate bandwidth or a few per cent of the transmission frequency. However, for UWB resolution, two techniques needed development, as shown in Figure 2.2. First, an impulse waveform was used for obtaining very fine resolution, when the pulse width was under a nanosecond in duration. This required very high power and a higher PRF to obtain reasonable signal-to-noise ratio (SNR) on the targets. A second innovation was to use short pulses of medium bandwidth and a higher PRF, and a change in the center frequency between pulses to get an overall improvement of the waveform bandwidth. This frequency jump burst (FJB) was instrumental in early UWB systems, which only covered medium-range extent, limited by the effective PRF. Each of these waveforms has been used in early UWB radars and will be covered in the following sections.

The velocity of the platform and the wavelength determine the surface clutter spectrum using the following relationship:

$$f_{\mathrm{d}} = -\frac{2v_{\mathrm{P}}}{\lambda} \sin \phi_{\mathrm{s}} \cos \psi_{\mathrm{g}} \tag{2.11}$$

The angles are the azimuth angle off of the velocity vector ϕ_s and the incident angle ψ_g with respect to the local terrain. Figure 2.3a shows the surface clutter spectrum from an ultrahigh-frequency (UHF) radar for the ground reference coverage area (GRCA), and the scenario parameters given in Table 2.1. Figure 2.3b plots only the iso-range and iso-Doppler contours for the range and Doppler ambiguities, respectively. As a comparison, Figure 2.4 shows the clutter, iso-range and iso-Doppler contours at X-band with a PRF of 1,200 Hz. The importance of these characteristics will be explored below.

There are several important points for this illustrated scenario, which determine the representation of the target and clutter with respect to the waveform parameters. The Doppler spectrum around the platform is spread between ±PRF, corresponding to the aircraft velocity and wavelength from (2.11). A range ambiguity will be shown by a solid iso-range circular contour. Because the range ambiguity is beyond the horizon at the 300 Hz PRF, an ambiguity is not shown in Figure 2.3, but it appears in Figure 2.4 at the higher PRF of 1,200 Hz. The dashed radial lines illustrate the Doppler

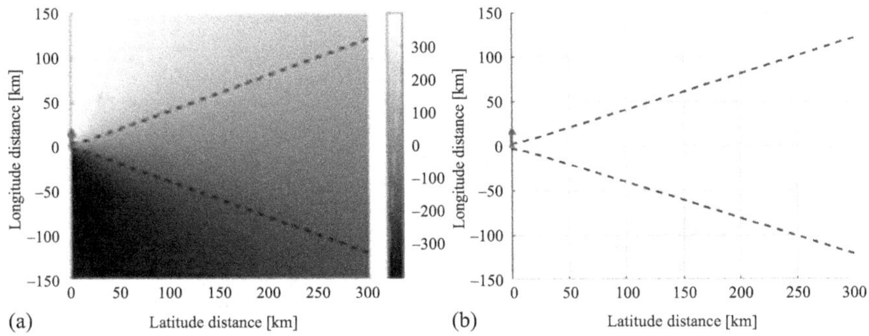

Figure 2.3 *UHF ground clutter Doppler spectrum (PRF 300 Hz). (a) Surface Doppler spectrum on GRCA and (b) iso-range and iso-Doppler contours*

ambiguity boundaries at ±(N_{amb} +½) PRF, where N_{amb} is the ambiguity number. The region between the two radial lines is Doppler or surface clutter unambiguous. So, there will be no folding of any angular clutter returns within the wedge. However, the antenna beamwidth needs to be narrower than the PRF/2 boundary, or cross-range Doppler will be folded into the imaging area. This performance is in contrast to X-band at a PRF of 1,200 Hz, shown in Figure 2.4a. Now the range ambiguities are shown at a radius of 125 km from the platform, and the multiple Doppler ambiguities are shown as the radial lines separated by the angle of ambiguous Dopplers.

The low PRF waveform (300 Hz) has the same FJB range eclipsing and ambiguities an LFM waveform with the same velocity and wavelength. Thus, there is no range folding of clutter returns into the region to obscure target detection in either the synthetic aperture radar (SAR) or ground moving target indication (GMTI) modes. However, the Doppler ambiguities are separated by very small angles. This separation can be determined by

$$\sin\left(\theta_{\text{amb}}\right) = \frac{N_{\text{amb}} v_{\text{P}}}{\lambda \text{PRF}} \tag{2.12}$$

where N_{amb} is the number of Doppler ambiguities and v_P is the platform. As a result, the angular separation between Doppler ambiguities will require a sufficiently large antenna to provide the attenuation of Doppler ambiguities from either GMTI or SAR modes.

Table 2.1 *Parameters for LFM waveform analysis*

Platform altitude	7.5	km	Cases	Freq (MHz)	PRF (Hz)
Platform speed	150	m/s	UHF	400	300
BW	30	MHz	X-band	9,500	1,200
Duty factor	10	%	GRCA	300 × 300	km²

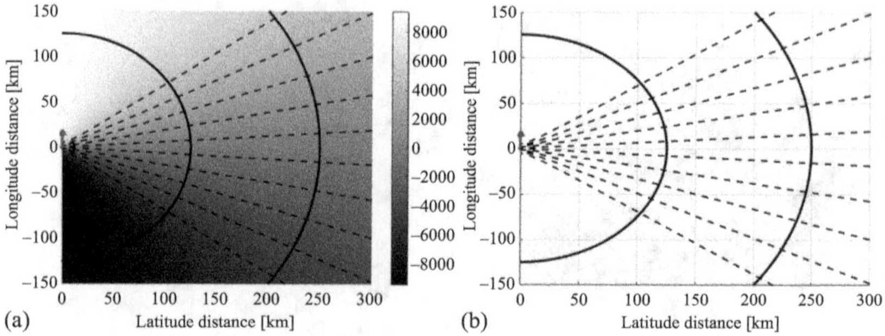

Figure 2.4 X-band ground clutter Doppler spectrum (PRF 1,200 Hz). (a) Surface Doppler spectrum on GRCA and (b) iso-range and iso-Doppler contours

The most important aspect of providing LFM range resolution from matched filtering of each of these waveforms is the simplicity of the waveform and the ease of matched filtering. The modulation of the waveform is provided by the change of frequency over the transmit time:

$$\begin{aligned}\omega &= 2\pi f = 2\pi f_c + \pi\gamma_p\left(t - nT\right)\\ &= \omega_c + \pi\gamma_p\hat{t}\end{aligned} \tag{2.13}$$

The improvement in range resolution is obtained by match filtering of the returned signal with a replica of the transmit waveform, given by [1]

$$s\left(n,t\right) = \mathrm{rect}\left(\frac{\hat{t} - 2R/c}{T}\right) e^{-j\frac{4\pi\left(f_c + \gamma\hat{t}\right)}{c}\left(R_t - R_o\right)} * e^{j\frac{4\pi\gamma}{c^2}\left(R_t^2 - R_o^2\right)} \tag{2.14}$$

Figure 2.5 illustrates the processing of an LFM waveform. A measure of the match filter effectiveness is the ambiguity function (AF). The AF shown provides an important aspect of LFM for surveillance imaging. For a resolution cell in the image, the range resolution can be obtained with the desired fine range resolution. If the target or platform is moving, the peak of the match-filtered (or pulse-compressed) signal will be displaced in the Doppler dimension of τ/T.

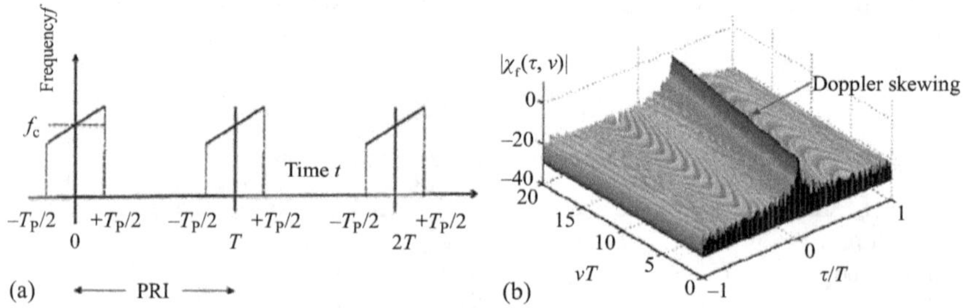

Figure 2.5 LFM waveform [2]. (a) Time waveform and (b) Ambiguity function

One of the fortunate characteristics of the LFM AF is the cross-coupling of range and Doppler, which makes the signal processing efficient in many applications. Other modulation techniques on the pulse, such as binary or m-ary phase coding, require significant increase in signal processing, essentially a different matched filter for each Doppler cell in the scene [2].

The AF is important in quantifying the ability of a matched filter to detect a radar signal. The function is expressed by $\chi(\tau,\nu)$

$$|\chi(\tau,\nu)| = \left| \int_{-\infty}^{\infty} u(t)\ u^*(t-\tau)\ \mathrm{e}^{2\pi\nu t}\, dt \right| \tag{2.15}$$

This function is the Fourier transform of a function u with a time-delayed version of itself. The maximum of the AF occurs at the origin $\chi(0,0)$ where the delay Doppler of the signal is matched. It should be noted that this assumes an infinite duration signal, which does not occur in radar systems.

When the radar waveform is periodic, it will repeat at integers of the PRI T [3]:

$$u(t) = u(t + nT) \tag{2.16}$$

For a pulsed signal, the equivalent function is the periodic AF $\chi_{NT}(\tau,\nu)$

$$|\chi_{NT}(\tau,\nu)| = \left| \frac{1}{NT} \int_{0}^{NT} u(t-\tau)\, u^*(t)\, \mathrm{e}^{\mathrm{j}2\pi\nu t} dt \right| \tag{2.17}$$

This AF can be related to the single-period AF $\chi_{\mathrm{T}}(\tau,\nu)$ by

$$|\chi_{NT}(\tau,\nu)| = |\chi_{T}(\tau,\nu)| \left| \frac{\sin(N\pi\nu T)}{N\sin(\pi\nu T)} \right| \tag{2.18}$$

where

$$|\chi_{T}(\tau,\nu)| = \left| \frac{1}{T} \int_{0}^{T} u(t-\tau)\, u^*(t)\, \mathrm{e}^{\mathrm{j}2\pi\nu\tau} dt \right| \tag{2.19}$$

The AF for a single pulse of an FJB waveform is shown in Figure 2.6. This illustrates the signal range Doppler ridge and the range resolution and soft decorrelation of the Doppler. The slope of the LFM waveform is a function of the speed of the radar platform and the PRF. It is evident, at this low platform velocity and relatively high PRF, that the slope is low. It would be very easy to detect and demodulate a signal by any reasonable compressive receiver.

The AF of eight pulses is shown in Figure 2.7. A compressive receiver that can determine the time–bandwidth characteristics readily detects both these functions. The objective of this new work is to evaluate the FJB waveform in more detail and determine the added techniques that need to be carried out to improve the noise-like features.

The AF is an important tool to evaluate the impact of Doppler on a match filter of the multimode radar waveforms. For a single pulse or a series of LFM pulses, there is a clear cross-coupling from the time to the Doppler domain. This will be shown as beneficial for image formation in Chapter 5. It is important in solving the long-standing objective of simultaneous SAR and GMTI, which will be covered in Chapter 9.

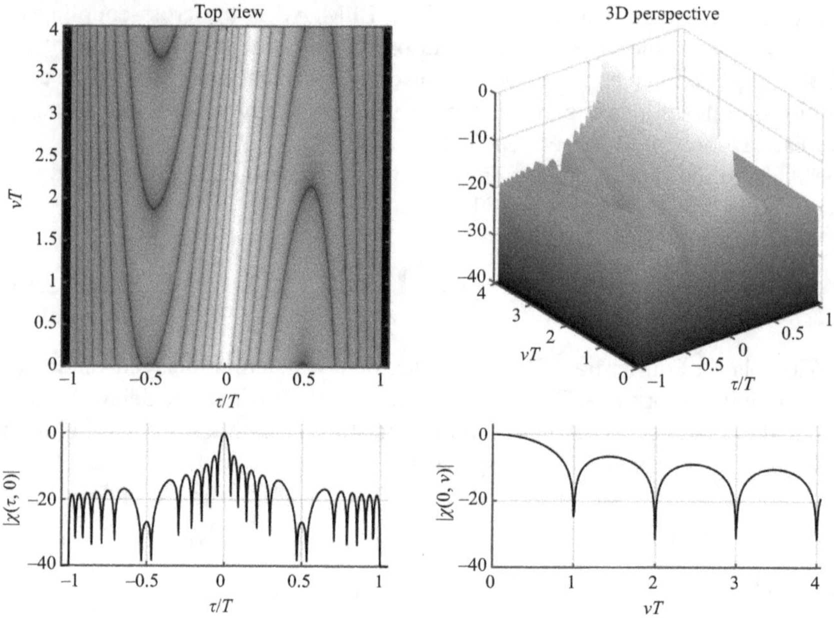

Figure 2.6 AF of a 30 MHz, 5-μs LFM subpulse [3]

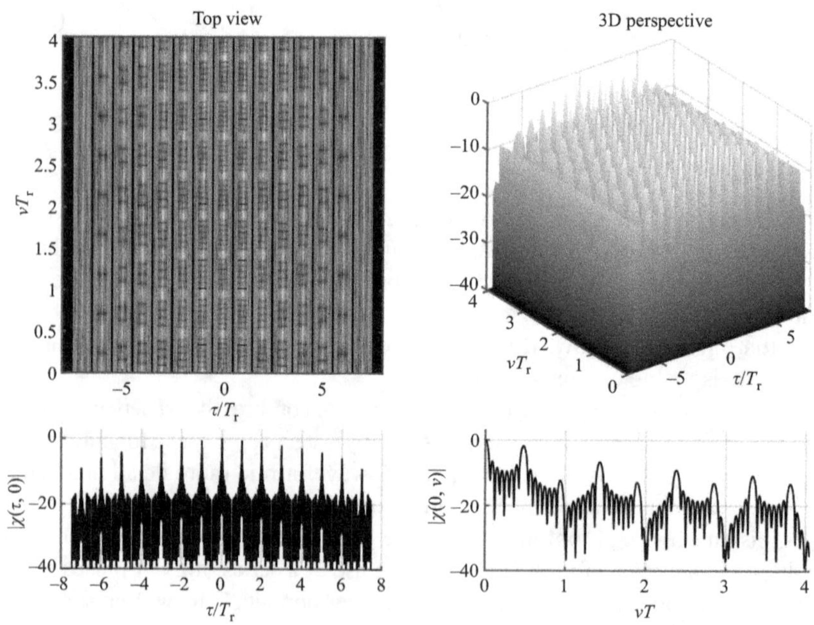

Figure 2.7 AF of eight LFM subpulses [3]

2.2 Signal to noise

The most important characterization of a surveillance radar is the received signal when compared to noise and clutter energy. An airborne radar operates at a significant distance from the region being surveyed; so, it needs to have sufficient signal strength to detect, characterize and track the object of importance. This section presents the basic factors affecting the radar's ability to meet the probability of detection and minimize the false alarm probability.

The radar measures a signal return based on system design factors and wave propagation. For a single pulse, the SNR received by the radar from a scatterer at range R is given by [4]

$$\mathrm{SNR} = \frac{P_\mathrm{T} G_\mathrm{T} A_\mathrm{R} \sigma}{(4\pi)^2 kBT_0 F_\mathrm{n} L_\mathrm{T} R^4} \tag{2.20}$$

where PI is the peak power of the transmitter; G_T the gain of the transmitter in the direction scatterer; λ the signal wavelength; A_R the area of the receiver, commonly expressed as $A_\mathrm{R} = G_\mathrm{T}\lambda^2/ 4\pi$; σ the radar cross-section of the scatterer; R the range to the scatterer; kT_0 the Boltzmann constant times the system reference temperature; B the signal bandwidth; F_n the noise figure of the receiver; and L_T i the total loss from the transmitter to the scattering cell and back to the receiver.

If there are multiple pulses N_p transmitted at a pulse repetition frequency PRF, the average power transmitted will be given as

$$P_\mathrm{ave} = P_\mathrm{T}\ \mathrm{PRF}\ \tau_\mathrm{p} n_\mathrm{p} \tag{2.21}$$

where n_p s the number of coherent pulses, PRF the pulse repetition frequency, and τ_p the uncompressed pulse width.

Furthermore, when the adar transmits a long, coded pulse and uses pulse compression to achieve improved average power, there is an integration gain approximated by the time–bandwidth product $B\tau_\mathrm{p}$. And, if multiple pulses n_p are coherently integrated, there is an additional increase in the SNR. Therefore, an improved SNR due to coherent integration and pulse compression SNR_CI can be approximated by [4]

$$\mathrm{SNR}_\mathrm{CI} = \frac{P_\mathrm{T} n_\mathrm{p} \mathrm{PRF}\, \tau_\mathrm{p} G_\mathrm{T} A_\mathrm{R} \sigma}{(4\pi)^2 kT_0 F_\mathrm{n} L_\mathrm{T} R^4} \tag{2.22}$$

It should be recognized that this equation is for a point target at range R. It is necessary to also evaluate the returns from the surface scatterers (clutter) to determine the interference in detecting any point targets. This surface return is also instrumental in measuring the characteristics of the ground for mapping or geoscience applications.

The return from distributed clutter is characterized by the backscatter coefficient σ^0 for the surface of scatterers. The radar clutter-to-noise ratio (CNR) is obtained by substituting the total radar scatter into (2.22)

$$\mathrm{CNR} = \frac{P_\mathrm{ave} G_\mathrm{T} A_\mathrm{R} \sigma^0 dA}{(4\pi)^2 kT_0 B F_\mathrm{n} L_\mathrm{c} R^4} \tag{2.23}$$

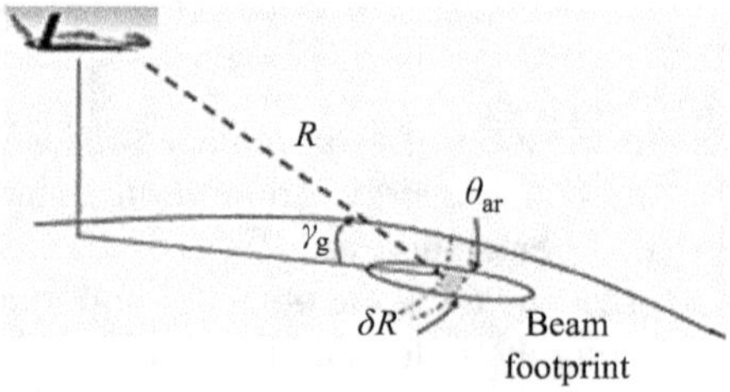

Figure 2.8 Clutter backscatter geometry

where σ^0 is the clutter backscatter coefficient, L_C the losses between radar and clutter patch and dA is the area of clutter in the resolution cell.

The area of the clutter cell is shown in Figure 2.8 and is given by

$$dA = \frac{\delta_R R \theta_{az}}{\sin \gamma_g} \tag{2.24}$$

where δ_R range resolution equals $c\tau/2$ and τ is the compressed pulse width; γ_g the grazing angle to the ground; and θ_{az} the azimuth beamwidth for the footprint on the ground.

For many applications, the cross-range resolution δ_{CR} can be enhanced by signal processing. Doppler processing is the most prevalent approach, where the Doppler cells across the beam can be used to limit the clutter content in the return. For pulse-Doppler GMTI, the clutter return is limited to a few Doppler cells, and the clutter-free region will be dominated by noise. For Doppler beam sharpening and SAR, there is a greater segregation of the target return from the distributed clutter through Doppler resolution. The ultimate goal is to have the range and Doppler resolution be finer than the size of the target. In this case, the target will be segregated from the clutter, and there will be a significant improvement in the signal to clutter. These effects will be covered in subsequent chapters.

The estimation of the signal-to-clutter ratio (SCR) is obtained by dividing (2.22) by (2.23) as

$$\mathrm{SCR} = \frac{\sigma}{\sigma^0 \delta_{CR} \delta_R \sin \gamma_g} \tag{2.25}$$

Once the SNR and CNR have been calculated, the radar design can be evaluated in terms of the standoff range to the search area. A common approach is to define a minimum SNR (SNR_{min}) needed for target detection or characterization. By specifying this factor, the maximum range (R_{max}) can be calculated [4] as

$$R_{max} = \left[\frac{P_{avg} G_T A_R \sigma}{(4\pi)^2 \, \mathrm{SNR}_{min} k T_0 F_n B L_t} \right]^{1/4} \tag{2.26}$$

Several signal-processing techniques have been developed to maximize the numerator of (2.26) and minimize the denominator. The following sections will detail the current knowledge on clutter characteristics and provide an introduction to the advanced airborne surveillance radar techniques.

2.2.1 Detection and false alarm probability

Surveillance radars require a specified probability of detection (P_d) for a target located at the desired detection range. The detection at the output of the receiver will include signal, noise and other interferences such as clutter. This requires a receiver that is a match filter to the target in terms of the waveform modulation and noise bandwidth. The noise and interference distribution need to be reduced in the receiver, such that the probability of false alarm (P_{fa}) is minimized. At this condition, the probability of detection and false alarms is determined by the probability density function (pdf) of the signal and the noise plus interference. In a pulsed radar, the ratio of single pulse signal energy to noise power per unit bandwidth is used to calculate the detection and false alarm probabilities.

Thermal noise is white in amplitude distribution and Gaussian in terms of its pdf. It is characterized by the input noise spectral density of kT_0. At the output of the receiver, the noise bandwidth is limited to $N = kT_0B$, where B is the receiver bandwidth after bandpass filtering. This is characterized as the band-limited Gaussian noise.

The noise statistics are characterized by a Gaussian pdf for the in-phase and quadrature-phase components (v_i and v_q), given by [5]

$$dP_{vi} = dP_{vq} = \frac{1}{\sqrt{2\pi kT_0B}} \exp\left(-\frac{v^2}{2kT_0B}\right) dv \tag{2.27}$$

Figure 2.9 depicts the noise spectrum with 3-dB bandwidth of $\sqrt{(kT_0B)}$, along with the signal spectrum centered at $\sqrt{2}E_s$ and having the same noise bandwidth. A threshold is shown at E_t, used in a receiver detector to limit the noise interference in the process. Given the pdf in (2.27), the detection probability P_d is obtained by integrating the target pdf dP_s from the threshold E_t to infinity ∞:

$$P_d = \int_{E_t}^{\infty} dP_s = \frac{1}{\sqrt{2\pi kT_0B}} \int_{E_t}^{\infty} \exp\left(-\frac{(v+E_s)^2}{2kT_0B}\right) dv \tag{2.28}$$

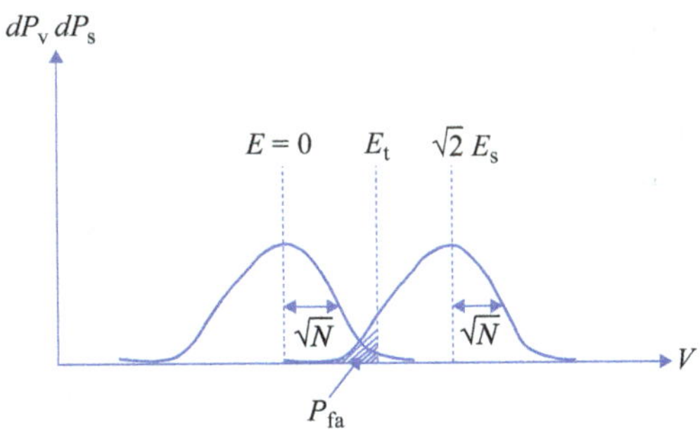

Figure 2.9 Pdf for coherent integration

It follows that the false alarm density is obtained by the overlap between the noise spectrum and the target spectrum and is given by the integral of the noise pdf dP_v from E_t to ∞:

$$P_{\mathrm{fa}} = \int_{E_t}^{\infty} dP_v = \frac{1}{\sqrt{2\pi kT_0B}} \int_{E_t}^{\infty} \exp\left(-\frac{v^2}{2kT_0B}\right) dv \tag{2.29}$$

2.2.2 *Signal to noise for detecting nonfluctuating target*

Calculating the required SNR to detect a target and minimize false alarms is a major part of designing a surveillance radar system. However, P_d and P_{fa} require integration of (2.28) and (2.29). Fortunately, for nonfluctuating targets, an approximation has been made by Albersheim, which is sufficiently accurate for estimation of radar systems with Gaussian interference. The relationship between SNR and P_d, P_{fa} is empirically derived for detecting nonfluctuating targets given the number of pulses N and is given by [6]

$$\mathrm{SNR}\,(\mathrm{dB}) = -5\log_{10}N + \left(6.2 + \frac{4.54}{\sqrt{N+0.44}}\right)\log_{10}\left(A + 0.12AB + 1.7B\right) \tag{2.30}$$

where

$$A = \ln\left(\frac{0.62}{P_{\mathrm{fa}}}\right) \tag{2.31}$$

and

$$B = \ln\left(\frac{P_{\mathrm{d}}}{1-P_{\mathrm{d}}}\right) \tag{2.32}$$

2.3 Clutter characteristics

Clutter is historically defined as scattering regions around a moving or stationary target within the resolution cells, which limits the detection probability of desired objects. These objects are typically point targets or groups of point targets that are associated with an "objective target." However, in many applications, the task for the radar is to characterize the clutter for scientific or model enhancement objectives. As such, the radar range equations in the previous sections will vary from signal-to-noise to clutter-to-noise approaches.

Clutter is often due to the surface being illuminated, referred to as spatial clutter. However, if the clutter is distributed between the radar and the objective target, then it can be characterized as volumetric clutter. The most common volumetric clutter is weather or chaff, which will attenuate and scatter the radar signal. In the UWB radar community, a common application is detecting objects under forests. In this case, the characteristics of clutter need to be quantified as it will obscure the objective target and directly affect the detection probability.

The following sections will provide some analytics developed to design UWB surveillance systems and factor in the clutter backscatter coefficient (σ^0) in the radar range equations.

2.3.1 Radar surface clutter return

Radar returns due to distributed natural and man-made cultural returns are commonly called clutter. For GMTI operation, these returns mask the moving targets and have to be removed from the data. For SAR operation, these returns are stationary returns that are important for characterizing any object or target in the surveillance images.

There are several clutter models that have been created over the past 40 years. They typically originate for a particular radar design and have resulted from parametric fitting of clutter reflectivity data to a model. For air to ground radar designs, there are three that are relevant to this book. The first is the Barton model, which has a physics-based analysis of the propagation of radar waves from the airborne radar to the Earth's surface and back [7]. The second model was started by the US Air Force when an airborne radar was being considered for detecting and tracking small targets at low grazing angles over land [8]. And the third model was developed by professors from Kansas University and University of Michigan, which proposed a five-parameter model at several frequencies [9]. The later model is generally called the Ulaby–Dobson model, although other variants have been developed.

These three principal models will be discussed below along with an analytic curve of the clutter backscatter coefficient at UHF, L-band and X-band frequencies.

Barton clutter model

The most common model for clutter is the constant-gamma model, where the clutter backscatter coefficient σ^0 is given by

$$\sigma^0 = \gamma \sin \gamma_g \tag{2.33}$$

where γ is the normalized reflectivity parameter and γ_g the grazing angle from the radar to the ground surface. The value of gamma is defined for a frequency and clutter type, e.g., farm lands, wooded hills, urban areas and forests. For farmlands, $\gamma = -15$ dB, whereas for forests, $\gamma = -10$ dB. However, the models for surface scattering over several classes of terrain and broad frequency ranges have been greatly expanded over the past couple of decades. This is especially true for airborne surveillance radars with heterogeneous terrains [7].

The constant gamma model breaks down into two regions: low grazing angles of $\phi_c < 10°$, and near incidence angles of $\phi_D > 60°$. This is illustrated in Figure 2.10, where the constant gamma model is located in the plateau region for $10° < \phi < 60°$. In the low-grazing-angle region, there is significant shadowing of the terrain, along with diffraction of the radar energy over ridges and undulating terrain. On the other hand, near-vertical incidence exhibits significant direct bounce of the radar energy off of the surface under the airborne radar. Surveillance radar systems required knowledge of all three regions and several classes of terrain scattering to correctly predict the radar's ability to detect targets.

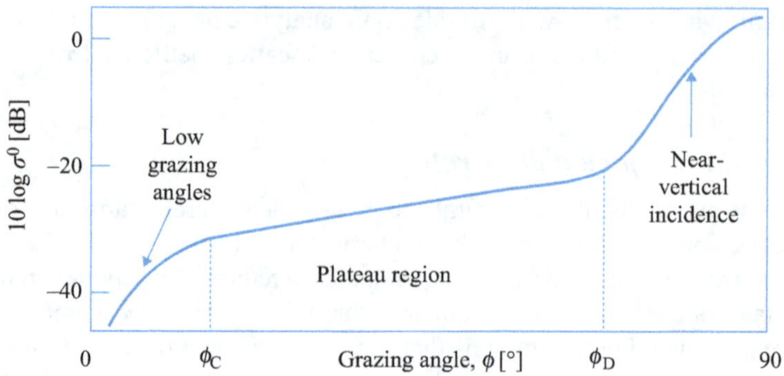

Figure 2.10 Regions of clutter backscatter coefficient [7]

Terrain clutter is the most complicated to model with several factors. First, terrain has both volume scatterers and surface scatterers. Second, there are a wide variety of "terrain types" and vegetation, each of which can provide significant variations in backscatter seasonally or even diurnally. Third, many terrain regions contain an increasing number of man-made scatterers or man-made modifications to the terrain and vegetation, which dominate the scatter statistics. Fourth, the characterization of clutter from a "terrain type" is complicated by its dependence on grazing angle, polarization, radar frequency, wind and moisture conditions, radar resolution cell size and even aspect. Finally, the roughness of the terrain prevents accurate knowledge of the power density on the surface and vegetation, especially at low grazing angles.

The propagation factor F_c for paths between the radar antenna and the clutter source on an undulating random Earth can be written, or low grazing angles, as

$$F_c \approx \frac{\psi}{\psi_c} = \psi \frac{4\pi\sigma_h}{\lambda} \approx \frac{R_1}{R} \quad (\psi < \psi_c) \tag{2.34}$$

where σ_h is the rms surface height deviation from the average height, R isthe range, R_1 the range at which the critical grazing angle ψ_c is reached, and h the height of the platform above the surface.

$$R_1 = \frac{4\pi\sigma_h h}{\lambda} \tag{2.35}$$

The reflectivity of land clutter is difficult to characterize because of the variability of structures and clutter types in the region. The constant-γ model applies to grazing angles above a critical angle ψ_c, with values of γ shown in Table 2.2. These values are almost always independent of wavelength.

The horizon range R_h depends on the radar antenna altitude h_r above the sea level, and the effective Earth's radius ka. This is estimated by

$$R_h \approx \sqrt{2kah_r} \tag{2.36}$$

where $a = 6{,}378$ km and $k = 4/3$. However, the radar horizon will depend on the local terrain altitude and height of any mountains within the radar field of view.

Table 2.2 Land surface parameters [7]

Surface description	**Mean reflectivity γ**	**RMS height deviation**	**Slope β_0**
Mountains	−5	100	0.1
Urban	−5	10	0.1
Wooded hills	−10	10	0.05
Rolling hills	−12	10	0.05
Farmland, desert	−15	3	0.03
Flatland	−20	1	0.02
Smooth surface	−25	0.3	0.01

There are three distinct clutter regions in range from the radar, depending on the propagation mode. The constant-γ model is the plateau region with $R_1 < R < R_h$. For ranges with $R < R_1$, the free space propagation holds to the surface with $F_c \approx 1$. However, the contribution of multiple scatterers in a resolution cell gives rise to interference and fading of the clutter. The critical grazing angle ψ_c is seen with the breaks in the reflectivity. And beyond R_h the effects of propagation over variation in the terrain height and knife edge effects contribute to the backscatter coefficient. The only consistent model exists in the constant-γ regime, where free space propagation is valid.

The terrain types required for the preliminary evaluation of early airborne surveillance radar system are mountains, rural, urban and farmland. However, "rural" and "urban" are vague definitions of "terrain." In this section, mountains, forest and farmland will be addressed. "Urban terrain" can then be constructed by the superposition of a high density of man-made structures on a low relief terrain such as farmland. Likewise, "rural" terrain can be the superposition of a low density of man-made structures in forest, farmland or mountains.

Air Force clutter model

The Air Force clutter model analysis was started for airborne surveillance radars to detect small targets in dense clutter. This was particularly important at low grazing angles over mountainous and forested regions. The three types of distributed terrain are modeled: farmland, forests and mountains. The expected value of the mean reflectivity from each type varies as a function of grazing angle with a Gaussian shape near vertical incidence and a constant gamma model in the plateau region and lower angles [8].

The near-incidence factor $C(\phi_i)$ can be given by

$$C(\phi_i) = 10\log\left(\frac{C_1}{\lambda}\right)\frac{1}{\sqrt{2\pi s_c}}\exp\left(\frac{\phi_i}{2\sigma_c^2}\right) \tag{2.37}$$

where $\phi_i = 90 - \phi_g$; $\sigma_c = 10$, scale factor C_1 is 50 for farmland and woodland and 10 for mountainous regions, and λ is the wavelength in meters. The proposed constant gamma values are 0.2, 0.1 and 0.032 for mountains, forest and farmland, respectively.

The variation of the mean reflectivity about the expected value is defined to be the gamma function, resulting in a K distribution for the instantaneous reflectivity. The shape parameter k is a function of both resolution cell size and grazing angle and is given by

$$k \cong \left(\frac{R\theta_a}{D_r}\right)\frac{c\tau}{2D_r}\gamma_g^2 \tag{2.38}$$

for farmland and forest, and

$$k \cong \left(\frac{R\theta_a}{D_r}\right)\frac{c\tau}{2D_r}(\phi_d - \phi_h)^2 \tag{2.39}$$

for mountains, where ϕ_d is the depression angle, γ_g the grazing angle and ϕ_h is the depression angle to the horizon.

The observed correlation lengths in range and cross-range can be obtained by convolving the matched filter length in each dimension with the function given in Figure 2.11. For the ARSR-4 radar collection system for these data, the matched filter range and azimuth resolutions were comparable or exceeded D_r. Therefore, a simple shape for the covariance function was used. For air defense radar waveforms with bandwidths less than 1 or 2 MHz and short coherent integration times, the correlation length will be approximately D_r for most terrains. As the system resolution increases, the effective correlation length will approach the width of the narrow peak.

For air defense search waveforms, k has a value much greater than 1. This leads to a narrow density function for $p(\sigma_o)$ about σ_g and the observed instantaneous density function $p(\sigma_o)$ is exponential. The use of a high resolution waveform such as SAR for some special purpose such as target classification may lead to nonexponential distributions at angles up to 20°.

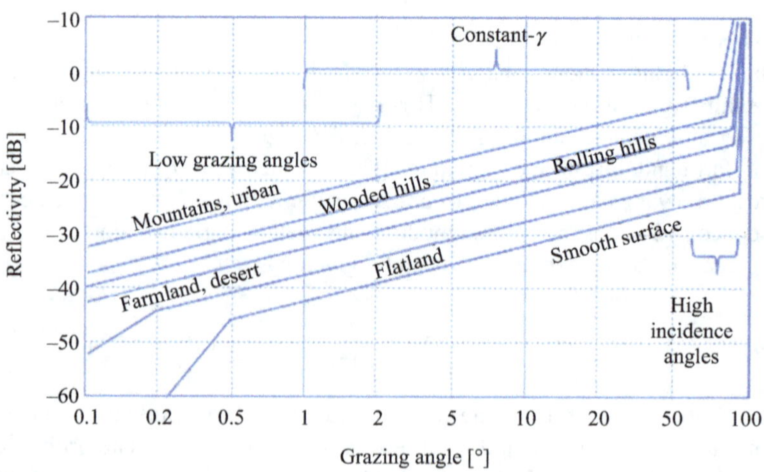

Figure 2.11 Constant gamma clutter model for six clutter types [7]

The σ_g performance varies significantly with grazing angle. As has been experienced with sea clutter, the range of possible grazing angles can be divided into three reasonably distinct regions: near-vertical incidence, a plateau region, and near-grazing incidence. Correspondingly, the value of σ_g is related to grazing angle by

$$\sigma_g = B(\gamma_g) + C(\gamma_g) \tag{2.40}$$

where $B(\gamma_g)$ is the plateau factor and $C(\gamma_g)$ is the near-incidence factor.

The most important region for an airborne system is the plateau region with grazing angles ranging from 10 to 60°. The return in the plateau region is often reasonably approximated by a constant gamma model $B(\gamma_g) = 7\ \sin(\gamma_g)$. The recommended values for the ADI model are 0.2 −7 dB) for mountains, 0.1 (−10 dB) for forest, and 0.032 (−15 dB) for farmland. In this region, there is still very little shadowing and the value for k as given in (2.39) will be equal to or greater than 10 for typical ADI search waveforms.

Clutter from near vertical incidence (grazing angles greater than 60°) is important only for ambiguous waveforms. The smoother terrain provides a higher backscatter than rougher terrain, and the S-band reflectivity is generally higher than the UHF data. The variation of σ_g with angle can be reasonably approximated by a Gaussian shape and the near-incidence factor $C(\phi_g)$ can be given by

$$c(\phi_g) = 10\log\left[\left(\frac{c_1}{\lambda}\right)\frac{1}{\sqrt{2\pi}s_c} * \exp(-(w90 - \gamma_{g)}/(2s_c^2)\right] \tag{2.41}$$

where σ_c =10, scale factor C_1 is 50 for farmland and woodland, and 10 for mountainous regions and λ is the wavelength in meters. In this region, there is essentially no shadowing and for typical ADI waveforms, k assumes large values much greater than 10. As for the plateau region, this leads to a narrow density function for $p(\sigma_0)$ about σ_g, and the observed instantaneous density function $p(\sigma_0)$ is exponential.

It is instrumental to have an analytic model for clutter in order to design and characterize long-range surveillance systems. There are several instances of models that are efficient for designing multichannel surveillance systems. One of those models is the Air Force five-parameter model that was developed in the 1990s, based on a wide range of data collections. One instance of this was the clutter model developed by Technology Services Corporation and incorporated in the Rome Laboratory Space Time Adaptive Processing (RLSTAP) analysis tool. Many versions of similar parametric tools have been published.

The model, similar to the Barton model above, is developed by fitting the clutter returns from many sites and clutter types, with three elements: plateau region, constant gamma and exponential incident angle [8]:

$$\sigma^0 = C1 + \text{GCA} * \sin\psi + C2\ \exp^{-\left[C3\left(\frac{\pi}{2}-\psi\right)\right]^{C4}} \tag{2.42}$$

where $C1$ is the plateau region of the backscatter coefficient, $C2$ the exponential multiplier, $C3$ the incidence angle multiplier, $C4$ exponent of the high incident angle factor and GCA gamma model coefficient.

Table 2.3 Air Force five-parameter clutter model (UHF) [8]

Clutter type	Plateau	Gamma coefficient σc	Exponential multiplier	Incidence angle multiplier	Exponent
Desert/roads	0.435E-05	0.2E-3	5.06	30.54	2
Range land	0.128E-3	0.631E-3	1.59	15.28	2
Agriculture	0.12E-3	0.6E-3	1.5	15.00	2
Forest	0.21E-2	0.631E-2	0.77	15.37	2
Urban	0.1E-1	0.251	2.82	19.09	2

The parameters at UHF from RLSTAP are given in Table 2.3. The plot of σ is given in Figure 2.12. Note that the range land and agriculture has little difference in shape.

Ulaby–Dobson clutter model

A second clutter model was explored for the surveillance radar community by adapting the Ulaby–Dobson clutter models for L-band, S-band and X-band. There were extensive data collections in the 1974 timeframe at the University of Kansas [9]. These collections entailed radiometric measurements of several clutter regions, over four seasons and with both horizontal and vertical polarizations. This chapter summarizes the three frequency data for the backscatter characteristics. The models were incorporated in the Air Force Research Laboratory RLSTAP and has been used for many radar system design evaluations [11]. The Ulaby–Dobson model is given by [9]

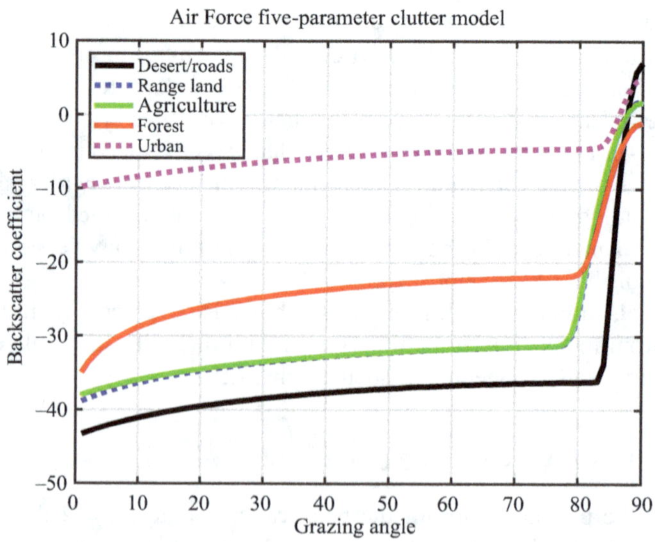

Figure 2.12 Air Force five-parameter UHF clutter backscatter coefficient [10]

Table 2.4 Ulaby–Dobson L-band clutter model coefficients [10]

L-band	Desert/roads	Range land	Agriculture	Forest	Urban
P1	–191.9	–127.1	–73.5	–95.8	–49.95
P2	0.131	15.61	23.67	27.95	27.41
P3	–4.112	–1.486	–0.896	–0.857	–0.386
P4	175.99	66.254	19.561	38.407	34.868
P5	0.855	1.469	–1.509	1.53	–0.049
P6	–0.653	–0.454	–0.022	–0.163	–1.427

$$\sigma^0_{db} = P1 + P2*e^{-P3\,\psi} + P4 * \cos(P5 * \psi + P6) \tag{2.43}$$

There are six parameters corresponding to regression of data over five clutter types, four seasons and two polarization. For this effort, we are only treating Spring Horizontal. The coefficients are all in decibels and have been used in (2.41).

For L-band, the coefficients are summarized in Table 2.4, and their corresponding clutter backscatter characteristics are given in Figure 2.13. These results exhibit similar characteristics in the three regions when compared to UHF in Figure 2.12.

For S-band, the Ulaby–Dobson coefficients are listed in Table 2.5, and their, clutter backscatter characteristics are given in Figure 2.14. There are only modest differences between the L-band and S-band clutter backscatter characteristics, predominantly in the range land and agriculture types. Finally, the model was applied to X-band clutter data, and the coefficients are given in Table 2.6, with the backscatter

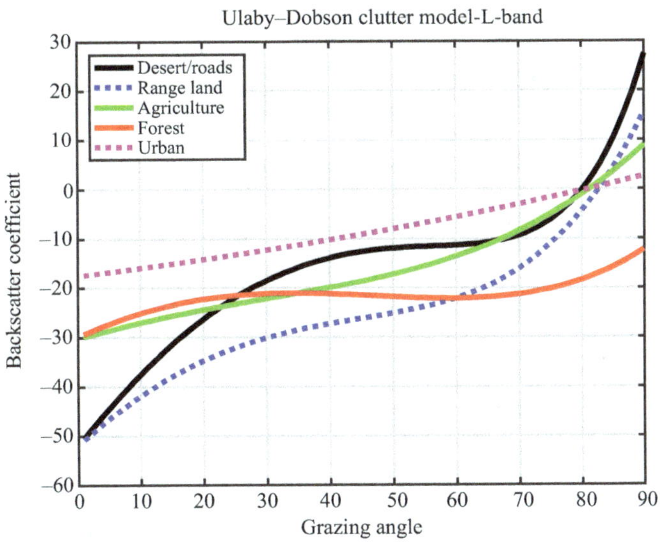

Figure 2.13 L-band Ulaby–Dobson backscatter coefficient [11]

Table 2.5 Ulaby–Dobson S-band clutter model coefficients [10]

L-Band	Desert/roads	Range land	Agriculture	Forest	Urban
P1	–135.17	–56.9	–52.394	–96.04	–44.261
P2	7.81e-04	3.235	24.784	28.118	57.393
P3	–7.033	–2.093	–0.669	–0.836	0.994
P4	123.734	24.602	11.434	38.33	92.24
P5	0.722	2.479	–0.666	1.5	0.548
P6	–0.738	–1.162	–1.782	–0.216	–1.94

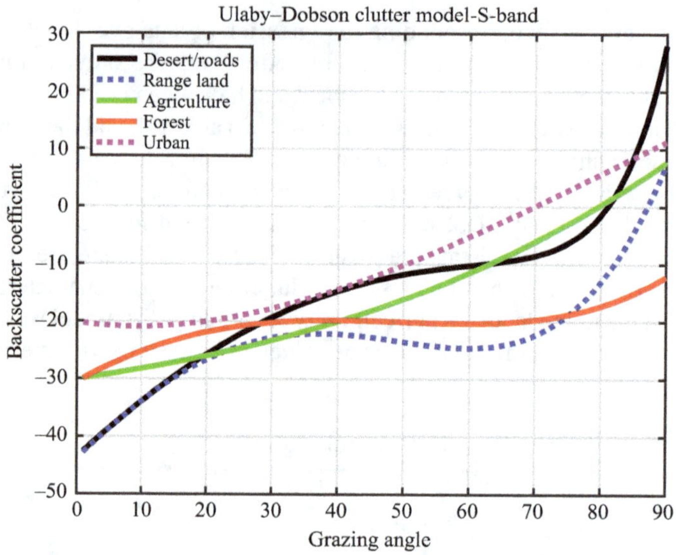

Figure 2.14 S-band Ulaby–Dobson backscatter coefficients [11]

Table 2.6 Ulaby–Dobson X-band clutter model coefficients [10]

L-Band	Desert/roads	Range land	Agriculture	Forest	Urban
P1	–71.248	–106.565	–65.021	–150.63	–185.98
P2	12.783	17.679	–11.306	63.671	7.069
P3	–1.435	–1.116	–0.875	–0.566	–1.864
P4	29.326	59.345	83.325	55.941	169.639
P5	–2.151	1.194	0.17	1.239	0.98
P6	0.689	–0.436	–1.002	–0.083	–0.351

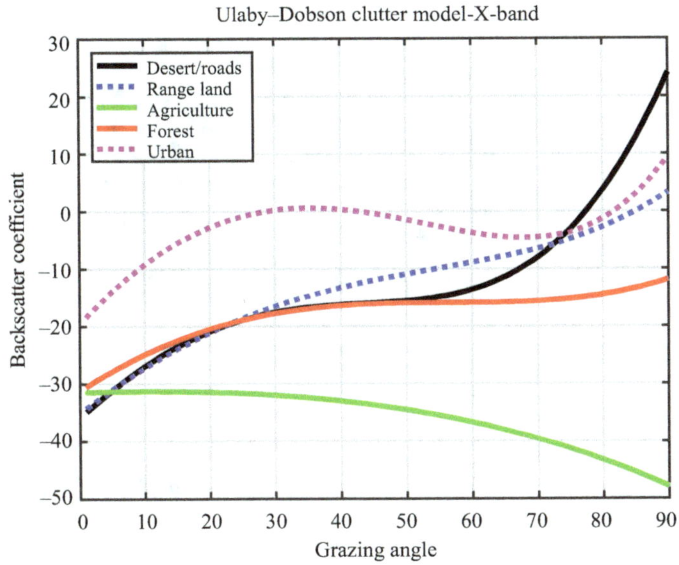

Figure 2.15 X-band Ulaby–Dobson backscatter coefficient [11]

plots given in Figure 2.15. These three models have been used extensively for airborne fire-control and surveillance radars.

Radar modeling of scenarios

The benefit of having these clutter models as a function of clutter type and frequency have been utilized by many radar deveopments for detailed modeling of multiple channel radar systems. These results will be shown in later chapters. However, it is instrumental to show how these models are used. As an example, Figure 2.16 shows an area in upstate New York that has rich terrain features and altitude changes. Such details are important in evaluating radar returns where the clutter cross-section has significant changes in radar cross-section returns. There are important repositories of the Earth's surface available from several geographic sources. The most prevalent one is the US Geographical Service (USGS) database of terrain height (a digital elevation terrain data or DTED) and land-use land cover (LULC). Figure 2.17 shows the local terrain height over the same area, and Figure 2.18 shows the LULC over the same area.

For a detailed modeling of radar performance ifor potential operation in specific geographic regions, the backscatter characteristics are used in a radar model and simulation from the range, grazing angle, bandwidth and power details in a candidate radar. These models will be used in Chapters 7 and 8 where modeling of multiple modes are illustrated for consideration in new radar applications. The modes include multiple channel GMTI for STAP, along-track interferometry (ATI) for high-resolution models of slow-moving vehicles, and a common mode for both

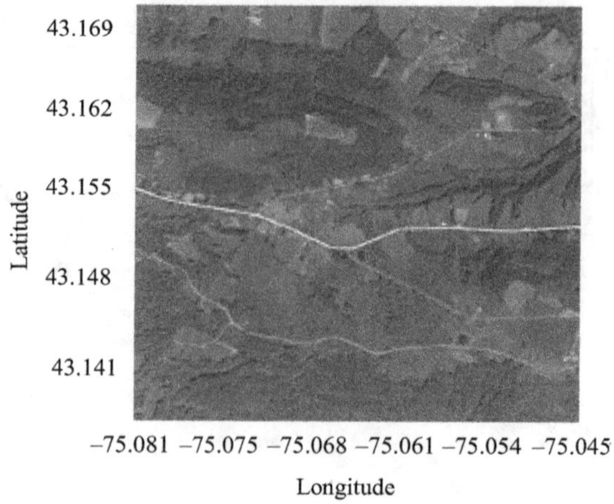

Figure 2.16 Google Earth image of Newport, NY [11]

SAR and GMTI using the FJB waveform. Because of the dynamics of both the radar platform and multiple targets on the ground, a detailed time base simulation is used.

2.3.2 Internal clutter motion

Internal clutter motion (ICM) is a problem for both SAR and GMTI when the clutter is moving due to wind or other mechanisms, such as rotating windmills and motion of power lines. In the case of SAR, any ICM is manifest as a defocusing or

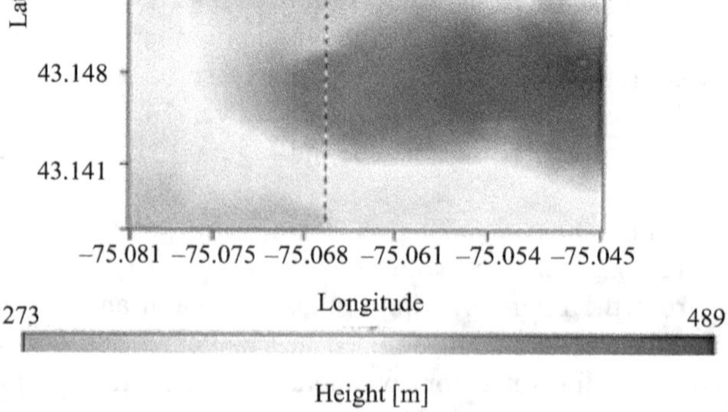

Figure 2.17 USGS terrain height data over Newport, NY [11]

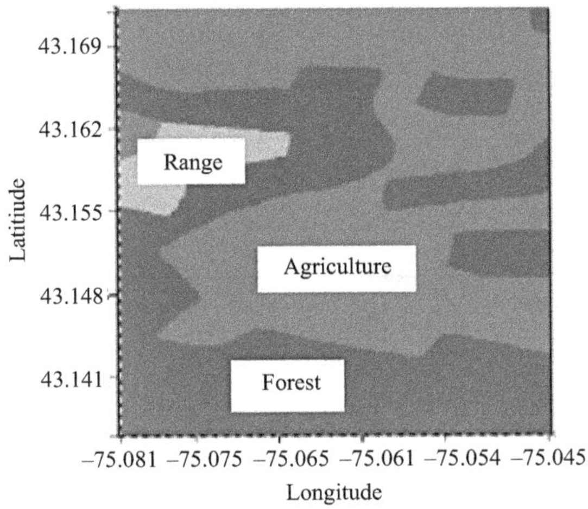

Figure 2.18 LULC data over Newport, NY [11]

smearing of the image and often obscures an object in the region of the tree line. For GMTI radar, the ICM is a Doppler effect that broadens the clutter line and affects the minimum detectable velocity (MDV). It is important to understand the effects of windblown clutter, especially tree lines and forests, on the radar system design.

An extensive collection of radar low-angle propagation data was obtained in the late 1980s from VHF to Ku-band. These data are well documented by the Billingsley model and published in a significant treatise on the cause of the internal motion and physics of the scattering. Billingsley showed that a reasonable model for the wind-blown clutter spectrum pdf $P(V_c)$ would be a delta-function DC-term plus an exponential AC term [12]:

$$P(V_c) = \frac{r}{1+r}\delta(V_c) + \frac{1}{1+r}\frac{\beta}{2}e^{(-\beta|V_c|)} \tag{2.44}$$

where V_c is the velocity of clutter Doppler in meters per second, r the ratio of the DC to AC power, β the shape parameter of the exponential spectrum and $\delta(V_c)$ the Dirac delta function centered a zero Doppler.

Based on extensive data with variation in terrain type and seasonal variations, Billingsley generated empirical models for r and β. The DC/AC ratio model is given as [12]

$$r_{dB} = 632 - 121\log_{10}(f_{MHz}) - 155\log_{10}(w_{mph}) \tag{2.45}$$

where f_{MHz} is the radar frequency in megahertz and w_{mph} the wind speed in miles per hour.

The exponential shape parameter β depends on the wind speed and is given in Table 2.7 for conditions from light air to gale force winds.

The model has found wide applications in radar even at higher grazing angles, as long as the parameters of the model can be appropriately modified for the grazing

Table 2.7 Billingsley exponential shape parameter beta versus wind condition

Wind condition (m/s)	β (s/m)
Light air (0.4–3.1)	12
Breezy (3.1–6.7)	8
Windy (6.7–13.4)	5.7
Gale force (>13.4)	4.3

Source: MIT Lincoln Laboratory [12].

angle. Since the DC term in (3.35) is not a Dirac delta function in nature, but exhibits a finite Doppler spread under 0.25 m/s, it is commonly referred to as a quasi-DC term.

To characterize ICM at foliage penetration (FOPEN) frequencies, a limited set of data collections were taken at the Redstone Arsenal in Huntsville, AL, from a 100-m high tower. The data were collected with both horizontal and vertical polarizations against a tree line and under varying meteorological conditions. As part of the calibration of environmental conditions, the winds were measured above the tree line to determine the effects on the foliage ICM at varying wind conditions [12].

The clutter spread for light wind conditions (i.e., 5.5 mph wind velocity) is shown in Figure 2.19 and is examined in homogeneous forest clutter at 19.4° and 22.3° incidence angles. The exponential spreading factor β for clutter on each side of the quasi-DC term is comparable with the model in (2.43). However, in both the cases, the Billingsley model overestimates the DC/AC factor.

The heuristic explanation of the performance is based on the penetration of the radar scattering into the tops of the forest. The large stationary scattering is representative of the tree trunks and large branches, which are somewhat sheltered from the wind. However, the large branches in the crown of the tree will have some modest motion, giving rise to the departure from the quasi-DC function of 0.25 m/s. The ratio of AC/DC returns is attributed to the amount of biomass that is contributing to

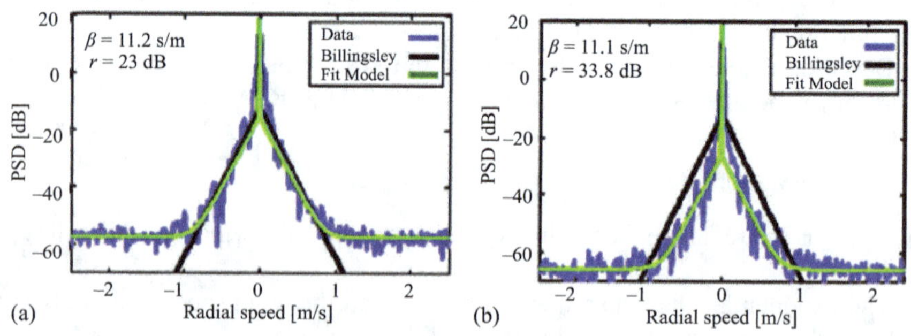

Figure 2.19 Forest clutter measured spectral spread for 5-m/s wind velocity [12]. (a) 19.4° incident angle and (b) 22.3° incident angle

the higher Doppler spectrum. For light wind, the AC component will be low, since the motion is from smaller branches close to the surface of the forest. At higher wind speed, the motion will be transferred to the lower parts of the tree where more attenuation and scattering will take place.

The ratio of DC to AC power in the FOPEN ICM was found to depend on the incidence angle to the tops of the forest. Data across a variation in incidence angles from 12 to 30° were collected. Based on a curve fit of these data, a simple empirical model of the DC/AC ratio dependence on incidence angle is given by [12]

$$r_{\mathrm{dB}} = r_{\mathrm{dB}}^{\mathrm{Billingsley}} + \max(22.7 - 5.8 \csc \psi_{\mathrm{I}}, 0) \tag{2.46}$$

where ψ_{I} is the incidence angle from the radar to the foliage. The ratio can then be examined as a function of the incidence angle, with wind speed as shown in Figure 2.20.

The preceding discussion of ICM should be examined in terms of the limitations of the data collected and the model of (2.44). First, the forests were from a single geographic location. Sections 3.4 and 3.5 illustrated the wide variation in loss and clutter return between the single and double canopy forests. The model also assumed that the loss in the foliage followed the csc θ model from (2.44). Again, this was derived from penetration in large stands of homogeneous trees. At tree lines or with spatial variations in forest density, an enhanced scattering from lower branches and trunks can be anticipated. More importantly, the wind velocity will affect the motion of tree branches and limbs farther from the tops. These mechanisms will affect the absolute spectrum and the ratio of DC/AC as a function of wind. The advantage of an empirical curve is when it is employed in early system analysis. For GMTI radar

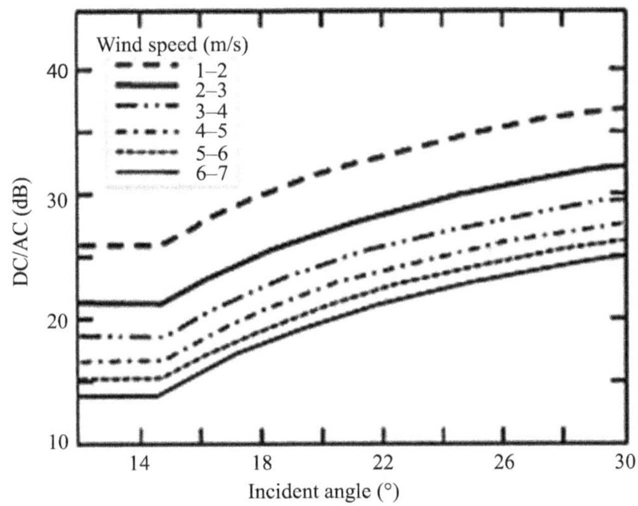

Figure 2.20 Empirical model of UHF internal clutter spread DC/AC ratio© 2006 IEEE [12]

systems, it is known that ICM will affect the Doppler processing and the MDV of any slow-moving targets in the clutter resolution cell.

2.3.3 Radar foliage volumetric clutter return

Volumetric clutter occupies the space around a target, as opposed to the surface clutter lying behind the target. As a result, there are two factors to be considered. First, the clutter will be prevalent to the radar waveform and compete with the signal reflected from the target. Second, the foliage will attenuate the radar waveform, reducing the measurement of the target. There had been little research and understanding of the physics of the foliage volumetric clutter.

There were a few SAR data collections that provided comparison of collections on the same terrain and objects. One of those collections is illustrated here. The Defense Advanced Research Projects Agency flew several systems over an area of Camp Roberts California to evaluate both foliage penetration and digital elevation map (DEM) interferometric SAR image generation. These collections were carried out at X-band and P-band to compare the characteristics of the DEM generation as well as the clutter and propagation losses.

Figure 2.21 illustrates the SAR images obtained in a side-by-side collection of X-band and UHF SAR images. The X-band image was obtained by the IFSARE system built by the Environmental Research Institute of Michigan (ERIM). This radar was developed to provide high ACR DEM with fine elevation accuracy. The illustration is a segment of the SAR image used to measure DEM for a large area of central California. It was collected the same day that several military vehicles had been placed under the foliage area known as "Sherwood Forest." It is apparent that only the tops of the trees were visible in these images [13].

The UHF image was collected with the UWB P3 FOPEN SAR, also built by ERIM. The three tactical targets under the foliage were revealed. However, it is clear that false alarm rates would be very high if only the horizontal polarization

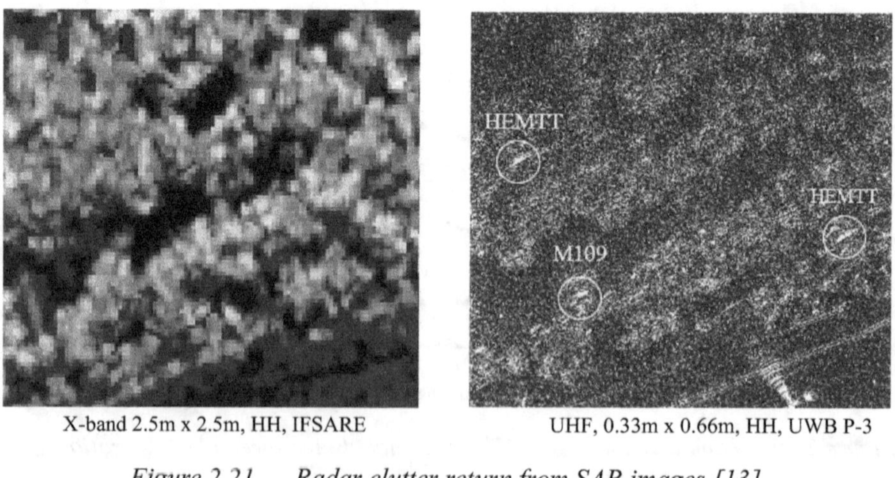

X-band 2.5m x 2.5m, HH, IFSARE

UHF, 0.33m x 0.66m, HH, UWB P-3

Figure 2.21 Radar clutter return from SAR images [13]

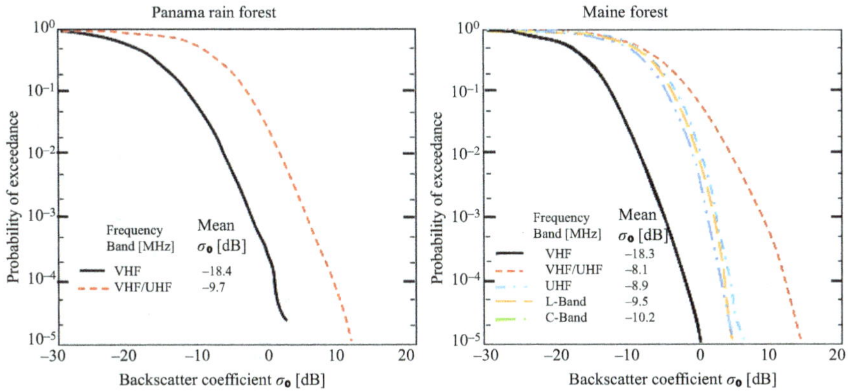

Figure 2.22 Clutter backscatter characteristics in different forest types [14]

image were to be used. Note that the strong return in the foreground was one of the calibration trihedrals used to calibrate the images for sensitivity at the multiple polarizations. Clutter backscatter is important in calculating the clutter-to-noise characteristics of the surveillance area as well as determining the signal-to-clutter characteristics of an object in the scene.

The curves shown in Figure 2.22 were conducted for determining the backscatter from the forest scene, a most difficult aspect of finding targets near or under forested areas. The data were collected from VHF to C-band, using the NASA JPL AirSAR, Navy UWB P-3 and Swedish FOA Carabas I platforms. There are three important aspects of the scattering:

- The tops of the forest are the predominant scattering for L-band and higher frequencies. This is because the propagation does not fully penetrate the trees.
- At VHF and UHF frequencies, there is a significant penetration of the forest, and the backscatter is due to the ground trunk interaction.
- At VHF, the ground trunk interaction is significantly less. In fact, the backscatter at these long wavelengths is approximately 10 dB below the UHF to microwave regions.

2.3.4 Sea clutter backscatter

The clutter return from large bodies of water is significantly more complex than that of land. This is due to the motion of the waves in water and the effects of wind over the surface [15]. The detailed treatment of these effects is beyond the scope of the UWB surveillance radar clutter models in this book. However, it is important to have a clutter model of water that can be used in surveillance over wide areas. These areas can include rivers, ponds and, to some extent, larger bodies of water.

When the sea surface reflectivity is averaged over polarization and wind direction, the clutter reflection coefficient is approximately given in terms of sea state, the

Table 2.8 Sea state clutter performance parameters for Sea States 0–5 [5]

Sea state	Wind scale K_B	Wind velocity v_w (m/s)	rms height deviation σ_s (m)	Slope β_0 (m)
0	1	1.5	0.01	0.055
1	2	2.6	0.03	0.063
2	3	4.6	0.10	0.073
3	4	6.7	0.24	0.080
4	5	8.2	0.38	0.085
5	6	10.8	0.57	0.091

Beaufort wind scale and the values of the local wind root mean square height deviation and slope. The following approximations give the surface parameters:

$$v_w \approx 1.8K_B + [0.193\,(K_B - 1)]^6 \tag{2.47}$$

$$\sigma_h \approx [0.158\,(K_B - 1)]^2 + [0.11\,(K_B - 1)]^6 + [0.13\,(K_B - 1)]^{12} \tag{2.48}$$

$$\beta_0 \approx 0.062 + 0.007K_B \tag{2.49}$$

Using the surface reflectivity model in (2.45), the constant gamma factor is given by

$$10 \log \gamma = (6\,SS - 10 \log \lambda - 58) = 6\,K_B - 10 \log \lambda - 64 \tag{2.50}$$

The sea surface parameters are given in Table 2.8.

The sea clutter backscatter coefficients are evaluated using (2.48) in Figure 2.23 for four frequencies: UHF 450 MHz, L-band 1.4 GHz, X-band 9.5 GHz, and Ku-band 16.0 GHz. Two cases are run: Sea State 1 in Figure 2.23a and Sea State 5 in Figure 2.23b. The cases were evaluated for grazing angles from 2° to 80°. When grazing angles are below 2°, the model is suspect due to the shadowing of the sea surface in higher sea states. And when the grazing angles are above 60° (or high incidence angles), the model is similarly suspect due to the variability of the sea surface.

It is clear that the Sea State 1 is significantly lower than Sea State 5. Sea State 1 represents a glistening case where the radar energy has little or no backscatter mechanism. And the higher sea states have strong backscatter due to the wave tops and a periodic scattering toward the radar. Sea States 0 and 1 are representative of calm rivers and ponds in most surveillance applications over large land masses.

2.4 Target characteristics

Man-made targets are composed of many scattering points over the volume of a vehicle, building or lines of wires (e.g., power lines, communications lines and towers). From a basic surveillance radar view, the radar returns depend on the range and azimuthal returns. Early surveillance radar had coarse resolution, so these targets

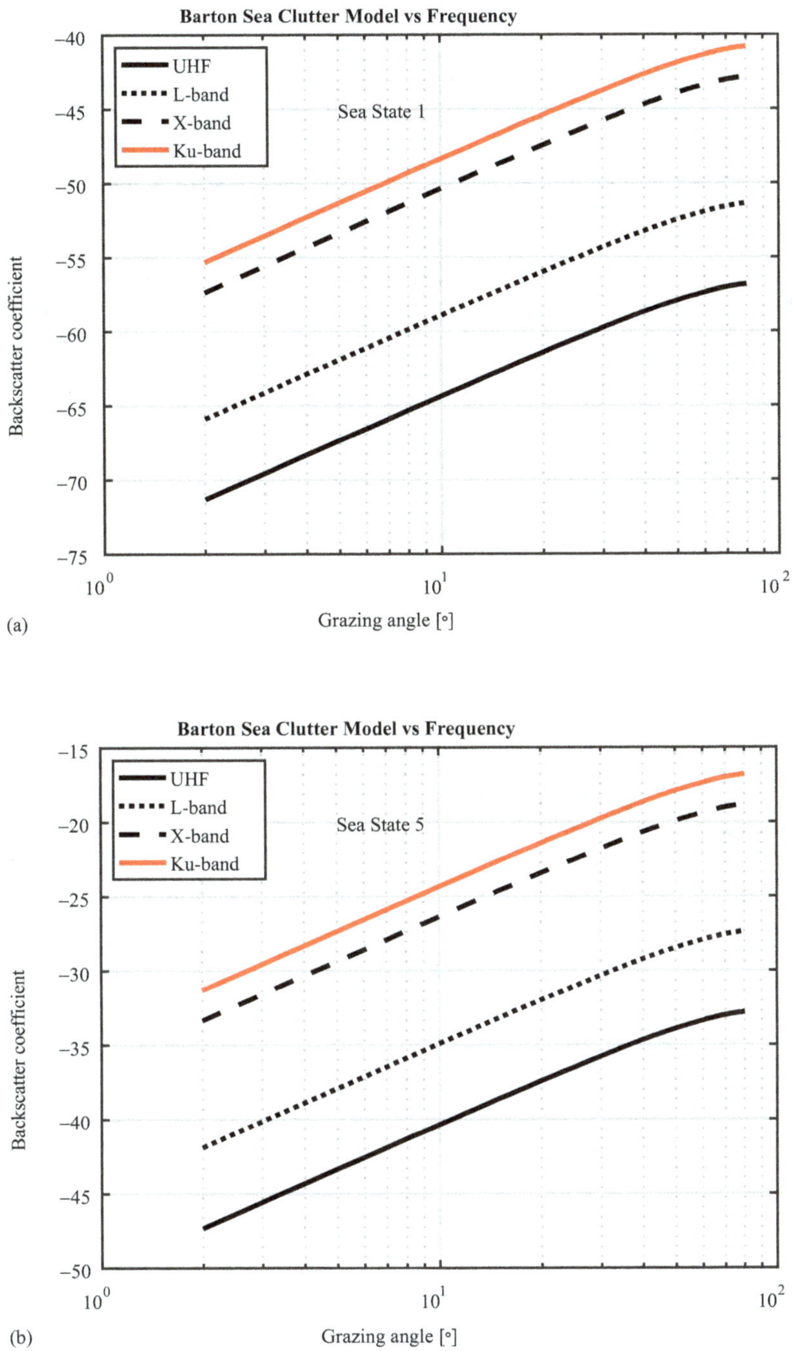

Figure 2.23 *Barton sea state backscatter coefficient versus grazing angle. (a) Sea State 1 and (b) Sea State 5*

would provide a return within one or a few resolution cells. As such, there would be several scattering cells within the radar return. As the radar changed viewing angle, or the target moved through range and Doppler cells, the radar cross-section (RCS) would vary significantly presenting a fading or scintillation of the return. This scintillation could be on a pulse-by-pulse or a dwell-by-dwell basis depending on the distribution of many scattering cells. This scattering was studied extensively by Swerling, and the statistical returns have been generally described as Swerling models [16].

Cultural clutter has similar characteristics. Grass lands, forests and urban areas all present scintillation to the radar. Early ground surveillance radars developed the frequency agility technique to combat this scintillation on their radar screens. This scintillation is less of a problem with the wide band or SARs. Nevertheless, it is instrumental to look at the effects.

UWB radars may have less of a scintillation effect if the scattering cells are comparable with the range and cross-range resolution. However, there are techniques adopted by these systems to reduce the scintillation. These techniques are generally referred to as pre-summing and area prefiltering. The rapid scanning of the radar frequency for high range resolution partially alleviates the scintillation. However, if the return is to provide knowledge of the distributed scattering for terrain characterization or interferometric processing for DTED or GMTI/SAR, then the statistics of the terrain need to be mitigated with either frequency agility or averaging over multiple looks.

2.4.1 Swerling models

Targets, especially man-made targets, have a variation in the RCS due to constructive and destructive interferences due to multiple scatterers in a resolution cell. These are referred to as the Swerling models [16]. These models characterize four cases of the statistical variation of a target return to the radar, as summarized in Table 2.9.

The first two cases are based on an exponential pdf, otherwise known as a chi-square distribution with two degrees of freedom:

$$dP_{\mathrm{a}} = \frac{1}{\bar{\sigma}} \exp\left[-\frac{\sigma}{\bar{\sigma}}\right] d\sigma \tag{2.51}$$

$$dP_{\mathrm{e}} = \frac{E_{\mathrm{s}}}{S} \exp\left[-\frac{E_{\mathrm{s}}^2}{2S}\right] dE_{\mathrm{s}} \tag{2.52}$$

Table 2.9 Swerling model decorrelation statistics [4]

Pdf of RCS	**Decorrelation**	
	Scan to scan	**Pulse to pulse**
Exponential	Case 1	Case 2
Chi-square. Degree 4	Case 3	Case 4

The first equation is the probability that the target RCS lies between s and $s + ds$, and $\bar{\sigma}$ is the average RCS from the target. The second equation represents the probability of the detected envelope of the signal voltage lies between E_s and $E_s + dE_s$, and the average signal power is $S = \overline{E_s^2}$.

The next two cases follow the chi-squared pdf with four degrees of freedom for the RCS. The pdf is given as [5]

$$dP_a = \frac{4\sigma}{\bar{\sigma}^2} \exp\left[-\frac{2\sigma}{\bar{\sigma}}\right] d\sigma \tag{2.53}$$

$$dP_e = \frac{2E_s^3}{S^2} \exp\left[-\frac{E_s^2}{S}\right] dE_s \tag{2.54}$$

Cases 1 and 3 correspond to slowly varying fluctuation of the RCS where the radar return remains constant over a coherent integration time. This is compared to cases 2 and 4 where the radar return changes amplitude and phase on a pulse-by-pulse basis.

2.4.2 *Frequency agile waveform*

Frequency agile (F/A) radars have been operated for several decades. They were originally developed to remove the fluctuation of targets versus clutter and improve the ability to detect objects in noncoherent systems. Historically, frequency agile waveforms were used to improve ground imaging resolution, thereby reducing target fluctuations.

Target fluctuations for man-made targets is a response due to multiple scatterers within the resolution cell. This fluctuating target requires higher SNR than non fluctuating targets in order to circumvent the fading of targets during a dwell. Frequency agility has demonstrated improved detection probability. When there is agility within a dwell, experiments have determined agility somewhat reduces target and clutter scintillation. However, the frequency agility dwell-by-dwell generally enables improved Swerling-2 target detection. The analysis of this important capability is covered below [17].

A frequency agile waveform consists of a series of pulses transmitted at a different frequency. For each frequency, the relative spacing of the scatterers on the target appear different. The radar frequency can be changed in microseconds or milliseconds depending on the type of target to be seen. There are two distinct F/A approaches as shown in Figure 2.24:

- Frequency agile operation within a dwell by changing the center frequency on a pulse-by-pulse basis.
- Frequency agility between coherent dwells can be used where severable pulses were integrated and the center frequency is changed before the next dwell. This approach has been shown to convert a Swerling-2 target to a nonfluctuating target, or Swerling-0.

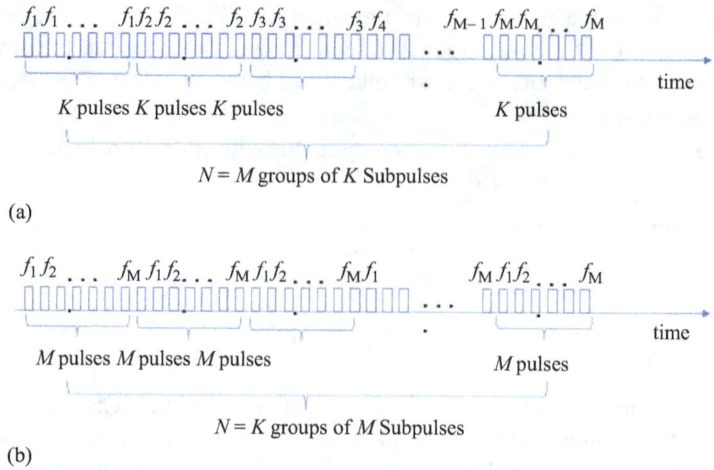

Figure 2.24 Frequency agile surveillance modes. (a) Frequency agility within A dwell and (b) frequency agility dwell to dwell

If the target is relatively stationary in aspect angle with respect to the radar, the coherent integration of M pulses with frequency shifts within a dwell may be beneficial. However, if the objective is to coherently integrate the target to increase the SNR, the frequency will be shifted dwell to dwell.

For detection and tracking radars, frequency agility can:

- Reduce the probability that a target will have an aspect angle with a null in the RCS toward the radar.
- Reduce range and angle tracking errors caused by finite target extent (glint).
- Allow improvement in the target signal to clutter fluctuations.

The most difficult targets to detect with high probability are those whose echoes fluctuate slowly. Hence, the use of frequency agility will cause the null in the RCS to come out of a minimum condition.

An assessment of the impact of frequency agility can be obtained by dimension of the target. If the size of the target is comparable or larger than the wavelength, the correlation coefficient ρ of the detected echoes can be expressed as [17]

$$\rho(\Delta f) = \frac{\sin^2\left[2\pi\Delta f\left(L\sin\vartheta\right)/c\right]}{\left[2\pi\Delta f\left(L\sin\vartheta\right)/c\right]^2} \tag{2.55}$$

where Δf is the carrier frequency shift, $L \sin \vartheta$ the projection of the target length onto the radial dimension and c the velocity of light.

This equation assumes that the pulse length $c\tau/2$ is greater than the projection of the target extent to the radar line of sight, $L \sin \vartheta$. There is a critical frequency shift (Δf_c) that yields the first zero of $\rho(\Delta f) = 0$, as the definition of "decorrelation":

$$\Delta f_c = \frac{c}{2L \sin \vartheta} = \frac{150 \text{ MHz}}{\text{tgt. length in meters}} \tag{2.56}$$

The frequency correlation function relates to the temporal correlation function for a line target undergoing a uniform rotation rate $\dot{\psi}$:

$$\rho(\tau; \Delta f) = \rho(\tau - \tau'; 0) \tag{2.57}$$

where

$$\tau' = \left(\frac{\Delta f \tan \dot{\psi}}{f_0 \dot{\psi}} \right) \tag{2.58}$$

And ψ is the angle between the plane perpendicular to the radar line of sight and the target. This relationship shows that a certain degree of decorrelation may be obtained by either shifting the frequency or allowing the target to change its aspect.

The operation of a surveillance radar in the presence of clutter has been widely considered in the early development of air-to-ground surveillance systems. A non-fluctuation target requires a smaller SNR (or SCR) than a fluctuating target. This is shown in the next slide. As a result, a lower required SNR will enable a longer range detection or a shorter revisit rate. Figure 2.25 illustrates the concept of separation of the target from clutter. Glint or scintillation exists for both targets and distributed clutter. The objective is to reduce the spatial or spectral distribution of the returns through the use of frequency agility [18].

The advantage of frequency agility in detection of targets has been developed. The objective is to reduce the tails in the probability curve through the use of frequency diversity. As the number of frequencies used in a dwell increase, the required SNR is reduced.

Swerling target models give an indication of the amount of fluctuation during a dwell. A Swerling-1 target is characterized by a target that remains constant during a dwell, but it demonstrates fluctuations from dwell-to-dwell. Conversely, in a Swerling-2 model, the target fluctuates from pulse to pulse during a dwell. As the number of frequencies M is increased, the required SNR is reduced. A burst of three

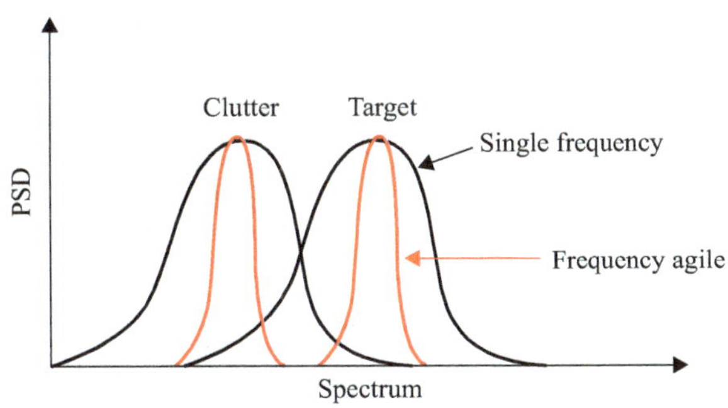

Figure 2.25 Frequency agile reduction of clutter spread [18]

different frequencies will reduce the SNR by 5 dB. The detection probability P_d for a burst of N pulses, with K sub-bursts of M pulses is given by [18]

$$P_d = 1 - \frac{a^M}{\Gamma(M)\,\Gamma(N-M)} \int_0^T dy \int_0^1 dt\, G(y,t) \tag{2.59}$$

$$G(y,t) = \exp(-y[1-(1-\alpha)t]\,y^{N-1}\,\tau^{M-1}\,(1-\tau)^{N-M-1} \tag{2.60}$$

where the term α and the sum of pulses y are related to the SNR

$$a = \frac{1}{1+K\,(S/N)} \tag{2.61}$$

and

$$y = \sum_{n=1}^{N} y_n \tag{2.62}$$

These results indicate a general expression for the probability of detecting a slowly fluctuating target, i.e., for a given frequency, there are target fluctuations from scan to scan but not from pulse to pulse. The target is detected after using multi-frequency, noncoherently integrated pulse burst. On the basis of numerical results from several data collections, there is a strong indication that for an SNR greater than 0 dB, frequency diversity does improve P_d, as illustrated in Figure 2.26. Results from the HOWLS tests confirmed these theories with frequency agile returns from the detection of military targets in distributed clutter and are summarized in Section 3.2.1.

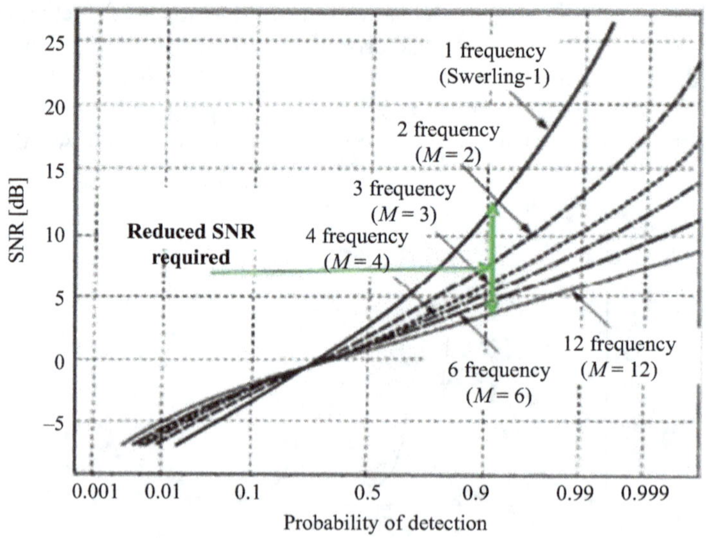

Figure 2.26 Frequency agile waveform improvement in target detection [18]

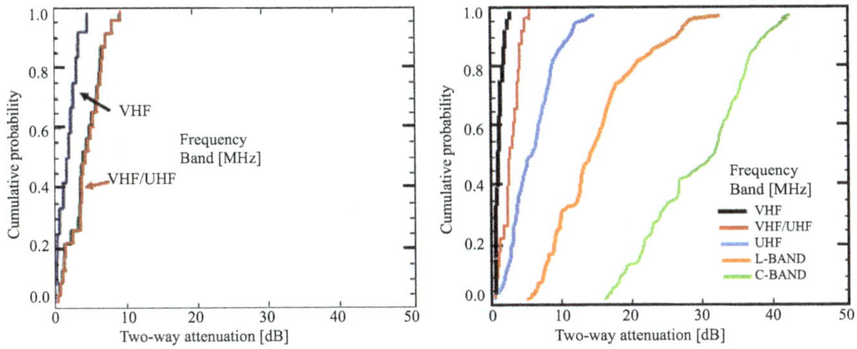

Figure 2.27 Attenuation of radar through forests versus frequency [19]

2.5 Propagation losses

A database of attenuation statistics was published for the Panama Rain Forest and the Maine collections in 1993. These data were for horizontal polarization and are shown in Figure 2.27. Significant conclusions were that low VHF (Carabas I) attenuation was significantly less than any other frequency. This is explained by the very long wavelength compared to the size of the tree branches and trunks. Rain forest attenuation was higher than the northern latitude forests. Again, the low VHF was lower than the VHF/UHF data from the SRI FOLPEN II system. No higher frequency data were collected in Panama. Microwave frequencies had such high attenuation that future FOPEN system development was concentrated at UHF and below. These data were collected over several grazing angles. It is important to look at the loss for various grazing angles for application to a system design [19].

Loretta Bessette (MIT Lincoln Labs) has characterized the losses from all of the FOPEN data collections. A statistical analysis of the median two-way foliage loss $Loss_{two\text{-}way}$ versus grazing angle θ and frequency freq shows regular trends. A two-parameter model has been constructed for polarization and forest type (single or double canopy). Note that there are ranges of parameters in Table 2.10. These ranges can be indicative of foliage type (deciduous versus conifer) or variations within a particular forest (heavy old growth versus light new growth) [19].

Table 2.10 Parameters for foliage loss [18]

Median foliage attenuation model for single and double canopies				
Polarization	**Single canopy**		**Double canopy**	
	α	β	α	β
HH	0.18 ± 0.045	0.53	0.34 ± 0.085	0.53
VV	0.30 ± 0.075	0.47	0.71 ± 0.1775	0.47

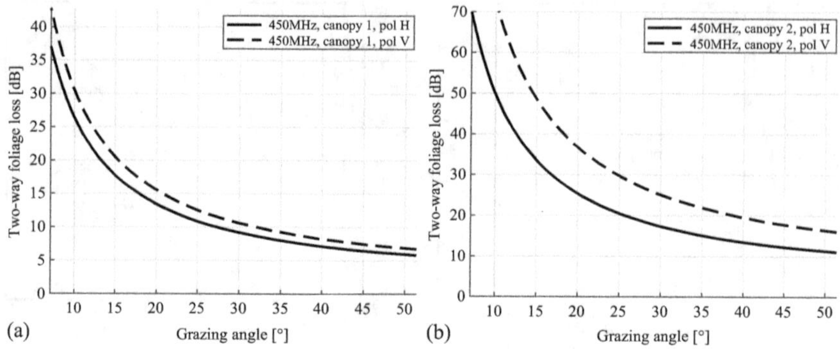

Figure 2.28 Analytic model UHF foliage penetration loss. (a) Single canopy and (b) double canopy

$$\text{Loss}_{\text{two-way}} = \frac{(-\alpha)\,(\text{freq})^{\beta}}{\sin(\theta)} \tag{2.63}$$

Figure 2.28 shows the loss at 450 MHz for the horizontal and vertical frequencies, for both single and double canopies. There is a slight penalty at V-pol for single canopy over H-pol, which is not significant with the variation of the foliage type in an area. However, for double canopy, there is almost 15 dB of two-way loss for the V-pol. It should be noted that in both cases, the loss grows significantly at low grazing angles. The design of UHF foliage penetration surveillance radars consider the median grazing angle for characterization of targets under trees to be 30°.

When the model is used for assessing higher frequencies, the foliage loss rises significantly. Figure 2.29 shows the loss for L-band and X-band in single canopy operation. The losses at L-band correspond roughly with the double canopy losses at UHF. There are many cases where the L-band operation is considered. But it is only where there are sparse trees and the losses are limited, and the operational benefit of

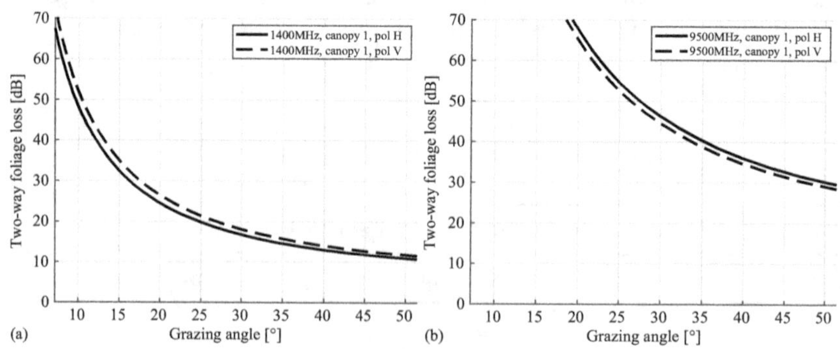

Figure 2.29 Analytic model L-band and X-band single canopy foliage loss. (a) L-band frequency and (b) X-band frequency

a smaller payload is considered [20]. The X-band losses in Figure 2.29b are excessive and verify the historical benefit of microwave radar operation for targets in the open. In conducting system design, these factors should be considered to bound the expected performance on detecting tactical targets within the foliage.

2.6 Area coverage rate

Surveillance radars have been measured by ACR considerations. For ground surveillance radars, ACR is defined as the amount of ground area that is surveyed per unit time. With the advent of unmanned air vehicle platforms that can fly long hours, and dwell over multiple areas, a new measure of ACR is needed. Furthermore, multimode radar processing facilitates the simultaneous operation of multiple radar modes such as GMTI, SAR and ATI. This section discusses the ACR for each of these modes.

The primary measure of effectiveness for a surveillance radar is volumetric ACR, measured in square radians per second [21]:

$$\mathrm{ACR} = \frac{\Omega}{t_1} = \frac{N_{\mathrm{B}} P_{\mathrm{T}} A_{\mathrm{R}} \sigma_{\mathrm{o}}}{4\pi k T_{\mathrm{o}} L \left(S/N\right)_{\mathrm{min}} R^4} \tag{2.64}$$

where Ω is the angular coverage of the illumination, t_I the dwell time, and N_{B} the number of receive beams used to cover the WAS volume. For a conventional SAR or GMTI radar, a single beam is often used, and the surface area processed for the target detection determines the area. It is common to consider the number of individual, sequential search beams needed to cover a large area. In this case, the total search time is $N_{\mathrm{B}} * t_I$. For example, if there are 10 beams required for surveillance and 2 s for target detection, then the revisit time will be 20 s.

One measure of ACR is to take the radar beam footprint as shown in Figure 2.30, determine its area, and divide by the coherent processing interval (CPI). This approach is an "instantaneous ACR," since it only considers the footprint on the ground for a single dwell. CPI times vary depending on the mode (SAR versus GMTI). It is assumed that $\mathrm{CPI}_{\mathrm{GMTI}} \ll \mathrm{CPI}_{\mathrm{SAR}}$. CPI for GMTI mode is given directly by the dwell time. CPI time for stripmap SAR mode is calculated by

$$\mathrm{CPI}_{\mathrm{SAR}} = \frac{R\phi_{\mathrm{SAR}}}{V_{\mathrm{p}}} \tag{2.65}$$

Let us assume that the radar has a beamwidth of ϕ radians. It is also assumed that the beamwidth of the SAR mode is greater than the beamwidth of the GMTI mode ($\phi_{\mathrm{GMTI}} < \phi_{\mathrm{SAR}}$). Given a swath extent of $(R_2 - R_1)$, and a CPI time in seconds, the instantaneous ACR can be given by

$$\mathrm{ACR}_{\mathrm{i}} = \frac{\left(R_2^2 - R_1^2\right)\phi_{\mathrm{i}}}{2CPI_{\mathrm{i}}} \tag{2.66}$$

where $R_1 < R < R_2$ and i represents either GMTI or SAR mode. For a large area, the platform dynamics need to be taken into consideration in evaluating the ACR.

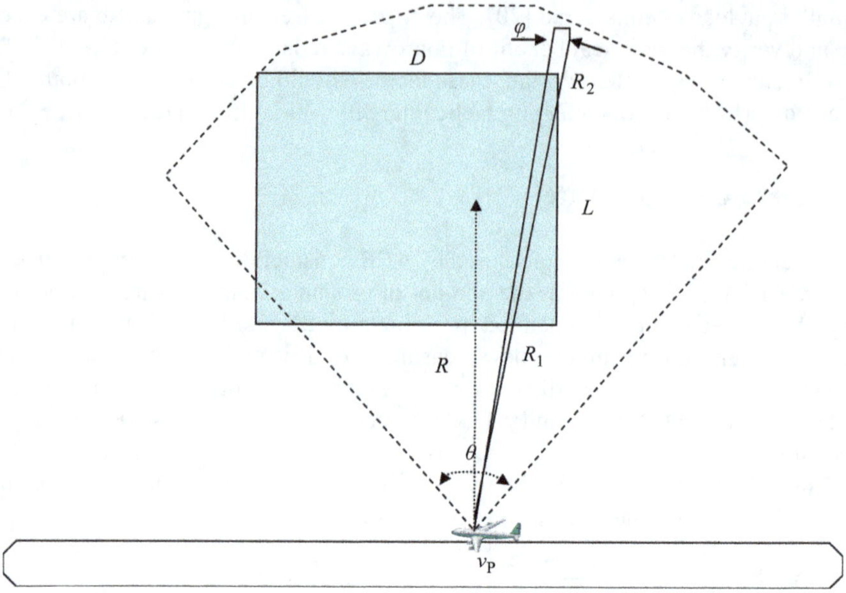

Figure 2.30 Scenario for wide-area surveillance from a standoff platform

Figure 2.30 depicts a normal standoff, WAS scenario. The platform will move in an orbit a sufficient distance (R_1) from the ground referenced coverage area ($D \times L$). The instantaneous field for the area from radar will subtend an angle of θ, and the electronic scan will provide coverage of the area with N beams, where N is determined by

$$N = \frac{\sin(\theta)}{\sin(\phi)} \tag{2.67}$$

The sine function is used for electronic scanned arrays, since they exhibit beam broadening with scan angle.

If one is looking at the ACR associated with the WAS scenario, a different formulation should be used. As the platform flies by an area of width D at a velocity of V_p, it can observe a point in the area for a time CPI. If it takes M beams to cover the area, and each beam has a CPI_i in seconds, the WAS ACR can be given by:

$$\mathrm{ACR_{WAS}} = \frac{\mathrm{DL}}{\mathrm{M\,CPI_i}} \tag{2.68}$$

Tasking of the radar is dynamic, with the number of beams sequentially selected to cover the area of interest. If the search pattern angle θ_s is used instead of the instantaneous FOV, M is determined by substituting for θ in (2.67).

The duty factor for this WAS mission is defined as the total time the radar is looking at a point on the ground during its total mission. This is defined as

$$\mathrm{DF_{WAS}} = \frac{\text{time observing target on ground}}{\text{total mission time}} \tag{2.69}$$

In Figure 2.30, $\mathrm{DF_{WAS}}$ would be the time that the GRCA is imaged by the radar in both legs of the platform path. Typically, the whole area would not be observed, so the ratio of area visible to total area would be used to weight the time. In addition, the turnaround time would be subtracted.

In practice, an SCR that provides suitable correlation for detection and phase for velocity estimates is achieved by adjusting the resolution of the SAR image. In the unresolved target case, adequate cell size will maximize SCR and therefore ATI detection probability. Based on the parameters given in Table 2.11, a comparison of the ACR for the three independent modes can be constructed as shown in Figure 2.31. The receive architecture is a modest four-beam receive pattern. And the ATI assumed a cross-range resolution four times the SAR. The GMTI is seen to have at least two orders of magnitude ACR higher than the SAR. And the combined

Table 2.11 Multimode UHF evaluation parameters

Center frequency	**400 MHz**	**Platform altitude**	**6.1 km**
Bandwidth	$M \times 15$ MHz	Platform speed	60 m/s
Sub-pulses (M)	8	Slant range	15 km
PRF	300 Hz	Antenna width	6 m
Sub-pulses width (T_{pm})	41.667 μs	Number azimuth channel (N_c)	16
SAR resolution (range, cross-range) [weighted]	2.3, 6.7 m	GMTI range resolution (weighed)	18 m

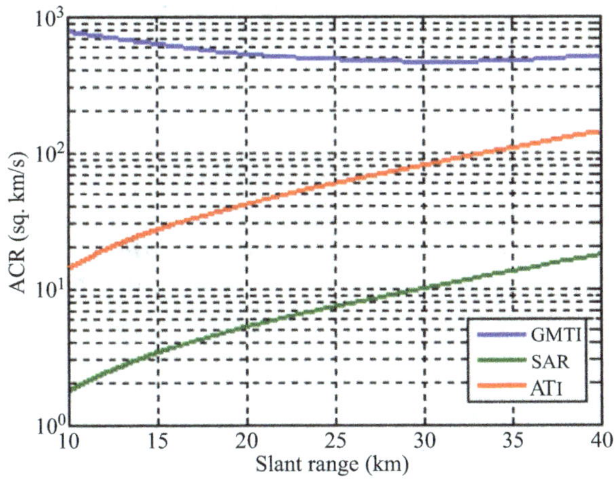

Figure 2.31 Area coverage rate for three simulated cases [21]

mode will enable the full GMTI ACR, even in the case of SAR or ATI processing within the transmit beamwidth.

Chapter 9 will expand on these ACR discussions when a common SAR and GMTI waveform is presented. This new mode has the potential for improving the ACR when the surveillance radar system needs to switch between SAR and GMTI mode. The long integration time for SAR or ATI will significantly reduce the GMTI area coverage.

2.7 Summary

Surveillance radars for wide-area coverage have been designed for both military and civilian applications. The waveform has consistently employed a linear FM pulse to achieve the high-range resolution and for cross-coupling of the returns between range and Doppler. This approach has been questioned in recent years because of its limitation in achieving UWB operation in the shrinking radio frequency spectrum. The FJB waveform has been introduced in this chapter to set the stage for multifunction operation in later chapters.

The surface clutter (both ground and sea) have provided a challenge in detecting small targets. Models of surface clutter are presented with reasonable analytic techniques. These models continue to be used in designing wideband surveillance radar systems. In particular, the consideration of the clutter backscatter coefficient is a reasonable starting point in the design of a modern radar. However, a consideration of wind-blown and ocean clutter motion need to be analyzed. And when targets or surface features are under foliage, the clutter models must consider the volumetric backscatter of the direct path along with the attenuation of the radar return.

A requirement for ACR is a major consideration for battlefield surveillance radars. When there is a need for multimode search for a mix of fixed and moving targets, a search for common SAR and GMTI waveform is needed. A first look at ACR is presented, and it will be expanded on in Chapter 9 for a future multimode radar mode.

References

[1] Carrara W.G., Goodman R.S., Majewski R.M. *Spotlight Synthetic Aperture Radar*. Norwood, MA: Artech House; 1995. pp. 61–9.

[2] Mozeson E., Levanon N. 'Matlab code for plotting ambiguity functions'. *IEEE Trans. Aerosp. Electron. Syst.* 2002;**38**(N3):1064–8.

[3] Pillai U.S. *Ambiguity Functions and Waveform Design for Simultaneous SAR and GMTI*. C&P Technologies Inc., Report for DARPA under Phase II SBIR Contract D11PC20007; 2011.

[4] Scheer J.A. 'Radar search and overview of detection and interference' in Richard M.A. (ed.). *Principles of Modern Radar, Chapter 3, Basic Principles*. Vol. 1. Edison, NJ: Scitech Publishing (IET); 2010. pp. 66–111.

[5] Barton D.K. *Radar System Analysis and Modeling*. Norwood, MA: Artech House; 2005. pp. 41–99.
[6] Alberhseim W.J. 'A closed-form approximation to Robertson's detection characteristics'. *Proc. IEEE*. 1981;**69**(7):839.
[7] Barton D.K. 'Land clutter models for radar design and analysis'. *Proc. IEEE*. 1985;**73**(2):198–204.
[8] Simkins W.L. Report AFRL-SN-RS-TR-1998-95. Air Defense Initiative Clutter Model. New Hartford, NY: PAR Corporation; 1998.
[9] Ulaby F. 'Vegetation clutter model'. *IEEE Trans. Antennas Propag*. 1980;**28**(4):538–45.
[10] Lavery G. *RLSTAP Glyph Notes*. Unpublished report by. Utica, NY: Black River Systems Company; 2005.
[11] Hancock R. *RLSTAP User's Manual Contract F30602.97-C-003*. Rockwall, TX: CAESoft Inc.; 2002.
[12] Billingsley J.B. *Low Angle Radar Clutter – Measurements and Empirical Models*. Norwich, NY: SciTech Publishing; 2001. pp. 600–19.
[13] Davis M.E. *Foliage Penetration Radar – Detection and Characterization of Objects Under Trees*. Raleigh, NC: Scitech Publishing; 2011. pp. 78–91.
[14] Binder B.T., Toups M.F., Ayasli S., Adams E.M. *SAR foliage penetration phenomenology of tropical rain forest and northern US forest*. Proceedings of the 1995 IEEE International Radar Conference; Washington DC; 1995. pp. 158–63.
[15] Watts S., Rosenberg L. *Advances in radar performance modelling in sea clutter*. Proceedings of the 2015 IEEE International Radar Conference; Washington DC; 2015.
[16] Swerling P. 'Probability of detection for fluctuating targets'. *IEEE Trans. Inform. Theory*. 1960;**6**(2):269–308.
[17] Nathanson F.E., Reilly J.P. 'Frequency agility for radar target detection and tracking'. *APL Tech. Digest*. 1970:2–8.
[18] Fante R.L. *Multifrequency Detection of Slowly Fluctuating Targets*. Bedford MA: Unpublished MITRE Memo MTR 95B00021; 1995.
[19] Bessette L.A., Ayasli S. *Ultra wideband P-3 and CARABAS II foliage attenuation and backscatter analysis*. Proceedings of the 2001 IEEE Radar Conference; Atlanta, GA; 2001. pp. 357–62.
[20] Davis M.E. *FOPEN radar design for sparse forest surveillance*. Proceedings of the 2016 IEEE Radar Conference; Philadelphia, PA; 2016.
[21] Kapfer R.M., Davis M.E. *Ultra wideband, multi-mode radar processing*. Proceedings of the 2012 IEEE Radar Conference; Atlanta, GA; 2012.

Chapter 3
Surveillance radar modes

The development of coherent radars and digital processing provided several critical innovations to improve the detection and characterization of man-made targets in cultural clutter. The first aspect of mode development was having an understanding of the propagation of radar waveforms, along with the signal loss due to target scintillation and propagation through media. Propagation in free space was well understood. However, when there existed anisotropic media between the radar and the desired target or clutter return, significant measurements of these characteristics were undertaken. This is especially true for propagation through foliage, use of polarization and understanding the dependence of wavelength on the radar return.

Early radar operation was noncoherent due to the lack of components to generate a waveform and maintaining a stable reference phase during a processing interval. These radars employed frequency agility to remove target or clutter scintillation effects on the returns. It was well understood that a frequency agile, noncoherent dwell could counter the variation of target returns due to temporal or spatial effects within the detection cell. So, the first mode to be discussed is the benefit of frequency agility on improved target detection. This agility can be on a pulse-to-pulse basis or between coherent processing intervals. The former approach is most useful in detecting stationary targets in strong clutter. And when the frequency agility is used after a coherent processing interval, the moving target scintillation can be mitigated within a Doppler filter.

The next series of modes use either spatial (or channel) comparison of returns or using Doppler characteristics of moving targets for discrimination of the moving targets from background clutter. The displaced phase center antenna (DPCA) was a technique that provided excellent suppression of background clutter in the detection of moving targets. The clutter Doppler variation due to the platform motion could also be used to improve the cross-range resolution of the surface spatial return. This technique is called Doppler beam sharpening (DBS). These two techniques were precursors to space time adaptive processing (STAP) and synthetic aperture radar (SAR) image formation. The next two modes developed for ground surveillance radar were ground moving target indication (GMTI) and (SAR. The former employed Doppler processing for discriminating moving targets from background clutter. The second mode utilizes very long integration aperture, synthetically formed due to the platform motion, along with wide bandwidth waveforms to greatly improve the ground image resolution in both range and cross-range.

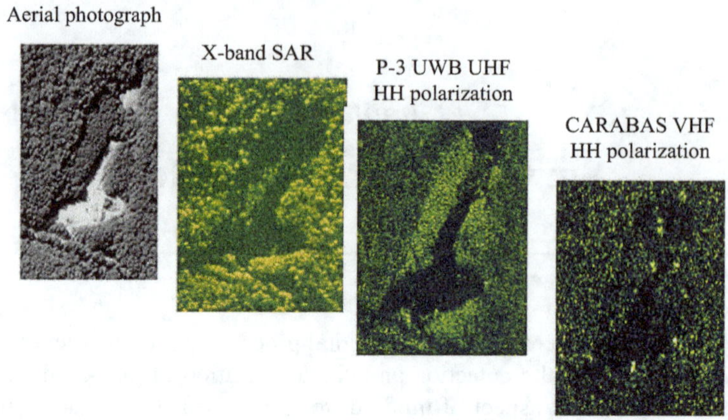

Figure 3.1 Image characteristics versus illumination wavelength [1]

Every new surveillance radar system must answer the question of which frequency to use. It should also answer the question of "why ultra-wideband (UWB) or fully polarimetric operation"? The first question is easy to answer and is illustrated in Figure 3.1. Optical photographs and microwave radars can reliably detect man-made objects that are in the open. However, if there is volumetric clutter (forest or underground), the targets are hidden in the dense forest cover. In this case, longer wavelength waveforms are essential due to their ability to penetrate the forest and some depth of the ground. The second and third questions are more important to the design detail of the future systems.

High-resolution imagery serves two purposes: to provide a better separation of the object scattering from the background clutter and to provide more detail of scattering of objects for characterization. In applications for foliage or terrain characterization, this factor is not as strong as motivation. But to find a small vehicle or a buried land mine, it is of major consideration. Hence, there was a strong motivation to develop radar systems that had both high spatial resolution and polarization diversity.

Polarization diversity has been evolving as a significant capability for both detection and characterization of terrain and man-made objects. If characterization is an important system objective, then polarization must be factored into the system waveform and processing approach from the start.

3.1 Surveillance radar testbeds

This section covers the history of UWB surveillance radar testbeds and their development and employment of critical radar modes. Two experimental radars were developed by MIT Lincoln Laboratory in the 1970s to explore the ability of detecting and characterizing tactical threats in the battlefield. The first was the multiple aperture surveillance radar (MASR), which developed the first phased array to employ DPCA to cancel mainbeam clutter and enable detection and tracking of slow-moving ground targets [2]. The second testbed was the Hostile Weapons Locator System (HOWLS), which was developed for DARPA and the US Army to demonstrate the utility of detecting slow-moving unmanned air vehicles in a tactical scenario with artillery and mechanized weapons [3]. The final testbed will demonstrate the capability of a multiple frequency radar to detect and characterize targets and terrain features in areas of mixed open and concealed environments. This geographic synthetic aperture radar (GeoSAR) system was developed by the Defense Advanced Research Projects Agency (DARPA) and the National Aeronautics and Space Administration (NASA) Jet Propulsion Laboratory, integrating both X-band and P-band interferometric radars onto a business jet platform [4].

This section is an introduction to the set of early battlefield radar modes. With the advent of coherent processing and phased array antennas, there was a significant realization of the capabilities for battlefield surveillance from a small tactical manned or unmanned platforms.

3.1.1 Multiple aperture surveillance radar

One of the earliest battlefield surveillance testbeds was the MIT Lincoln Laboratory's MASR. The concept of DPCA had been employed on air-to-air surveillance systems. With the tactical radar and target velocity applications, the mainbeam clutter spread and pulse repetition frequency (PRF) separation of mainbeam clutter would be the predominant issue and using two channels could be sufficient. However, for air-to-ground surveillance, especially from a low-altitude vehicle, the sidelobe clutter returns will be a limiting aspect. MASR was designed to demonstrate a nonadaptive approach to cancelling the ground clutter in both mainbeam and sidelobes.

The objective of this development was to test the limits of GMTI by employing the DPCA technique. DPCA had been used for detecting airborne targets from a moving platform. However, the clutter was not as much of a problem using PRF waveforms, where the airborne targets would be exo-clutter, *i.e.,* outside of the mainbeam of the antenna. For GMTI, it was difficult to cancel the clutter where the antenna mainbeam was intercepting the ground clutter, and the targets of interest were moving significantly slower than the radar platform. The antenna had a unique approach of "stopping the motion" by switching the antenna elements and compensating for the motion of the platform over a few PRIs. This made the antenna essentially stationary during this period, and the ground clutter was essentially cancelled by matching the element pattern between pulses. This lack of motion significantly cancelled the cutter to a narrow Doppler regime, whereby the slow-moving targets could be detected [2].

MASR was an L-band radar with a phased array antenna with 34 columns as seen in Figure 3.2. The completed antenna system was designed for frequency agile waveform operation over 5% bandwidth. It was important to manufacture the aperture to provide a match of the amplitude and phase of each column over the full bandwidth and to provide the element scan angle of ±45°. An excellent measured element match, as shown in Table 3.1, was obtained with 0.4° rms in phase and 0.1 dB rms in amplitude. This level of match was desired to have over 45 dB of clutter suppression [5].

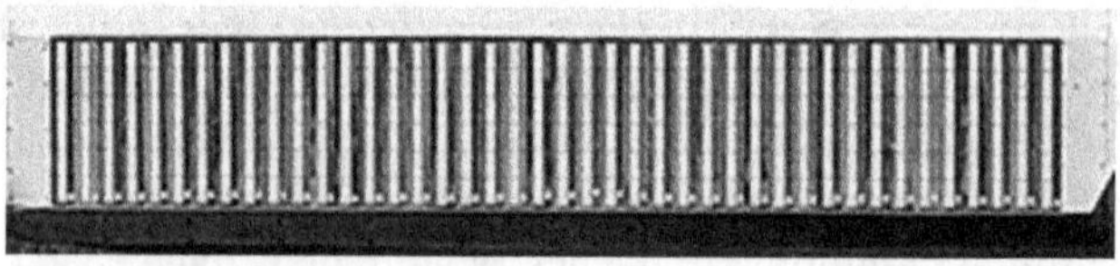

Figure 3.2 *MASR antenna [5]*

The MASR antenna was 15 feet in length and 1.5 feet in height to provide a relatively narrow beam in azimuth. This aperture size was designed to minimize the mainbeam Doppler spread of clutter and to provide sufficient angular resolution for target location. The phased array architecture consisted of 34 (active) vertical columns with six waveguide radiators each to provide the elevation beam, as shown in Figure 3.3. There were also four dummy columns on each end of the array, to provide the element match over the full 34 columns used for phase scanning. A DPCA phase center switching network selected the 32 active columns to form the beam and switch between along-track elements between pulses for the DPCA operation. The wide scan angle was provided by a six-bit diode stripline phase shifter assembly, one per active column. Finally, there was both a sum Σ and difference Δ power divider to provide the low sidelobe of –40 dB. Given this fine control of the column phasing, the sidelobes could be maintained over the ±45° in azimuth scan, as shown in Figure 3.4 [6].

The major innovation in this system was the switching matrix on each column to move the phase center on a pulse-to-pulse basis, as demonstrated in Figure 3.5. With very low azimuth sidelobes, and fine match on transmit and receive, the mainbeam clutter could be effectively cancelled. The electronic control in the antenna would enable the forward section of columns to transmit and receive the signal reflected off of the ground clutter and any moving targets within the mainbeam. After the antenna has moved a distance equal to an integer multiple of the column spaces, the antenna would be switched to transmit from a section of the antenna behind the first pulse. When this pulse was transmitted and received, the returns would be essentially stationary with the first pulse. This arrangement would be repeated over many pulses to form the range-Doppler map. However, the ground clutter would be essentially cancelled, and the moving targets (or nonstationary clutter) would be detected by the signal processing. The inertial navigation system onboard determined the radar

Table 3.1 *MASR antenna measurement results [6]*

Match (rms)	Goal	Measured	Measurement error
Phase (°)	0.7	0.4	<0.1
Amplitude (dB)	0.11	0.1	<0.01
Reflection	0.02	0.017	

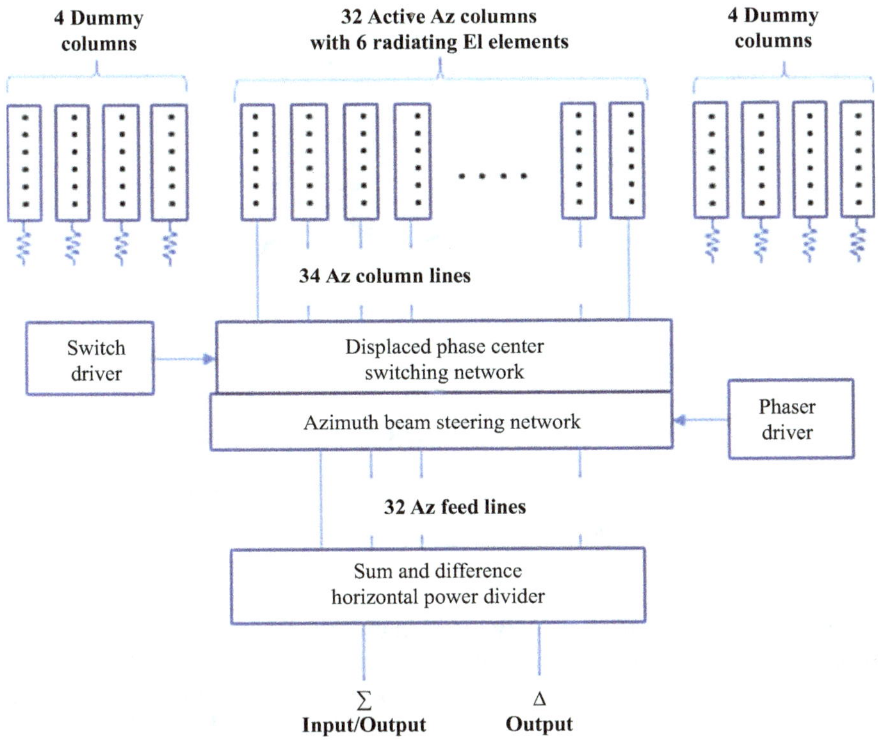

Figure 3.3 MASR phased array architecture [5]

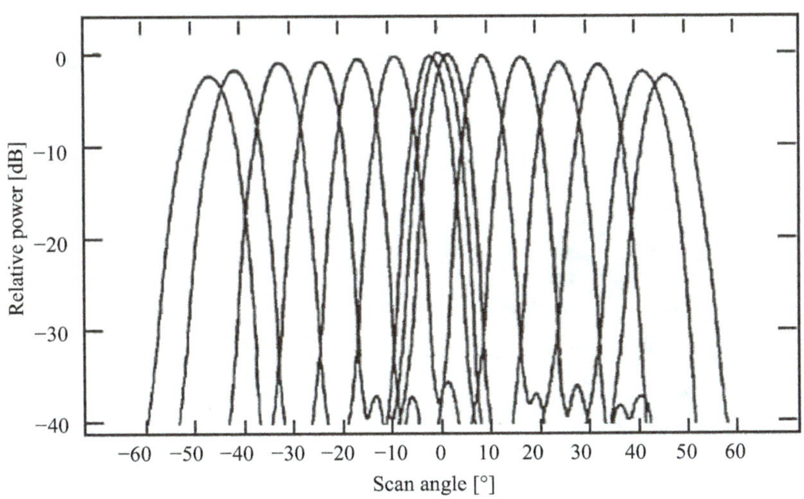

Figure 3.4 MASR antenna scan characteristics [6]

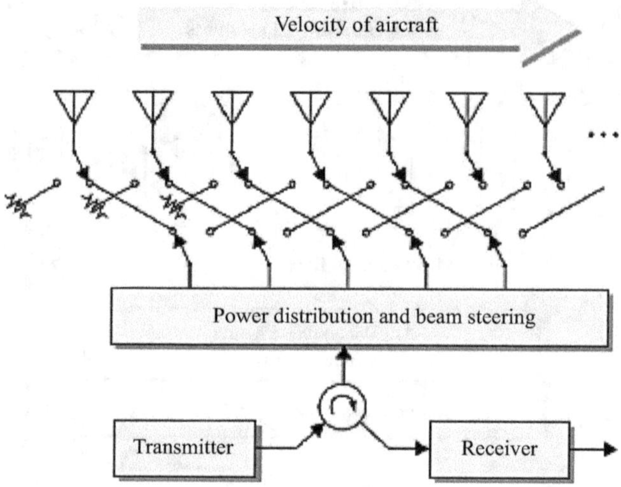

Figure 3.5 MASR DPCA two-channel beam switching matrix [2]

motion between pulses, and the array switching was synchronized to the pulse repetition interval (PRI) to enable the clutter cancellation [6].

The power distribution and beam steering control are shown in Figure 3.5. There is a double pole, double throw switch attached to each column of the antenna. The phase center switching occurred in front of the beam steering phase shifters, so that there was at least 60 dB isolation between each column. And the output of the switch was designed to move the array an integer of a half wavelength on each pulse. The PRI and switching times were designed so that the aircraft moved an integer of the element spacing on each pulse [5].

Figure 3.6 demonstrates the cancellation of the mainbeam clutter between the two-phase centers. The single slow-moving target can be clearly seen after DPCA

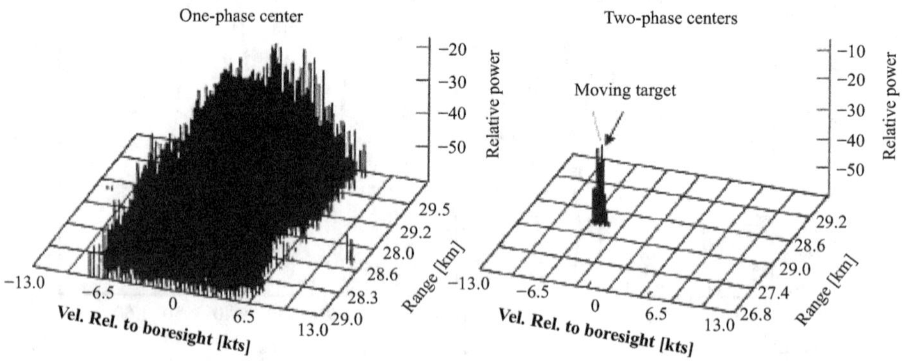

Figure 3.6 Comparison of MTI and DPCA detection of slow-moving targets [2]

processing at the array face. Detailed measurements of the subclutter visibility verified 46 dB of clutter cancellation from the moving platform [2].

It was also verified that the size of the L-band antenna combined with DPCA switching was sufficient to be operated off of a slow-moving civilian aircraft for tactical moving target data collection. That choice of frequency was based on the desire for nonadapted cancellation of clutter for wide-area surveillance of moving ground targets. The slow speed of the De Havilland Twin Otter provided a balance of beam switching at reasonable PRIs and meeting the constraints of antenna fabrication with excellent channel match. However, the size and weight of the antenna was a challenge for any future unmanned aircraft [6].

The performance analysis of MASR can be analyzed based on the antenna technology and the simulation of returned clutter and target to the antenna. The array performance is predicted based on the 32-element active array on an individual pulse, and the displacement of the array at the PRI. An errored antenna pattern U_{err} is simulated with:

$$U_{err} = \left(\cos\theta + \delta_{ae}\right)\sum_{i=1}^{Nel} A_i\, e^{j\frac{2\pi d}{\lambda}\left(\sin(\theta+\delta_{ph})+\sin\theta_s\right)} \tag{3.1}$$

where θ is the angle in the face of the array, θ_s is the steering angle off of the array normal, δ_{ae} and δ_{ph} are the element amplitude and phase errors, respectively, A_i represents the Taylor 40 dB and $N_{bar} = 5$ weighting of the array for sidelobe control. The multiplier $(\cos\theta + \delta_{ae})$ is an element amplitude function as the beam is steered. Applying the amplitude and phase error into (3.1), we obtained the antenna pattern, as shown in Figure 3.7, and a good match to the measured performance, as shown in Figure 3.4. This scan performance and sidelobe structure are qualitatively

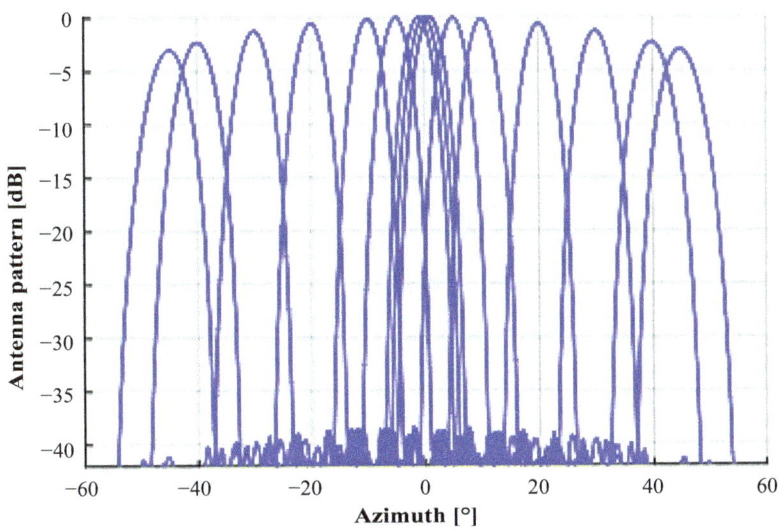

Figure 3.7 Simulation of MASR errored antenna pattern – steered ±45°

Table 3.2 DPCA radar parameters

Variable	Value	Units	Variable	Value	Units
Frequency	1,400	MHz	Bandwidth	5	MHz
PRF	3,226	Hz	Platform velocity	175	m/s
Range center	25	km	Column offset	1	$\lambda/2$
Peak power	2,500	W	Number of pulses	256	

an estimated pattern because the amplitude and phase errors were applied with a random number generator. However, the actual antenna with its specific distributions of the errors as shown in Figure 3.4 has a good match. It is this match that can be used to evaluate the DPCA capabilities below.

The simulation of DPCA can be carried out with a rather simplistic model with parameters given in Table 3.2, which has the distributed clutter, motion of the antenna with two DPCA subarrays (overlapped), a scan angle θ_s and slant range to the center of the analysis box [7]. These are shown in Figure 3.8, along with an inset with six targets of varying velocity to or away from the radar, and the center of the simulation at slant range R_s. The targets were simulated to illustrate the detection of slow-moving targets in both the mainbeam of the antenna and in the exo-clutter region. The simulation has 256 range bins referenced to the 25-Km distance from the radar, R_c. Each target is 20 dBsm, and the range and normalized Doppler data are given in Table 3.3. Hence, at a platform velocity of 175 m/s and a wavelength of 0.2143 m, the PRF will be limited to 3,226 Hz at a $\lambda/2$ spacing between columns.

The clutter and target return from the first 32 columns are illustrated in Figure 3.9. The moving targets are distributed in normalized Doppler at the slant ranges corresponding to Table 3.3. The strong mainbeam clutter can be seen at a compensated zero Doppler. And the moving targets can be seen as transitioning

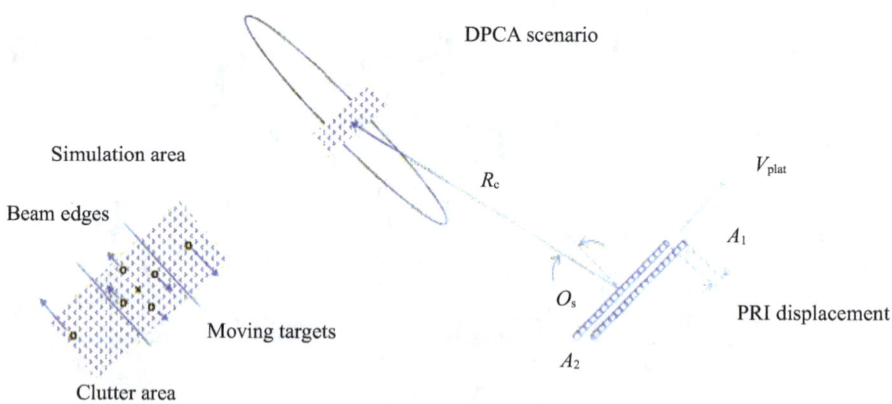

Figure 3.8 DPCA simulation to model clutter cancellation

Table 3.3 DPCA target characteristics

Target	Range bin offset from R_c	Radar cross-section (dBsm)	Normalized Doppler (m^{-1})
1	32	10	0.1
2	16	10	0.05
3	8	10	0.02
4	–8	10	–0.02
5	–16	10	–0.05
6	–32	10	–0.1

from exo- to endo-clutter. It is this condition that needs to be analyzed in terms of moving target detection and any attenuation due to the DPCA technique.

For DPCA to work, the PRF needs to be chosen such that the platform will move one column distance between each pulse. This can be observed by the clutter range-Doppler slope β_{DPCA} given by the distance traveled d, with the platform velocity v_{plat}, and the PRI [8]:

$$\beta_{DPCA} = \frac{2v_{plat}\,\mathrm{PRI}}{d} \tag{3.2}$$

When $\beta_{DPCA} = 1$, the cancellation of the clutter should be maximum. This clutter ridge is obtained by calculating the periodogram of the normalized Doppler and range, as shown in Figure 3.10. The clutter ridge exactly fills the normalized Doppler space [–0.5 0.5] and the clutter angle space of ϕ^a in the interval [–1 1]. The six targets are seen at zero Doppler in the interval [–0.1 0.1].

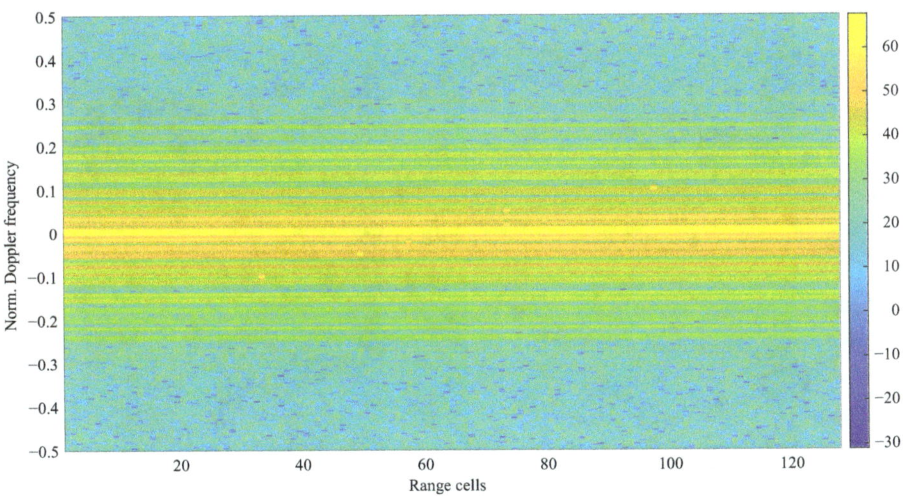

Figure 3.9 MASR simulated clutter with six moving targets

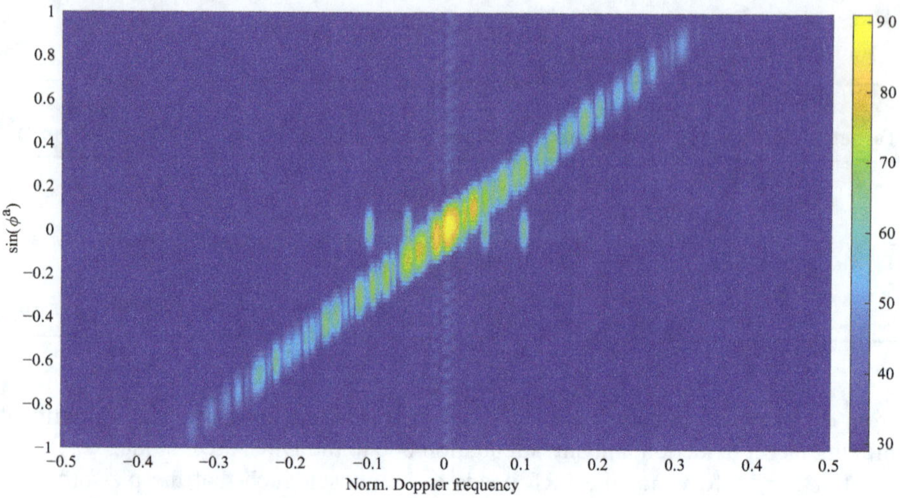

Figure 3.10 Periodogram of MASR simulation

The radar returns for both targets and clutter are calculated in a range-Doppler return, as shown in Figure 3.11. The range and Doppler returns are both weighted to Taylor 60 dB, $N_{bar} = 6$. The six targets are seen spread across the clutter return, and they are almost obscured by the mainbeam clutter. The image indicates the CNR of the peak of the beam as 42.7 dB.

The DPCA processing takes the form of subtracting the clutter and target returns from each array element between the pulses. This subtraction forms the difference between the measurements of array elements 1 to $N-1$ for pulse number m, and for array elements 0 to $N-2$ for pulses $m + 1$. This difference between the signal return for each pulse and to each element can be written in compact form [8] as

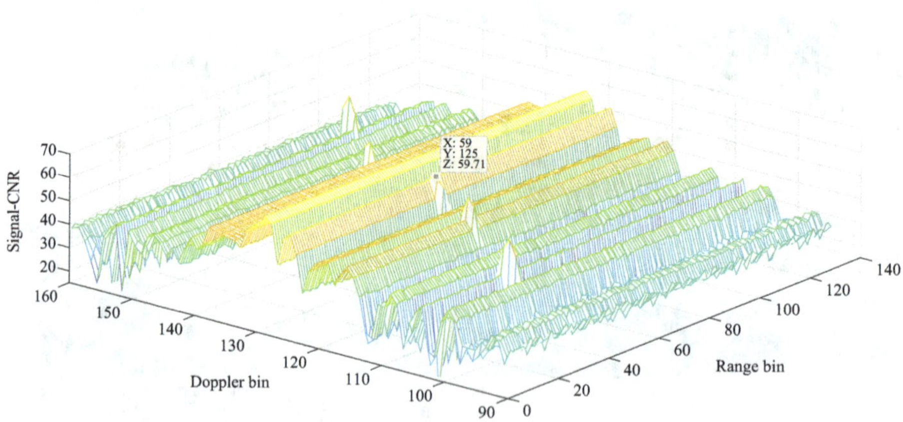

Figure 3.11 MASR simulated range-Doppler image, PRF = 3,226 Hz

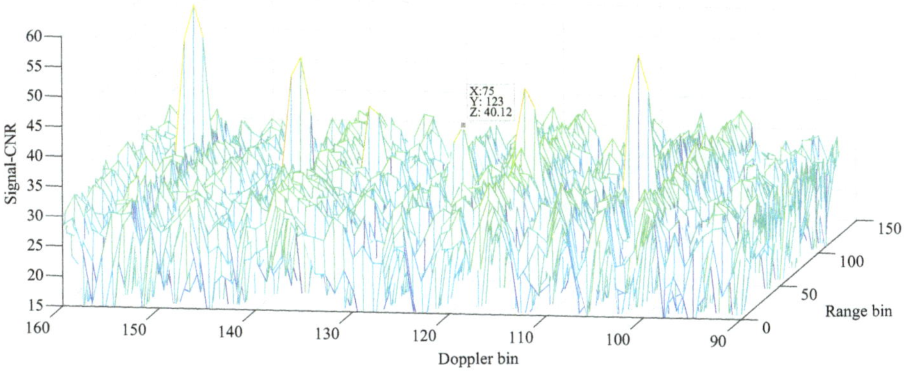

Figure 3.12 MASR simulated DPCA range-Doppler image, PRF = 3,226 Hz

$$\Delta x(m) = \begin{bmatrix} \mathbf{0} & I_{N-1} \end{bmatrix} \boldsymbol{x}(m) - \begin{bmatrix} \boldsymbol{I}_{N-1} & \mathbf{0} \end{bmatrix} \boldsymbol{x}(m+1), \quad m = 0, 1, \ldots, M-2 \tag{3.3}$$

where $\mathbf{0}$ is a zero vector of $[N-1\ 1]$ and $\boldsymbol{I}_{N-1}$ is an identity matrix of dimension $[N-1\ N-1]$.

The DPCA processing is then carried out on the two virtual 32-element antennas, as depicted in Figure 3.8. Processing was carried out by calculating the return of the two antennas, separated by one column, and delaying the data from Channel A to coincide for the return from Channel B. The DPCA range-Doppler image is shown in Figure 3.11. The peak clutter in the return is 51 dB, and the very large target is at 54 dB relative signal strength. Figure 3.12 shows the targets after DPCA processing, where two of the targets are exo-clutter at 50 dB, and the two targets nearest to the peak of mainbeam clutter are at 35 dB.

It should be noted that the variation in the clutter power in the mainbeam is an artifact of combining the strong targets with the clutter returns in the simulation. The floor of the DPCA plot is at 15 dB; so, the subclutter visibility is approximately 40 dB, corresponding to the MASR documentation. It is evident that the DPCA cancelled the mainbeam clutter. However, the subtraction process also reduces the target SNR.

This DPCA analysis was carried out at the PRF corresponding to the clutter ridge $\beta_{\text{DPCA}} = 1$. When the PRF or platform velocity is changed, the DPCA approach will be affected. The DPCA simulation was redone with PRF = 2,000 Hz, as illustrated in the periodogram in Figure 3.13. It can be clearly seen that with $\beta_{\text{DPCA}} = 0.62$, the clutter between pulses is not fully cancelled. Figure 3.14 shows the six targets after DPCA, with the same attenuation as before. However, the mainbeam clutter floor has increased by between 10 and 15 dB, reducing the clutter-to-noise ratio (CNR) for target detection. This departure from the DPCA criterion of $\beta_{\text{DPCA}} = 1$ was a difficulty that MASR did not fully address. However, the STAP approach of the

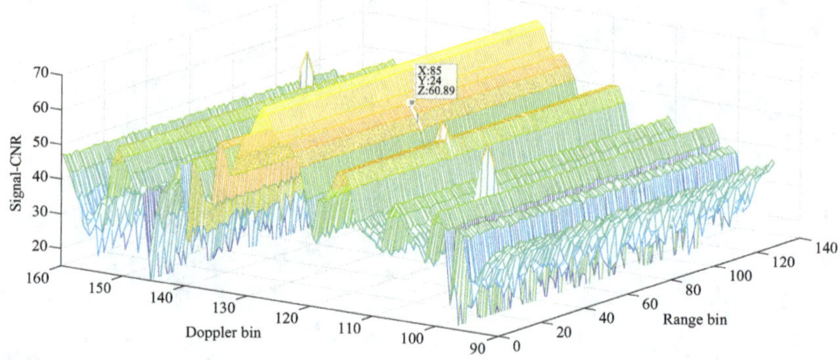

Figure 3.13 MASR simulated range-Doppler performance, PRF = 2,000 Hz

1990s provided a solution to this important surveillance technique. The benefit for STAP will be developed and shown in Chapter 7.

3.1.2 Hostile weapons locator systems

The HOWLS was funded in 1974 by DARPA Tactical Technology Office and managed by MIT Lincoln Laboratory. The main objective of this experimental radar was to develop technology for the army to detect and locate hostile combatants and associated weapon fires in a company-level operation. An emerging Amber class remotely piloted vehicle (RPV) was the intended airborne platform. However, Piper Navaho was chosen to provide the altitude and flight conditions representative of the RPVs in the 1970s. Furthermore, the frequency and size of the radar were designed for future development on medium-class RPVs. The radar system implementation was created to test the ability of electronic scan and coherent processing for

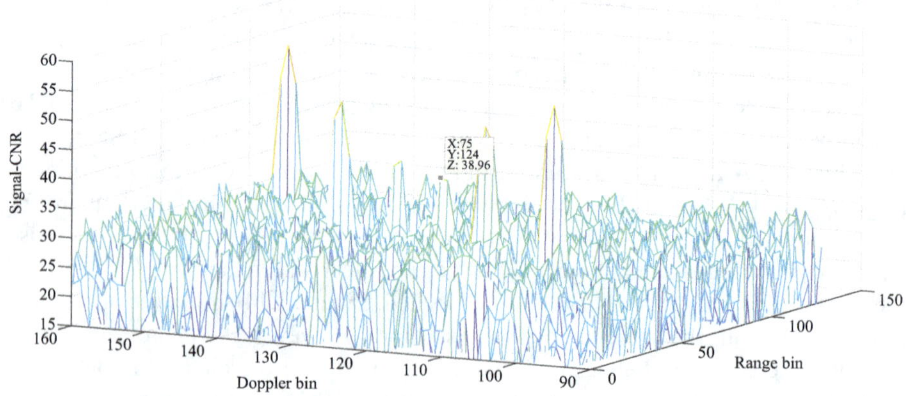

Figure 3.14 MASR DPCA simulation, PRF = 2,000 Hz

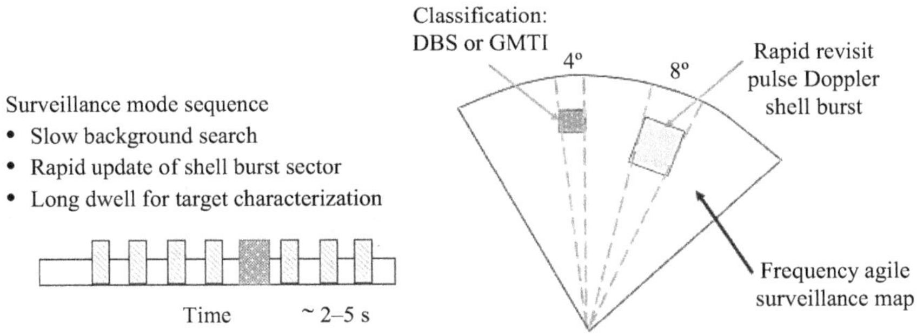

Figure 3.15 Mode interleave in HOWLS airborne radar

detecting and classifying tactical targets in a moderate clutter environment and in a medium-altitude flight regime [3].

This early battlefield surveillance radar for RPV installation was intended for use in registration of artillery and fixed targets. On a battlefield, there are several activities being conducted simultaneously. The first aspect of the HOWLS mode design was to determine the revisit rate for being able to collect the detection and tracking data and transmitting the data to a ground station for recording and initial signal processing. The three initial tenants were: conduct a slow background search to provide context for search and track; provide a rapid update over areas of interest to detect the weapon firing; and longer dwells for characterizing the event (either moving or stationary). The driving requirement was the ability to detect a weapon firing. And the timing of these events occurred in small multiple of seconds. As such, the interleave of modes, as illustrated in Figure 3.15, were designed to provide a revisit rate of less than 2 s, based on detections from areas determined to be appropriate firing sites or targeted areas.

The system was developed with the following major radar modes and features [9]:

- Ku-band electronically scanned array (ESA) was developed to provide the necessary stabilized pointing within the coherent multimode operation. This beam pointing was carried out with beam steering commands from the radar computer and stabilized by inputs from the inertial navigation system.
- GMTI mode against tracked and wheeled vehicles was carried out by a two- or three-pulse canceller, followed by a coherent processing interval to measure the target Doppler. The detection of the ground targets was carried out with a cell averaging constant false alarm rate algorithm. The tracking of the targets was accomplished by a short-time revisit rate, under the tasking from the ground. That revisit rate was typically between 1 and 3 s, depending on the speed of the targets.
- High PRF pulse Doppler mode was used for detection and characterization of artillery pieces. The speed of the artillery piece was at or near the speed of

sound. As a result of this high Doppler, the PRF had to be increased sufficiently that the target was not Doppler ambiguous.

- DBS mapping mode was used for detecting and characterizing stationary vehicles and structures of military relevance. DBS was chosen over SAR for HOWLS due to the long SAR dwells needed for obtaining ground resolutions of 3 m or less. The combination of GMTI for any moving targets and DBS for the region around the moving targets gave the context of interpretation of the threat with respect to the background surface features (e.g., roads, tree lines and buildings).
- Data link control and demonstration of multimode radar processing. The final technical demonstration of HOWLS and its radar coherent modes and phenomena exploitation was important in evaluating the utility of an RPV electronic steered antenna during several test campaigns. The small propeller aircraft flying at low altitudes was a risk at that time for practical employment. The combination of the ground processing center and direct data link to the airborne radar was instructive for future tactical multimode radars.

The ground referenced scan area was ±30° maximum angular sector, and the ground map needed to be updated over several minutes. This was the lowest priority task and employed a frequency agile noncoherent mapping to reduce the clutter glint. The radar was developed to store up to four "simultaneous" modes in the beam steering computer. The beam could be positioned and switched between modes at a 4-kHz rate or 250 ms. For detection of weapon firings in designated areas, a rapid revisit with high PRF dwells to detect artillery Doppler characteristics would be established with multiple scans within the basic revisit rate of between 2 and 5 s. And for these events, an alert confirm mode would consist of a selection of DBS for fixed targets or GMTI for moving targets and would detect and send the returns in a small angular sector and short-range segment.

The sizing of the buffer memory on the airborne radar enabled transmission of the radar detections averaged over the multiple scans and sub-modes. The downlink (for radar detections) was 12 million bits per second and the uplink was 20 kilobits per second (for mode scheduling). These values are not too different from today's tactical data links.

The mode characteristics are summarized in Table 3.4. These parameters were developed between the radar designer and the system integration team. Many of the details were early understanding of requirements for this experimental system. More critically were the synchronization and timing of the modes given the dynamics of the platform and the computational capabilities of computing microcircuits.

During the many experimental data collections, the airborne radar would digitize the radar returns and send them to the ground over a C-band data link. These data were stored for later algorithm development. But it also provided the ability to display the ground maps and detections of fixed and moving targets on the ground map. A live display provides the operator the relative radar returns and georeferenced for an assist in stabilizing the display during test runs. The display shows two

Table 3.4 HOWLS mode control characteristics [9]

Parameter	Ground map	Fixed target det.	Fixed target ID	GMTI	DBS	SAR spotlight
PRF (kHz)	4–11	4–11	4–40	4	4	3
Pulse res. (m)	3–10	3–10	10	10	3–10	5
Azimuth res. (mrad)	6	6	6	6	0.5–1	<0.1
Transmit mode	Non Coh. Freq. Agile	Non Coh. Freq. Agile	Coherent	Coherent	Coherent	Coherent
Independent samples	64	64	1	1	›2	>2
# Channels	1	1	2	2	2	2
Range window (km)	>1	1–4	0.1–1	>1	>1	0.5 Rng 0.5 Az
Range (km)	1–25	1–5	1–2.5	1–15	1–25	5–20
MDV	—	—	—	0.5–1.5	—	—
Blind vel (m/s)	—	—	—	37.2	—	—

rows of corner reflectors, the background radar map and several tactical targets with their return shown in color scale.

The need for collecting these multiple mode characteristics drove the technology for electronic scan of the antenna and interleaved multiple modes within the radar. Several design factors were considered for the HOWLS design in anticipation of demonstration test activities. The first factor was to provide electronic scan over a ±30° sector. This was instrumental in detecting activities or objects and then revisiting before the aircraft flew out of the region. In order to revisit these areas rapidly, the radar stored up to four simultaneous modes/beam positions in the beam steering computer located on the array subsystem. This proximity and the limited delay for getting the beam control messages to the array elements enabled a beam position and mode update at 4 kHz data rate. Finally, since there was insufficient computation onboard the aircraft, the data were sent to ground station over the C-band data link. The command and control of the radar was carried out from the ground processing system over the uplink. Figure 3.16 illustrates the images in the ground display with a photograph and tactical area overlay and the real-time display of HOWLS ground image and overlay of detection of targets. The white squares are the corner reflectors, and the red circles are the tracked and wheeled vehicles places in the area.

The HOWLS mode selection was extensive for its time period and not too much different than many of today's small UAV systems. The main difference is the employment of noncoherent modes in addition to coherent modes. A noncoherent ground map with frequency agility transmissions was well accepted for many fighter radars at the time. And frequency agile detection of fixed targets was efficient, as long as the frequency change between pulses caused any radar cross section (RCS) target nulls to be moved at the aspect angles consistent with the dwell times. There was an early version of SAR. But the long collection times for SAR was counter to the rapid revisit needed for artillery detection. As a result, the DBS mode was often used during the tactical detection trials. And high PRF modes were needed due to the projectile velocities and the detection of explosive debris during impact.

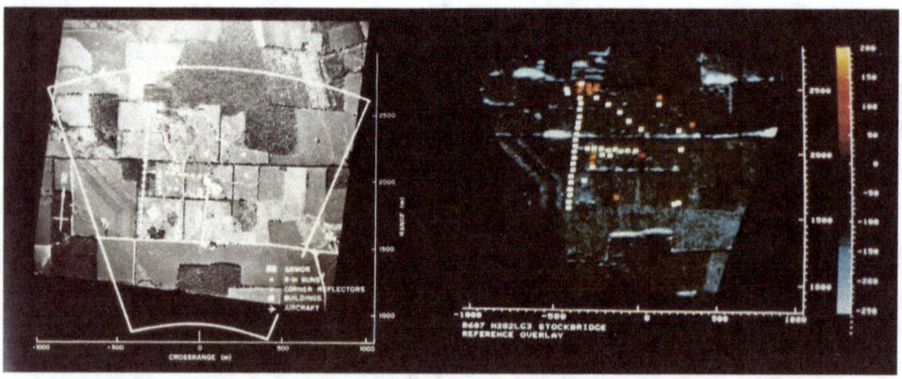

Photograph of test area — HOWLS ground display

Figure 3.16 Ground radar display from HOWLS Stockbridge NY collection [9]

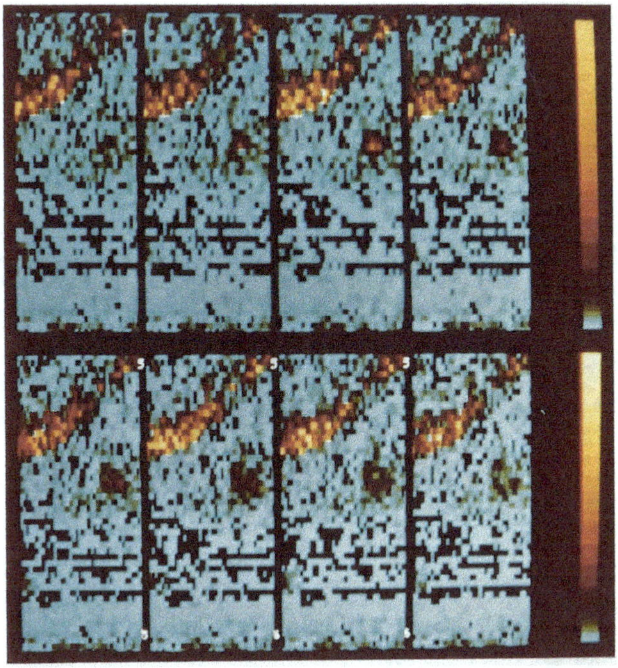

Figure 3.17 Detection of ground weapon impact with HOWLS [9]

The weapon locator mode of HOWLS was critical for the ability of a standoff sensor to detect and locate any weapons in the area. The target detection mode was using a high PRF waveform to obtain multiple detections in an area. Figure 3.17 shows a ground processing screen with multiple images approximately 0.2 s apart. The detection of the ground burst is pronounced in the series of images by the scattering of the surface debris in both range and Doppler. This shows that the shell impact is detected because of the large number of ground debris material along with the weapon particles are seen for 2 s. Hence, the weapon detection mode needs to have a revisit rate of approximately 1 s to ensure the detection and localization of the artillery burst.

HOWLS was taken to several operational test areas under the Netted Radar Program to demonstrate the utility of a small ESA radar for operational utility to the Army Division command and control. Figure 3.18 shows the display pictures from the tests at Ft. Sill Oklahoma, where two active displays and one passive display were located in the Target Integration Center. The left photo graph shows the road network as a background along with the HOWLS radar coverage lines. There are also various target types in track (walking personnel – targets 73 and 90, activity point – target 158, and helicopter – target 199). The picture on the right shows several tactical vehicles on a nearby highway. The upper number in the target tag is the track number, and the lower number is the target's current true speed in meters per second [10].

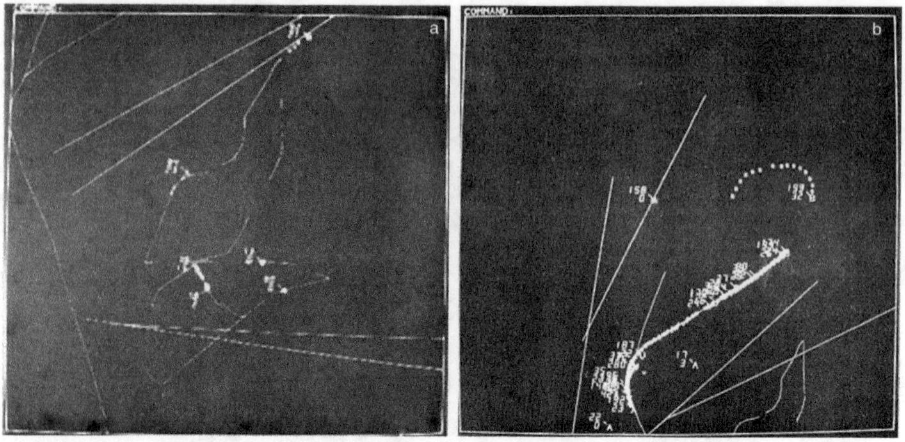

Figure 3.18 HOWLS tactical displays during Ft Sill demonstrations [10]

3.2 Battlefield surveillance radar modes

This section covers the history of battlefield surveillance radar modes up to 1985. Early radar operation was noncoherent due to the lack of waveform formation and maintaining phase during a processing interval. Early radars employed frequency agility to remove target or clutter scintillation effects on the returns. It was well understood that a frequency agile, noncoherent dwell could counter the scintillation of target returns due to temporal or spatial effects on the detection cell. So, the first mode to be discussed is the benefit of frequency agility on improved target detection. This agility can be on a pulse-to-pulse basis or from a between-coherent processing intervals.

The next series of modes use either spatial (or channel) comparison of returns or using Doppler characteristics of moving targets or radar platform for discrimination of the moving targets from background clutter. The DPCA mode was a technique that provided excellent suppression of background clutter in the detection of moving targets. The clutter Doppler variation due to the platform motion could also be used to improve the cross-range resolution of the surface spatial return. These two techniques were precursors to STAP processing and SAR image formation.

The next two modes developed for ground surveillance radar were GMTI and SAR. The former employs Doppler processing for discriminating moving targets from background clutter. However, the second mode utilizes very long integration aperture, synthetically formed due to the platform motion, to greatly improve the ground image resolution in both range and cross-range.

Finally, we will introduce interferometric processing between multiple channels to measure the terrain slope (cross-track interferometry) and measure the motion of slow-moving targets within the beam (along tract interferometry). Both these modes have been instrumental in both military target characterization and geophysical assessment of the scene.

3.2.1 Frequency agile target detection

Section 2.4.2 presented the theory behind using frequency agile waveforms for countering the scintillation of returns. This is primarily caused by multiple scattering centers within a resolution cell, providing fading of the returns as a function of the detection aspect angle. The Ku-band HOWLS radar featured a frequency agile (F/A) mode to determine the capabilities of detecting fixed targets in distributed clutter. A series of data collection runs at the Stockbridge NY test site to determine the benefit of using frequency agility of modest bandwidth coherent detection, but using the F/A on a pulse-by-pulse basis, as shown in Figure 2.24 [10].

The initial tests examined the improvement in ground mapping by using the F/A waveforms, where Figure 3.19 shows a radar map of the fielded area along with a progression of radar signal processing. The data from the frequency agile waveform are viewed with a range-pulse map in Figure 3.19b, and the scintillation is evident by all of the spikey returns. This is not an adequate surface surveillance mode. The first processing step was to process the pulses with an fast Fourier transform (FFT), as viewed in Figure 3.19c. However, the strong clutter at zero Doppler made it difficult to discriminate any targets or features. Finally, the FFT was carried out after two-pulse cancellation. The inherent clutter was cancelled, and the two tactical targets could be seen in the image.

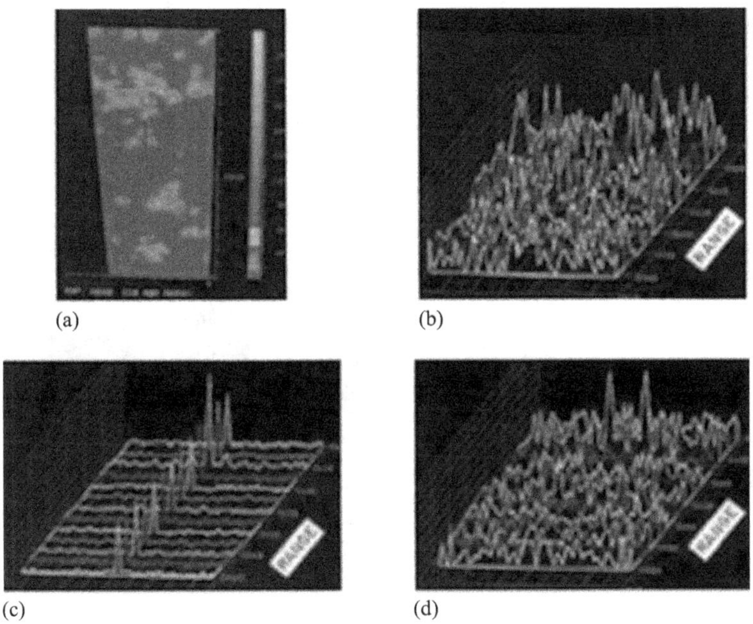

Figure 3.19 HOWLS frequency agile ground mapping mode [11]. (a) Radar map, (b) F/A data, (c) FFT of F/A data, (d) Range Doppler after MTI cancellation

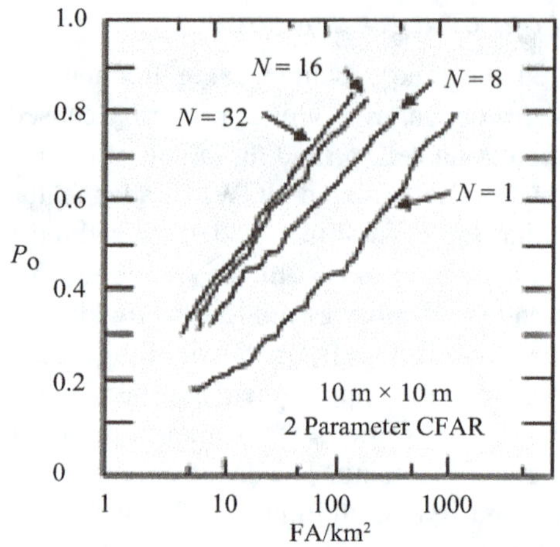

Figure 3.20 Target detection and false alarms for frequency agile waveform [11]

A series of data collections were used to test the statistics of a frequency agile waveform. Figure 3.20 provides the receiver-operating characteristics curve of detecting a calibrated target in clutter as a function of the number of frequencies. Empirically, this performance indicates that at least 16 different frequencies are needed to maximize the detection probability and minimize the false alarm density. Based on this evidence, a comparison of the image of Stockbridge NY test area with a single 15 MHz bandwidth waveform with the ground mapping waveform with a 32 frequency agile waveform. The results are shown in Figure 3.21. With a single 15 MHz bandwidth waveform, the resolution in the scene is partially obscured with scintillation of the background clutter. When the F/A waveform has 32 frequencies spread across 500 MHz bandwidth, the

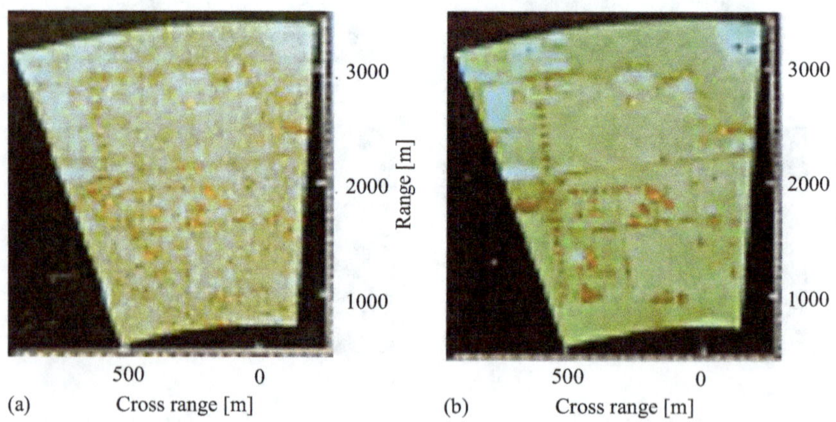

Figure 3.21 Ground map of Stockbridge with frequency agile waveform [11]. (a) 1 pulse [15 MHz], (b) 32 pulses [500 MHz]

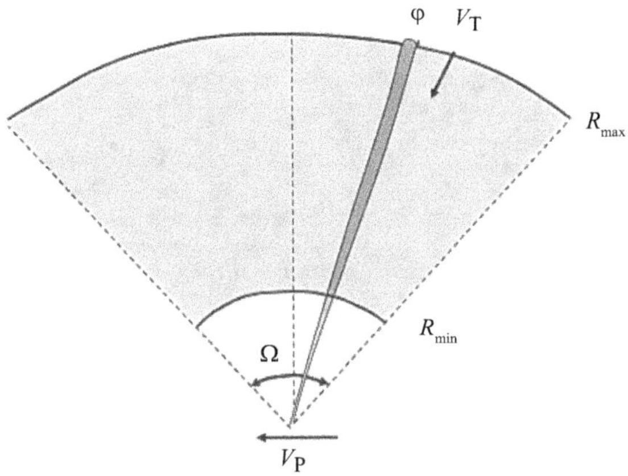

Figure 3.22 Wide-area GMTI surveillance

scintillation is significantly reduced. Figure 3.21b shows the details of the scene as well as all of the corner reflectors. The features match the photograph in Figure 3.16 with greatly reduced scintillation or false alarms [11].

3.2.2 Ground moving target indication

GMTI scans a narrow beam over an annular region, as shown in Figure 3.22, and dwells for a sufficient time that the moving target can be discriminated from the background clutter. The Doppler spread across the beam depends on the velocity of the radar platform and the beamwidth φ of the antenna. A target can be detected (assuming exo-clutter) when the radial velocity of the target is greater than the Doppler spread across the beam. This representation is approximate and assumes a rotating antenna. For electronic scanned antennas, the beamwidth will vary by the 1/sin of the scan angle, and this broadening of the beam gives a degradation of the MDV at the wider scan angles.

As previously discussed in Section 2.1.1, the moving platform causes clutter to spread across the antenna beamwidth. A moving target will have its own contribution to the radar return, as described in (2.11). And to be exo-clutter, the target radial velocity v_T needs to be greater than the ground clutter spread:

$$v_T \geq v_p \sin\left(\frac{k\lambda}{2D}\right) \tag{3.4}$$

where D is the width of the antenna, k the beam broadening factor from sidelobe control, and λ the wavelength.

In order to separate the moving target return from mainbeam clutter, moving target indication (MTI) processing is used. There are two stages of the processing:

- An FFT filter of N pulses is used as shown in Figure 3.23a within the PRF separation of the mainbeam clutter. The sidelobes of the mainbeam antenna is

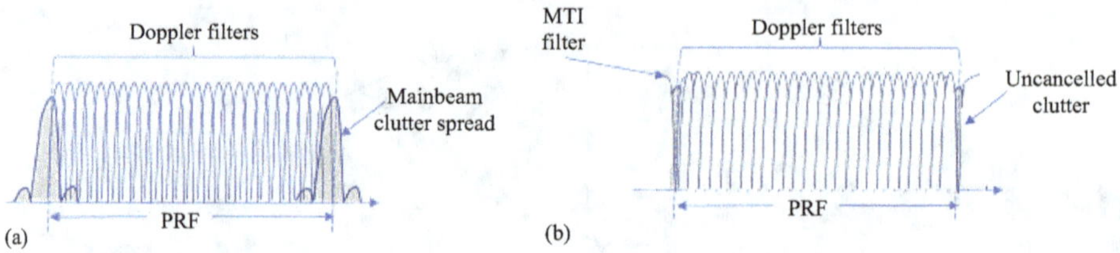

Figure 3.23 GMTI processing with Doppler filters. (a) Doppler processing, (b) MTI filter Plus Doppler processing

shown by the gray areas and contribute significant clutter to any potential target detections. This contribution is normally limited by applying sidelobe reduction in the FFT processing.

- Two-or three-pulse MTI canceller is used to suppress the mainbeam clutter, as shown in Figure 3.23b. This puts a null in the return and eliminates the clutter that is well correlated. The residual clutter after the MTI cancellation reduced the target signal to clutter for those filters separated from the Doppler null. The moving targets are then only competing with the sidelobe clutter from the adjacent FFT filter. It should be noted that the signal return should be motion compensated so that the mainbeam is at zero Doppler.

The normal application of processing steps is to use the MTI two- or three-pulse canceller to reduce the mainbeam clutter. Then the signal $z\{m\}$ is coherently integrated with FFT processing to minimize the CNR for slow targets.

The two-pulse (single delay) MTI canceller uses a filter given by the processing, as shown in Figure 3.24a [12]:

$$H(f_d) = 1 - e^{-j2\pi f_d T} \tag{3.5}$$

The processing entails taking the input digital signal $y\{m\}$ of I,Q data after the analog-to-digital conversion and passing it through the filter where each range bin is delayed by T and subtracted. And the three-pulse (double delay) MTI canceller is given by

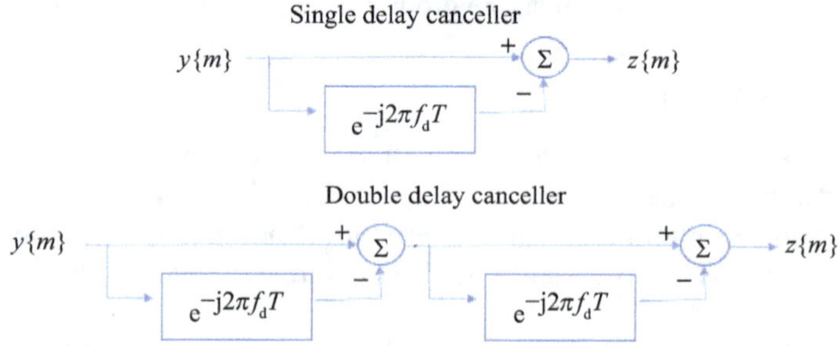

Figure 3.24 Cascaded MTI filters for mainbeam clutter suppression [12]

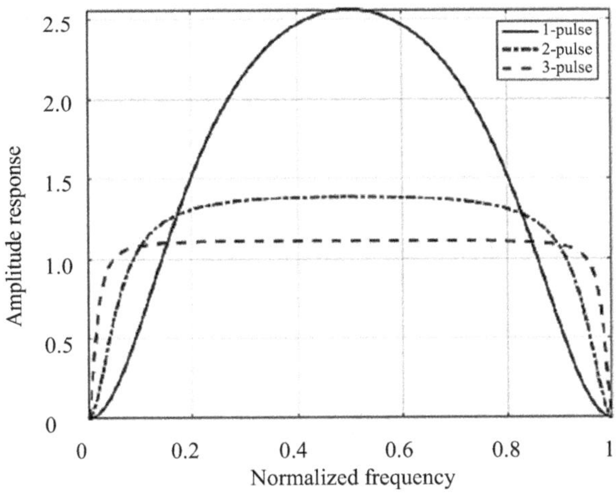

Figure 3.25 MTI filter for suppressing mainbeam clutter

$$H(f_d) = 1 - 2e^{-j2\pi f_d T} + e^{-j4\pi f_d T} \tag{3.6}$$

where $f_d T$ is the normalized Doppler frequency, as shown in Figure 3.24b.

Several delays can be included to provide the desired cancellation of the mainbeam clutter. The N-pulse canceller can be expressed by

$$H_N(z) = \left(1 - e^{-j2\pi f_d T}\right)^{N-1} \tag{3.7}$$

The response to this canceller is shown in Figure 3.25, where the two curves are for a one-pulse delay and a two-pulse delay for the subtraction.

It should be noted that the two- or three-pulse cancellers only remove the mainbeam clutter. An N-pulse FFT follows the MTI canceller, which provides the coherent integration of the N pulses and an attendant processing gain. This processing provides a Doppler bandwidth of approximately PRF/N for estimating the radial velocity to the targets. However, the Doppler bandwidth is broadened by any filter weighting for sidelobe control.

The minimum detectable velocity (MDV) is an important measure of clutter cancellation. A choice of acceptable loss in the signal to clutter is chosen to determine the MDV for a particular radar mode. This threshold could be as low as 5 dB of loss or as much as 10 dB depending on the clutter type and system consideration of types of slow-moving targets. Moreover, the MDV must also consider the radial velocity of the targets during the search and tracking as the platform moves along the ground reference collection area. If the target is moving in a straight line, its project to the radar radial line will change as the aircraft moves along. If the targets have a Doppler within the mainbeam (endo-clutter), then further adaptive processing is used. This technique of STAP has been developed extensively in the past 20 years and will be covered in Chapter 7.

A first-order analysis of the MDV as a function of antenna size and frequency can be obtained from (3.4). The residual clutter after the MTI cancellation has a reduction of CNR for the lower Doppler filters. The moving targets are then only competing with the sidelobe clutter from the adjacent FFT filter. Figure 3.26 illustrates the improvement in MDV of moving targets as a function of the antenna size and the platform velocity. The calculation is made for antenna pointing broadside of the platform motion. If the ESA scans off of broad side, the width of the antenna beam will increase with $1/\sin \theta$.

The size of the antenna is the most important factor in determining the MDV of the radar. Two examples of MDV versus antenna size for X-band (10 GHz) and UHF (400 MHz) are shown in Figure 3.26. It can be seen that an X-band radar can provide an MDV under 1 m/s, with an antenna below 3 m in length, and at a platform velocity of 200 m/s. This is sufficient for many applications. However, for UHF, the antenna needs to be over 5 m in length to detect a 10-m/s target. This very high MDV limits the utility of an UHF GMTI system for any appreciable velocity.

It is obvious why the microwave bands have been preferred for GMTI. More adaptive processing such as DPCA and STAP are needed to obtain reasonable MDV. However, both will cause some loss of detection due to errors in the correlation between the independent channels.

3.2.3 Doppler beam sharpening

Ground imaging had been limited by the cross-range resolution obtained by the antenna. Two techniques were developed in the 1970s for battlefield surveillance radar: DBS and SAR. This chart will show the benefit of DBS for improving the cross-range resolution for discriminating terrain features. The advantage of DBS is its considerably shorter integration times and processing complexity than SAR imaging mapping, while it provides significant cross-range resolution than a real beam ground map.

The approach for DBS is to place multiple Doppler filters across the beam width, as shown in Figure 3.27. This can be done for broadside and for squint angles. The ground Doppler return was shown to be based on the cone angle ψ_{cone} around the platform velocity vector v_{plat} [14]:

$$f_{\text{d}} = \frac{2\,v_{\text{plat}}}{\lambda} \cos \psi_{\text{cone}} \tag{3.8}$$

This Doppler effect is also related to the scan angle of the antenna broadside angle θ by

$$f_{\text{d}} \approx \frac{2\,v_{\text{plat}}}{\lambda} \sin \theta \tag{3.9}$$

Returns from different azimuth angles have different Doppler shifts. The projection of the Doppler centroid at the squint angle is used to select a number of filters across the beam. This will provide the cross-range resolution to discriminate terrain features and any landmarks for navigation and image processing.

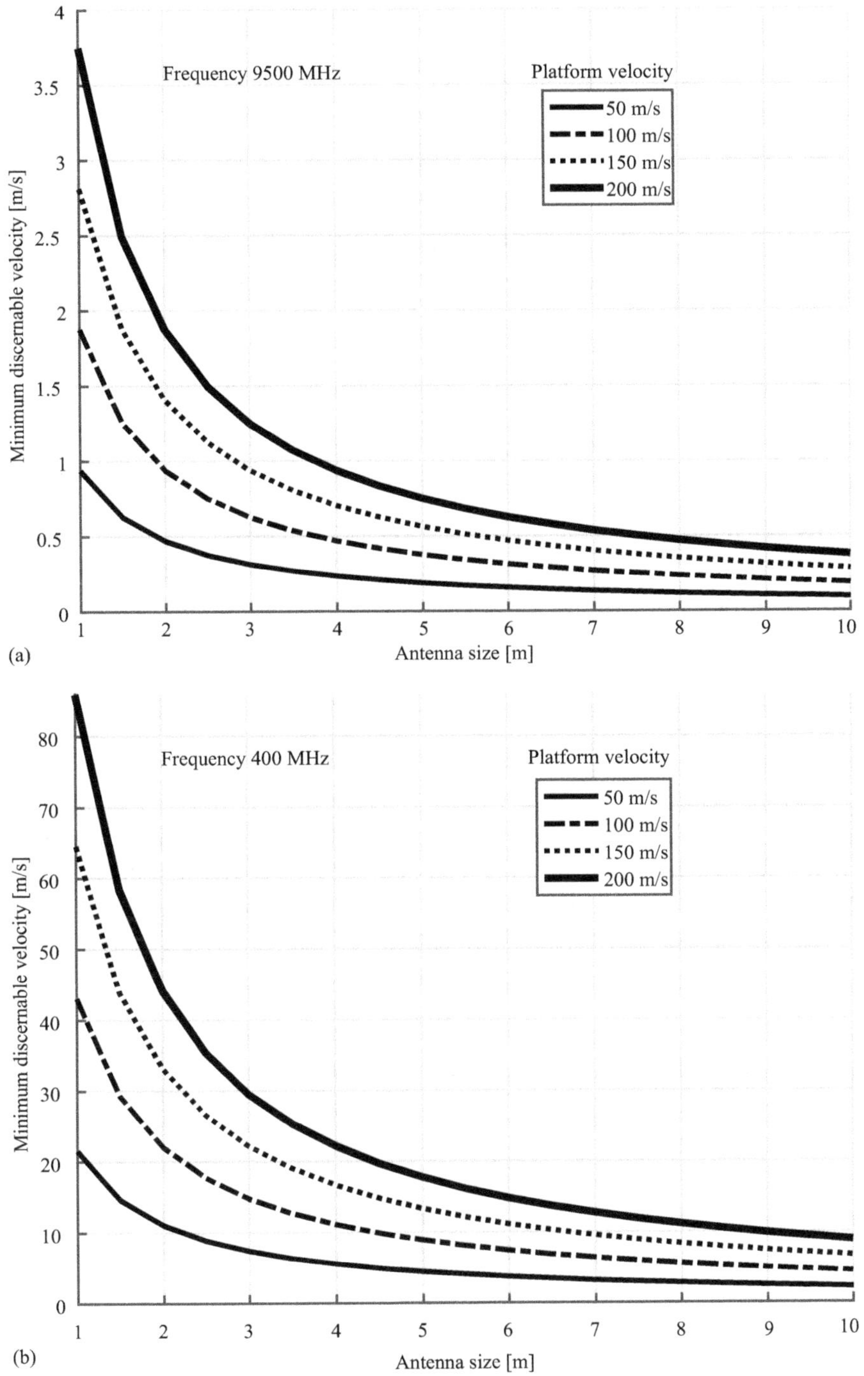

Figure 3.26 *MDV versus antenna size. (a) X-band GMTI and (b) UHF GMTI*

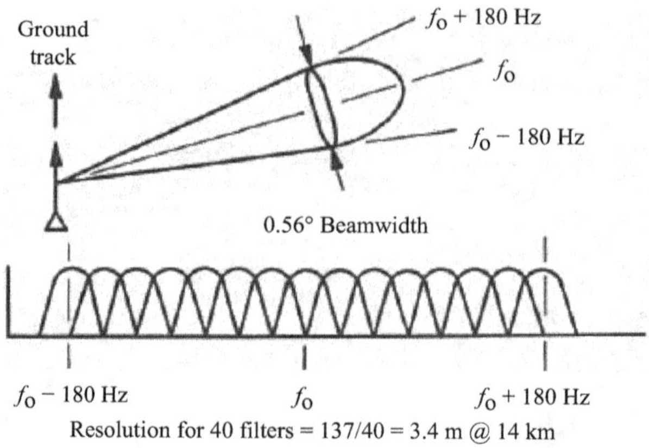

Figure 3.27 DBS with squint angle [13]

The image formation in DBS can also be considered from the point spread function of the image collection, as the aircraft collects data. This approach will be developed more completely in Chapter 7 with UWB SAR image formation. However, it is instrumental in covering it here where the collections for DBS are much shorter. The geometry from the radar to each point on the surface is given by the following simple relationship [12]:

$$t(u;x,y) = \frac{2}{c}\sqrt{(u-x)^2 + r^2} = \frac{2\,r}{c}\sqrt{1 + \frac{(u-x)^2}{r^2}} \tag{3.10}$$

where u is the location of the radar platform and (x,y) are the locations of the scatterers in the image area. When the along-track offset of the platform to the scatterer is very small compared to the slant range r to the scatterer, (3.10) can be simplified to

$$t(u;x,y) \approx \frac{2}{c}\left(r + \frac{u^2}{2r} - \frac{ux}{r} + \frac{x^2}{2r}\right) \tag{3.11}$$

Only the terms with u capture the platform motion. So, the time delay from the platform to the array of scatterers is given by

$$\Delta t\ (u;x,y) \approx \frac{2}{c}\left(\frac{u^2}{2r} - \frac{ux}{r}\right) \tag{3.12}$$

And the phase progression $\Delta\psi$ due to the time delay variations is given as

$$\Delta\psi \approx \frac{4\pi}{\lambda_c}\left(\frac{u^2}{2r} - \frac{ux}{r}\right) \tag{3.13}$$

The first term does not depend on the along-track position of the scatterer. It is the same for all scatterers at a given cross-track range and serves only to modulate the data by a quadratic phase. These quadratic phase functions are generally negligible

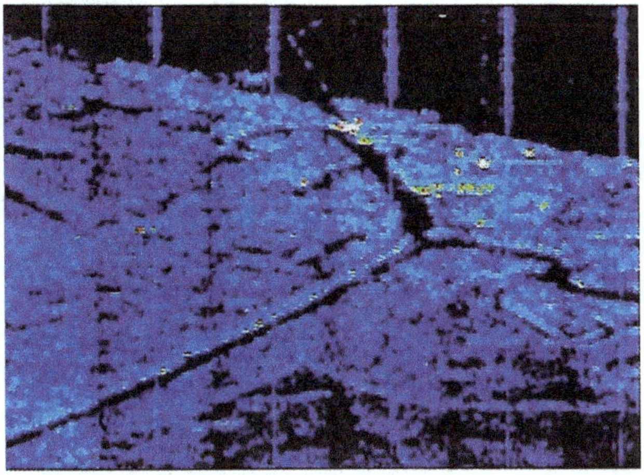

Figure 3.28 DBS image with multibeam surveillance radar [13]

if the angular deviation is less than a $\pi/2$ phase cycle. Baseline DBS imaging geometries meet this constraint, so the first term in (3.13) can be omitted. As a result, the DBS phase term can be approximated by

$$\Delta\psi = -\frac{4\pi}{\lambda_c}u\left(\frac{x}{r}\right) \tag{3.14}$$

This short derivation of the DBS phase function is important in explaining the linear position of scatterers within the narrow beam. The Doppler processing of these scatterers is uniformly distributed in the image, providing the improvement in the cross-range resolution. By Doppler filtering of the radar returns the cross-range resolution ground image can be improved by the Doppler filter width compared to the antenna beamwidth. The cross-range resolution improvement is obtained by the number of Doppler filters across the mainbeam extent of the radar illumination.

Figure 3.28 shows the DBS performance of the multibeam surveillance radar developed by General Electric for the Air Force and DARPA in 1985. Five simultaneous receive beams were formed in the aperture, in order to provide a wider image at the same cross-range resolution [13]. This technique will be explained in more detail in Chapter 9.

3.2.4 Synthetic aperture radar

SAR systems have been used for both battlefield surveillance and geoscience applications. The GeoSAR system incorporated both X-band and P-band interferometric radars on a platform. This provided the ability to collect multiple images for comparison of the returns provided by angle and wavelength diversity. Figure 3.29 shows both the two independent radar images each formed with a 160 MHz bandwidth. The system utilized 1-m post-spacing to provide DTED 3 estimation of the

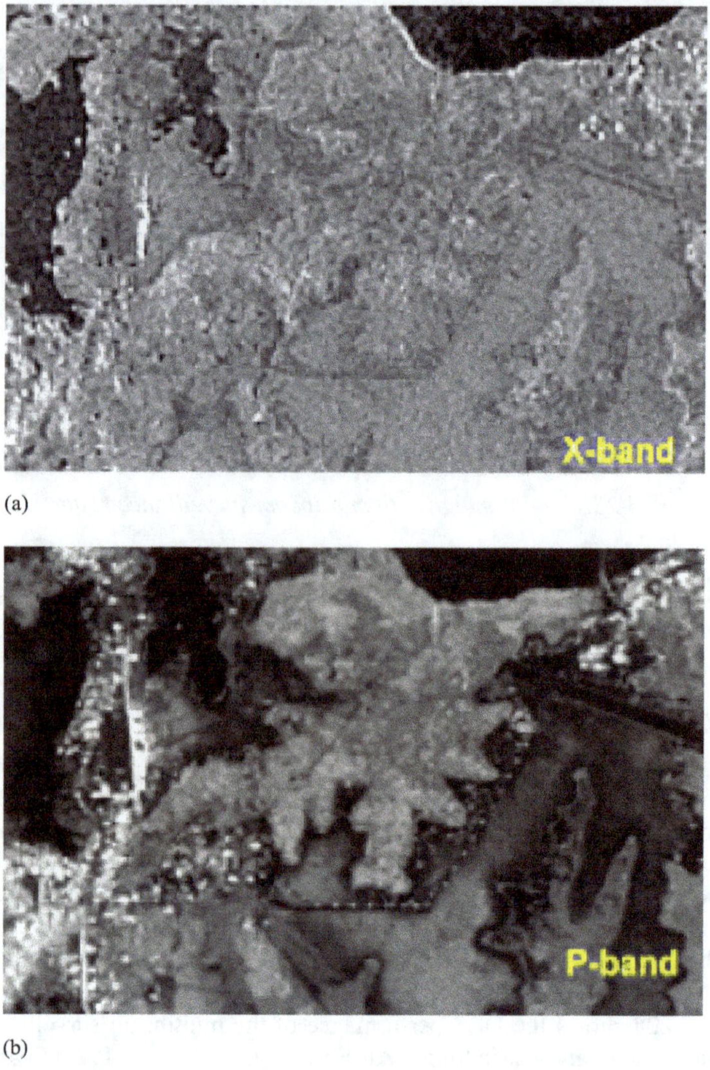

Figure 3.29 *SAR images from GeoSAR radar [15]*

terrain height. The images are from the Brazilian Amazon basin where the tests for terrain height measurements were conducted [4].

The X-band image clearly showed the tops of the jungle in a mountainous region and with underlying rivers. The P-band image shows the returns from below the tops of the trees. As was discussed in Chapter 2, the P-band radar can penetrate the forests, while the X-band is scattered by the tops of the trees. This is clearly seen in the picture. Another important physical aspect is that P-band polarization can give a strong indication of the types of scattering. This has been shown to be useful for characterizing the forests [1].

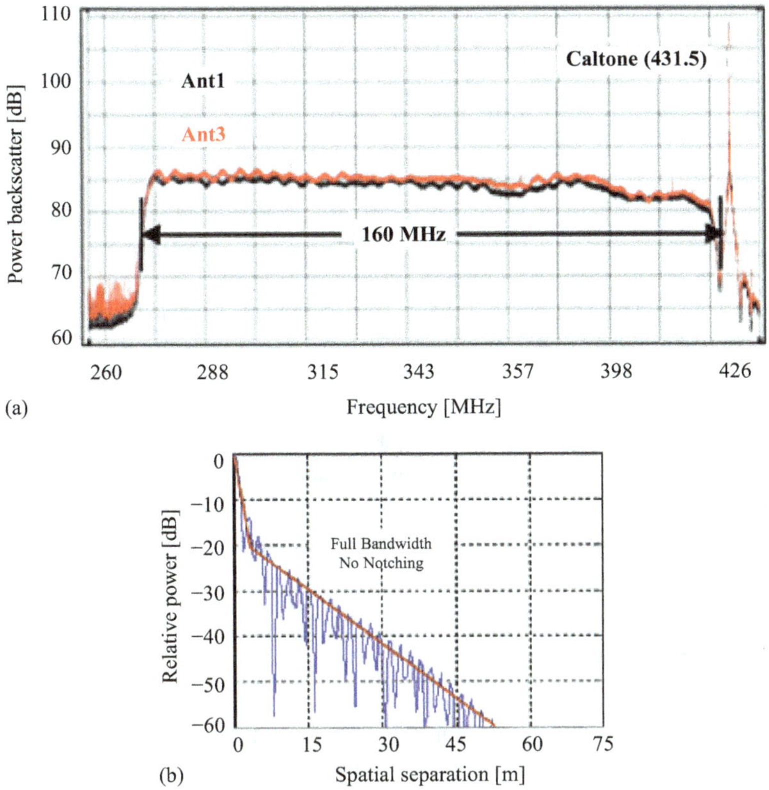

Figure 3.30 *UWB SAR waveform characteristics [15]. (a) UHF waveform spectrum and (b) UHF and X-band IPR*

GeoSAR was one of the early UWB systems with over 30% bandwidth. The UHF pulse spectrum is shown in Figure 3.30a. Also shown in the figure is the calibration tone (caltone) at 431.8 MHz, which is used for ensuring the coherence between the two receive channels. Figure 3.30b shows the resolution of the two frequency systems with the inherent roll off of a rectangular pulse. This is especially relevant at P-band where there are significant issues with obtaining the full bandwidth. The development of interferometric SAR modes will be given in Chapter 7, and the UWB mode and its difficult acceptance in the RF environment will be presented in Chapter 8.

There are several simplistic equations for SAR processing. The first is range resolution δ_R, which depends on the bandwidth B of the signal:

$$\delta_R = \frac{k_R c}{2B \cos \gamma_g} \tag{3.15}$$

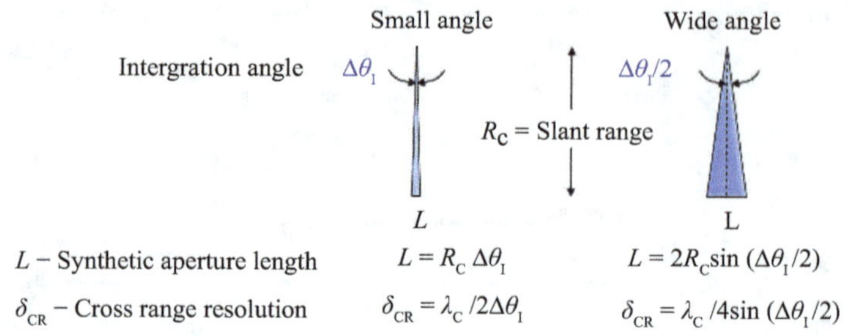

Figure 3.31 Integration approximation for small-angle SAR [1]

where the range spreading factor for sidelobe weighting is k_R. To estimate the resolution in the ground plane, one can divide by the cosine of the grazing angle ($\cos\gamma_g$) between the slant range and the local ground plane.

Chapter 1 has already discussed the fractional bandwidth, which is a critical parameter for understanding the UWB criterion. If ΔB is greater than 25%, the system is considered to be UWB.

$$\Delta B = \frac{2\,(f_H - f_L)}{\left(f_H + f_L\right)} \tag{3.16}$$

The cross-range resolution is estimated by the integration angle θ_I of a resolution cell in the scene. The synthetic aperture length is $R\tan(\theta/2)$, and the angle θ is approximated by L/R_a. Using this approximation, the cross-range resolution δ_{CR} is given by the following equation, valid for broadside and without a squint angle:

$$\delta_{CR} = \frac{k_{CR}\,\lambda C}{4\sin\left(\theta_I/2\right)} \tag{3.17}$$

The synthetic aperture length is given in the literature in several forms, depending on the SAR integration angle, as shown in Figure 3.31. For a small-angle approximation, the integration length is simply the slant range times the angle in radians. For wide angle, the integration length needs to use twice the sine of the half integration angle $\Delta\theta_I/2$. For small and wide angles, it is sufficient to use the slant range from the center of the SAR integration length L. The small-angle approximation is normally used for high frequencies and unweighted, coarse cross-range resolution.

However, for UWB operation and when the integration angle is much larger than a few degrees, a more exact interpretation of the geometry must be used. This is shown in Figure 3.32, where the slant range at the beginning and end of the data collection has to be used to correctly calculate the SAR length. The cross-range resolution is given by [1]

$$\delta_{CR} = \frac{\lambda_c\,k_{CR}\,R_C}{2\,L_{eff}\sin\alpha_{DC}\cos\left(\theta_I/2\right)} \tag{3.18}$$

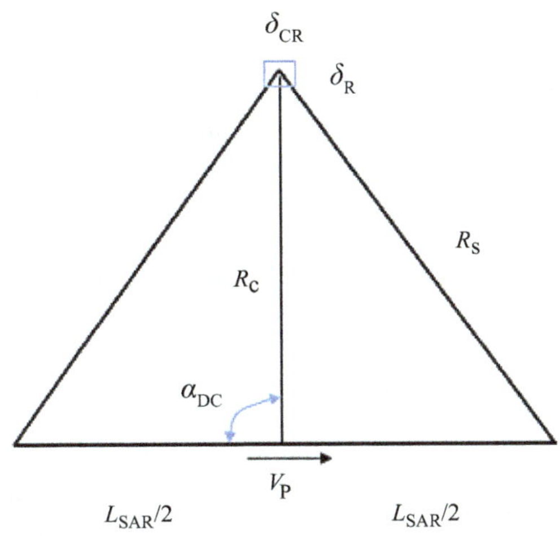

Figure 3.32 UWB SAR cross-range resolution geometry [1]

where the slant range to the center of the image collection is R_C and the squint angle is sin α_{DC}. The effective length of the SAR collection is given by L_{eff}, based on the desired cross-range resolution δ_{CR} and by expressing the $\cos(\theta_I/2)$ in terms of the cross-range resolution. This in effect calculates the effective SAR length based on the edges of the integration path and not at the center.

$$L_{eff} = \frac{\lambda_c\, k_{CR}\, R_C}{2\,\delta_{CR}\sqrt{1-\left(\dfrac{\lambda_c\, k_{CR}}{4\,\delta_{CR}}\right)^2}} \tag{3.19}$$

Once the effective SAR length is estimated, the integration time t_I is given based on the average velocity of the platform v_P:

$$t_I = \frac{L_{eff}}{v_P} \tag{3.20}$$

It should be noted that any squint angle α_{DC} will dilute the cross-range resolution. It should also be noted that these resolutions are expressed in the slant range. In the ground plane, the resolution is also diluted by the grazing angle projection into the ground plane, cos γ_g.

Figure 3.33 illustrates the required integration angle for two frequencies UHF – 450 MHz and L-band at 1,400 MHz, which are representative for penetrating SAR images – both foliage and ground. The UHF and L-band collections are evaluated at 1-m resolution. Whereas the X-band and Ku-band SAR collections are applicable to imaging in open areas, and they are both evaluated at 0.3-m resolution. The 3-m curve in Figure 3.33 indicates the representative integration angles for VHF SAR with its meter wavelength.

The synthetic aperture length has been shown in (3.19) as related to the integration angle and the standoff range. Figure 3.34 illustrates the integration times required for a 150-m/s platform at a variety of standoff ranges, and for the "penetrating" frequencies at UHF and L-band. For L-band, a 1-m resolution requires a collection times are over 50 s, which is contrasted to 100 s with the longer wavelength at UHF FOPEN SAR. The impact on the system is major with these very long SAR integration times. First, an entire pass is consumed in collecting a single UHF swath of data. In addition, the variation of the terrain can be large over this geographic area, requiring detailed estimation of the slant range to the surface. The demands on the motion measurement system and MOCOMP are amplified. As will be presented in Chapter 5 with the UWB SAR image formation processing, there are limitations on choice of algorithms for efficient and effective focus of the image.

Figure 3.35 shows the integration times for microwave frequencies versus imaging slant ranges. In these cases, a cross-range resolution below 0.5 m can be easily obtained under 4 s out to 25 km at X-band and 3 s at Ku-band. The integration times are shorter than the long-wavelength UHF and L-band, and the demands on the motion measurement and compensation are somewhat relaxed.

A very important system requirement for UWB SAR is the antenna coverage for these large angles, operating over a very-wide-frequency bandwidth. Two developmental FOPEN SAR antennas, as shown in Figure 3.36, were developed under DARPA's sponsorship in 1997. One was the GeoSAR antenna by Jet Propulsion Laboratory for the GeoSAR platform, and the second antenna by Ball Aerospace as a technology risk reduction to greatly improve UWB fully polarimetric antennas.

A notional gain objective was derived to provide for integration angles of 30–45° and operation over frequencies between 200 and 600 MHz. This is the line termed "specification" in the line plot in Figure 3.37. The GeoSAR system would operates only over a 160 MHz bandwidth centered at 350 MHz. But the cavity backed radiator was much wider bandwidth. Because these antennas could be installed in the wingtip pods of the Gulfstream aircraft, they gain was optimized broadside by

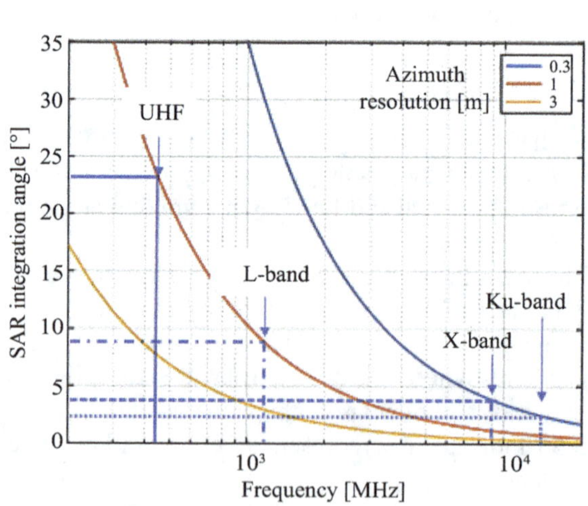

Figure 3.33 UWB SAR integration requirements

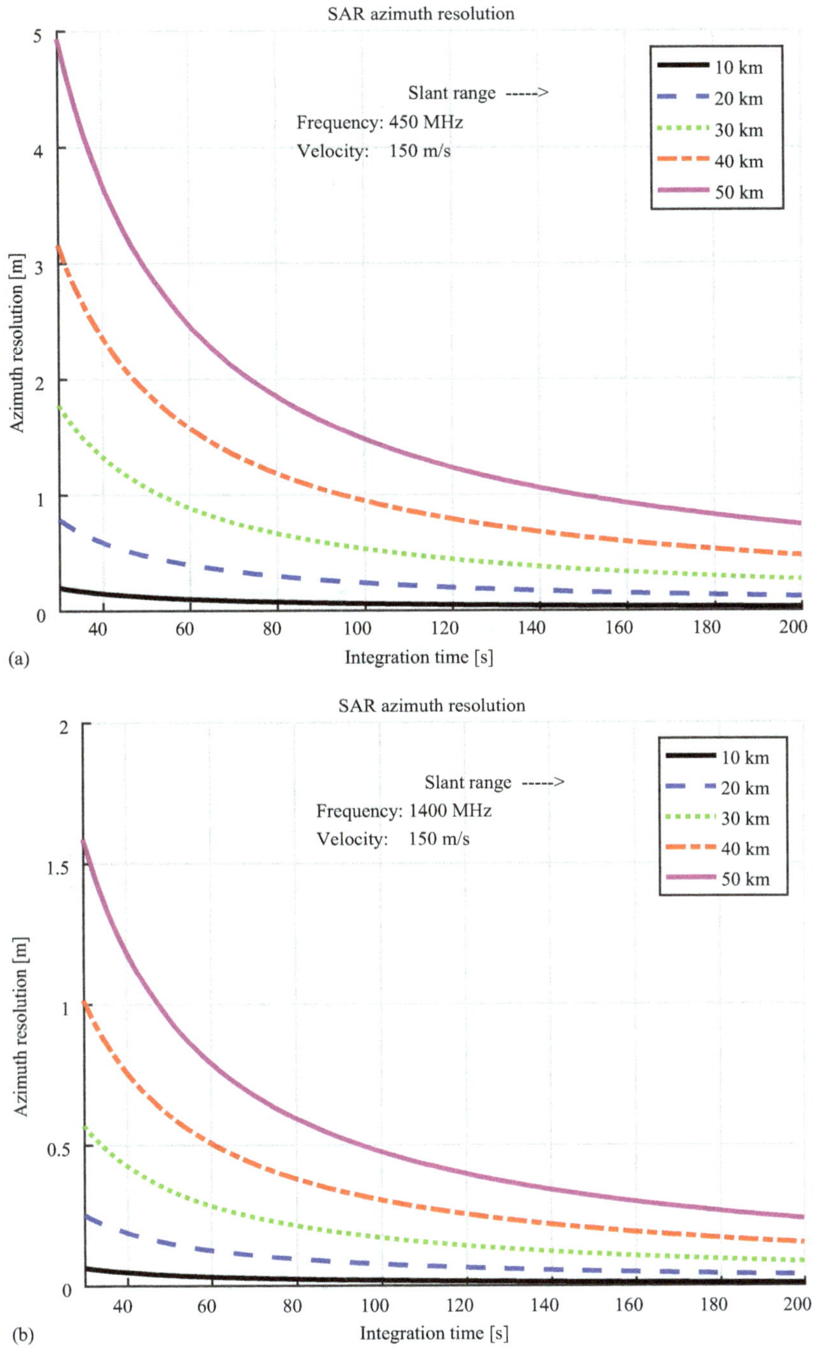

Figure 3.34 Penetrating frequency SAR resolution versus integration times at various slant ranges. (a) UHF and (b) L-band

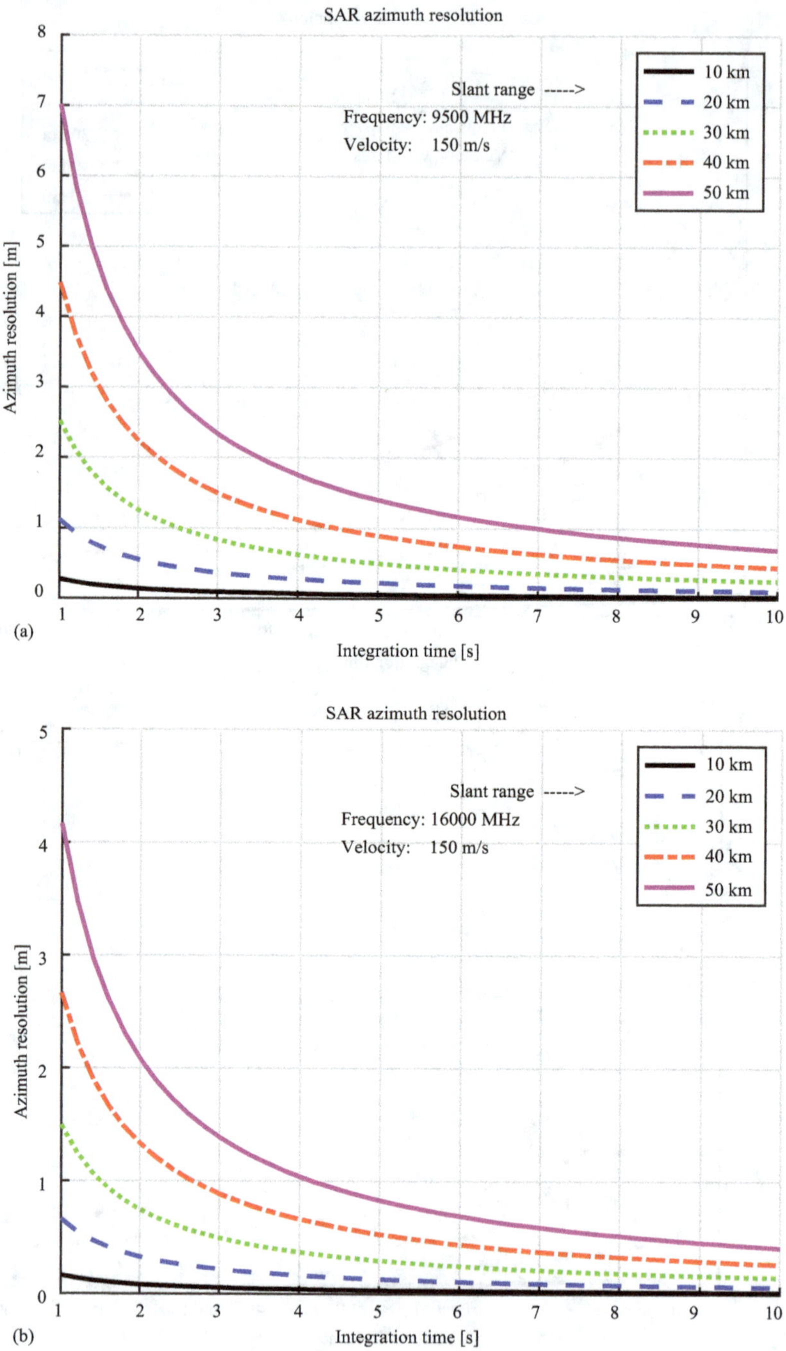

Figure 3.35 Microwave frequency SAR resolution versus integration times at various slant ranges. (a) X-band and (b) Ku-band

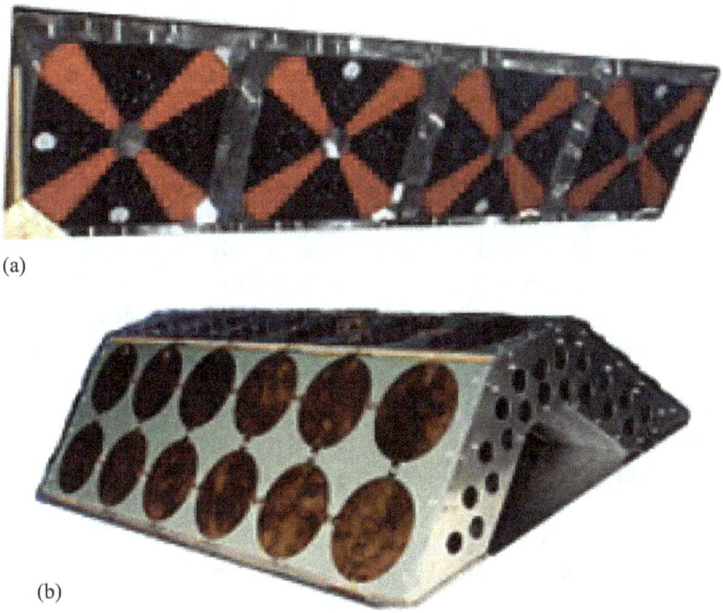

(a)

(b)

Figure 3.36 UWB SAR full polarimetric antennas for foliage penetration [16]. (a) GeoSAR and (b) Ball FOPEN

phyically tilting the antennas to the center of the image swath. It provided gain margin and met the 20 dB polarization isolation objectives.

The Ball antenna was to be installed in a centerline pod and could operate by switching feel points out of either side of the aircraft. As a result, the antenna was conformal in design, and the gain at the 35° depression angle suffered. However, the

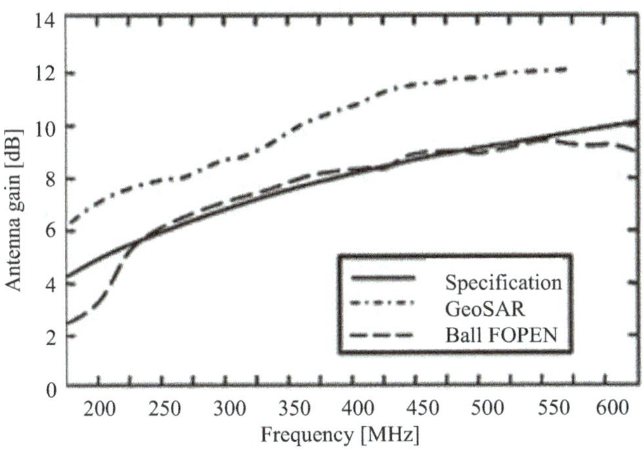

Figure 3.37 UWB antenna gain versus frequency [16]

antenna achieved the specified gain over 80% of the bandwidth. It too achieved the 20 dB polarization isolation.

3.3 Surveillance mode conclusions

Microwave battlefield surveillance radars are widely accepted in the radar systems that have been fielded in the past. However, the early UWB SAR systems had significant variation in waveform and processing techniques given the available RF and digital component capabilities. These limitations have hindered the adaptation of wideband and UWB waveforms to provide the spatial and spectral characteristics needed in the new operational capabilities.

These early radar systems were strongly dependent on early RF transmitter and antenna technology. The waveforms used for either fixed or moving targets were single modes tasked sequentially due to bandwidth or spectral interference constraints. Moreover, receiver technology required time/bandwidth control to avoid interference.

New technologies post-2000 enabled rapid improvements in the design and applications for area surveillance, along with commercial and geoscience requirements. These technologies include novel and small UWB antenna technologies, coherent and direct waveform synthesis, high dynamic range transmit and receive components, and high-performance computers for real-time processing. The antennas have been greatly reduced in volume and mass and provided fully polarimetric performance. It is important to design the power aperture system to coexist with the environment. Direct digital synthesis of waveforms, wideband conformal antennas and wideband signal processing are enabling new insights into system design.

The previous battlefield surveillance radar mode examples are instructive in solutions to a wide variety of system engineering problems for fielding an UWB radar system. The availability of components for both the antenna and transmitter design evolved significantly during the decade covered.

Because of the rapid growth of cellular communications, the problems of fielding an UWB radar have increased. However, the same technology that is present for cell phones and wireless communications can now be applied to radar systems. It is important to note that the rapid increase in the use of personal communications devices has further restricted the available bandwidth for UWB radar operation. The radar community needs to explore new techniques to greatly reduce the interference of these systems to the commercial radio community. Later chapters will explore these needed radar innovations.

References

[1] Davis M.E. *Foliage Penetration Radar – Detecting and Characterization of Objects Under Trees*. Raleigh, NC: Scitech Publishing; 2011. Chapter 2.

[2] Muehe C.E., Labitt M. 'Displaced phase center antenna technique'. *Lincoln Laboratory Journal*. 2000;**12**(2).

[3] Bryant T.G., Morse G.B., Novak L.M., Henry J.C. 'Tactical radars for ground surveillance'. *Lincoln Laboratory Journal*. 2000;**12**(2):341.

[4] Hensley S., Chapin E., Freedman A., *et al.* 'First P-band results using the GeoSAR mapping system'. *Proceedings 2001 IGARS Symposium, Sydney Australia*. May 2001:126–8.

[5] Tsandoulas G.N. 'Unidimensionally scanned phase arrays'. *IEEE Transactions on Antennas and Propagation*. 1980;**AP 2**(1):86–99.

[6] Tsandoulas G.N., Nader M.A. 'A high precision displace phase center array antenna'. *Antenna and Propagation Symposium Digest*; Amherst, MA; 1976. pp. 56–7.

[7] Zhang Y. *Signal Modeling and Simulation of Multichannel Bistatic System*. Macy, NY: Stiefvater Consultants; 2000.

[8] Dong Y. *Phased Array Radar Data Processing Using Adaptive Displaced Phase Centre Antenna Principle*. DSTO-RR-0334. Edinburgh, South Australia: Defence Science and Technology Organisation; 2007.

[9] Oot R.F. *"HOWLS experimental airborne radar". GE AESD Marketing Brochure*. Utica, NY: General Electric Aerospace; 1976.

[10] Knittel G.H. *Phase II Netted Radar Demonstration*. MIT Lincoln Laboratory Project Report TT-42, October 1980, DTIC A103684 (USAF publicly released September 1981); 1981.

[11] HOWLS experimental data pictures courtesy of MIT Lincoln Laboratory. 'HOWLS detection performance, tactical technology'. *Group 47, Battlefield Surveillance Program, 1989, DARPA publicly released*; 2020.

[12] Richards M.A. 'Doppler processing, Chapter 17'. *Principles of Modern Radar Basic Principles*. Edison, NJ: Scitech Publishing; 2010. pp. 625–76.

[13] Mancini H.J. A multimode battlefield surveillance radar. *Proceedings 1984 SAE Avionics, Guidance and Control Symposium*; Long Beach, CA; 1984.

[14] Showman G.A. 'An overview of radar imaging, Chapter 21' in Richards M.A., Scheer J.A., Holm W.A., Pub S., Edison N.J. (eds.). *Principles of Modern Radar – Basic Principles*. Vol. I. Scitech Publishing; 2010.

[15] Davis M.E., Reis J.J. "Operation of GeoSAR foliage penetration modes in spectrum regulated environment". *Proceedings 2015 IEEE Radar Conference*; Pretoria, South Africa; October 2015.

[16] Tomlinson P.G. *Implications of the Frequency Dependence of the P-3 Antenna Gain*. Arlington, VA: Unpublished Memo DSA Inc; 1994.

Chapter 4
UWB antennas

This chapter covers the technology development of ultra-wideband (UWB) antennas, including electronically scanned arrays (ESAs) and multichannel antennas for synthetic aperture radar (SAR) and ground moving target indication. Early battlefield surveillance radars employed passive ESAs to enable rapid scanning and multimode operation as a platform flew across the ground reference collection area. Passive arrays were employed because of the availability of reciprocal phase shifters that fit within the half-wavelength spacing between elements. There was also a need in the surveillance radar designs for radio frequency (RF) amplitude weighting of the array to control the sidelobes and limit effects of strong clutter and RF interference. Moreover, the ESA architectures developed the technology to form multiple beams or channels for additional spatial degrees of freedom. These degrees of freedom were used for detection and localization of the targets or exploitation of ground features.

Up until 1970, active electronically scanned arrays (AESAs) were only being used at frequencies at S-band and below. But they were prohibitively expensive for consideration above S-band. It was not until the development of monolithic microwave integrated circuits (MMICs) that active element arrays would be available in the late 1990s [1]. This revolutionary capability enabled extension of the bandwidth and improved advanced signal processing for extracting more information from the radar waveform returns. More importantly, the AESA could provide a significant increase in the effective power aperture due to reduction of losses between the final power amplifier and the radiating element and reduced noise figure on reception by eliminating the same loss.

The transition to UWB arrays was gradual, until the cost-effective development of wideband digital control processing and wide dynamic range analog-to-digital converters (ADCs). These technologies would enable effective control of the channel match over the ultra-wide bandwidths associated with fine range and cross-range resolutions. Moreover, a rapid technology development in the commercial communications world has been leveraged for both military and civilian radars.

4.1 Uniform linear array

The most basic ESA is a uniform linear array (ULA), as illustrated in Figure 4.1. There are N elements separated by a distance d, and a set of phase shifters φ_n, which can steer the beam. For a perfectly collimated linear array, the phase front requires

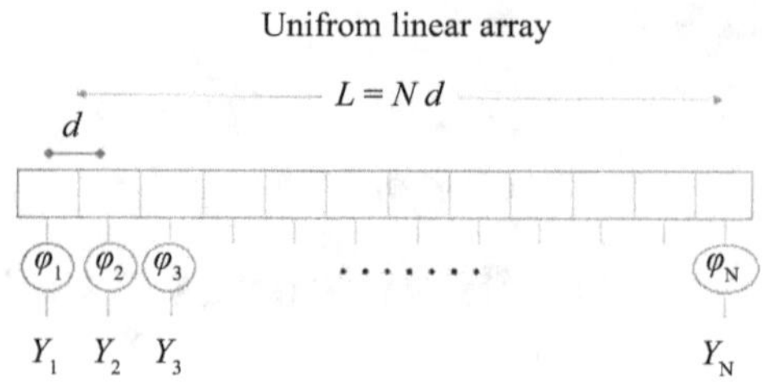

Figure 4.1 Uniform linear array

identical transmission distance from the input to the radiating element. The beam can be steered to an angle φ_S by the progression of phase shift across the array.

An ULA can form an RF beam Y by combining the output from each element ($n = 1{:}N$) and the use of a phase shift at each element to focus the beam [2].

$$Y = \sum_{n=1}^{N} A_n \exp\left(-j \frac{2\pi}{\lambda} nd\left(\sin\varphi - \sin\varphi_s\right)\right) \tag{4.1}$$

For uniform weighting $A_n = 1$, the antenna will have first sidelobes at –13 dB. For lower sidelobes on the array, a weighting is applied to each element in the feed structure on either or both transmit and receive. The beam is steered to angle φ_s, by providing a progressive phase shift θ_n:

$$\theta_n = \frac{2\pi}{\lambda} nd \sin\varphi_s \tag{4.2}$$

The beams are well behaved as long as the separation between the elements is a fraction of a wavelength. This separation determines the maximum scan angle without grating lobes. If the elemental separation is less than a half wavelength, then the beam can scan a hemisphere without having a grating lobe. If the separation is greater than $\lambda/2$, then the scan volume will be limited such that grating lobes do not appear in the beam space. Depending on the separation between the elements and when the phase θ_n is modulo 2π, there can be a grating lobe in the azimuth pattern affecting the array gain [2].

$$d_c \leq \frac{\lambda}{\left(1 + \left|\sin\varphi_s\right|\right)} \tag{4.3}$$

This critical element separation d_c between the elements is important for providing an azimuth scan sufficiently wide to meet the surveillance requirements. It should be noted that these developments are for narrow band operation using phase shifters with digitally controlled phase shift. For wideband arrays, the control will have to be a combination of time and phase, as presented in Section 4.4.

The early ESA designs were driven by having a control element small enough to fit within each element's cross-section, typically a half wavelength. At L-band, the wavelength is 30 cm, and at X-band, it is 3 cm. So, the microwave frequencies made it very difficult to achieve the element dimensions, without a complicated and heavy structure. A horizontal, linear array was the easiest to make, since the vertical dimension was not needed for 2D beam steering. For the sake of this analytical derivation, we will consider a simple linear array. When the elements are uniformly combined, the array has a theoretical peak sidelobe of −13 dB and a linear reduction of the sidelobe levels, as shown in Figure 4.2. This example array has 64 elements separated by half-wavelength spacing, and two scan examples are shown, 0° and 30°. Because the array gain has been normalized to unity, the anticipated fall off of gain as the beam is scanned is not shown. When the array pattern $A_n(\varphi)$ is factored without normalization, the antenna gain will vary with angle due to the projected array area and the decrease of the element gain A_n with angle φ.

The following sections will illustrate early surveillance radar linear arrays with passive element control technology. The most important aspect of this technology was that the amplitude control for sidelobe reduction was easily attained with waveguide feeds and power division for RF distribution. There are two types of array sidelobe control techniques that were readily obtained. The first is a class of Taylor weighting on the sum pattern of both transmit and receive [3], and the second was Bayliss weighting for difference patterns on receive [4].

4.1.1 Taylor weighting

Amplitude weighting for a mainbeam antenna requires low sidelobes to reduce the interferene sidelobes due to clutter or RF interference. The most common radar antenna weights are Taylor weights with specified sidelobe levels in dB [3]. The Taylor weighting is expressed as a finite Fourier series in the aperture variable p [5]:

$$g\left(p\right) = 1 + 2\ \sum_{n=1}^{N_{\text{bar}}} F\left(n, A, N_{\text{bar}}\right)\ \cos\left(n\pi p\right) \tag{4.4}$$

This distribution $g(p)$ is zero at the center of the beam and unity at the ends. The coefficient $F(n,A,N_{bar})$ is given by:

$$F\left(n, A, N_{\text{bar}}\right) = \frac{\left[\left(N_{\text{bar}} - 1\right)!\right]^2}{\left(\left(N_{\text{bar}} - 1 + n\right)!\right)} \prod_{m=1}^{N_{\text{bar}}-1} \left[1 - \frac{n^2}{z_m^2}\right] \tag{4.5}$$

This weighting function is constructed from the zeros of the array distribution, where the first few N_{bar} zeros are constrained to a constant amplitude. And the far-out zeros follow a $1/u$ decay of the sidelobe levels. The zeros z_{m} in (4.5) are given by

$$z_{\text{m}} = \pm\sigma\ \sqrt{A^2 + \left(m - 1/2\right)^2}, \qquad 1 \le m \le N_{\text{bar}} \tag{4.6}$$

$$z_{\text{m}} = \pm m, \qquad N_{\text{bar}} < m. \tag{4.7}$$

The dilution parameter in the sidelobes σ is expressed by

$$\sigma = \frac{N_{\text{bar}}}{\sqrt{A^2 + \left(N_{\text{bar}} - 1/2\right)^2}} \tag{4.8}$$

Finally, the parameter A is given in terms of the sidelobe ratio (SLR) of the sidelobes to the peak gain of the aperture:

$$A = \frac{1}{\pi} \ln\left(\text{SLR} + \sqrt{\text{SLR}^2 - 1}\right) \tag{4.9}$$

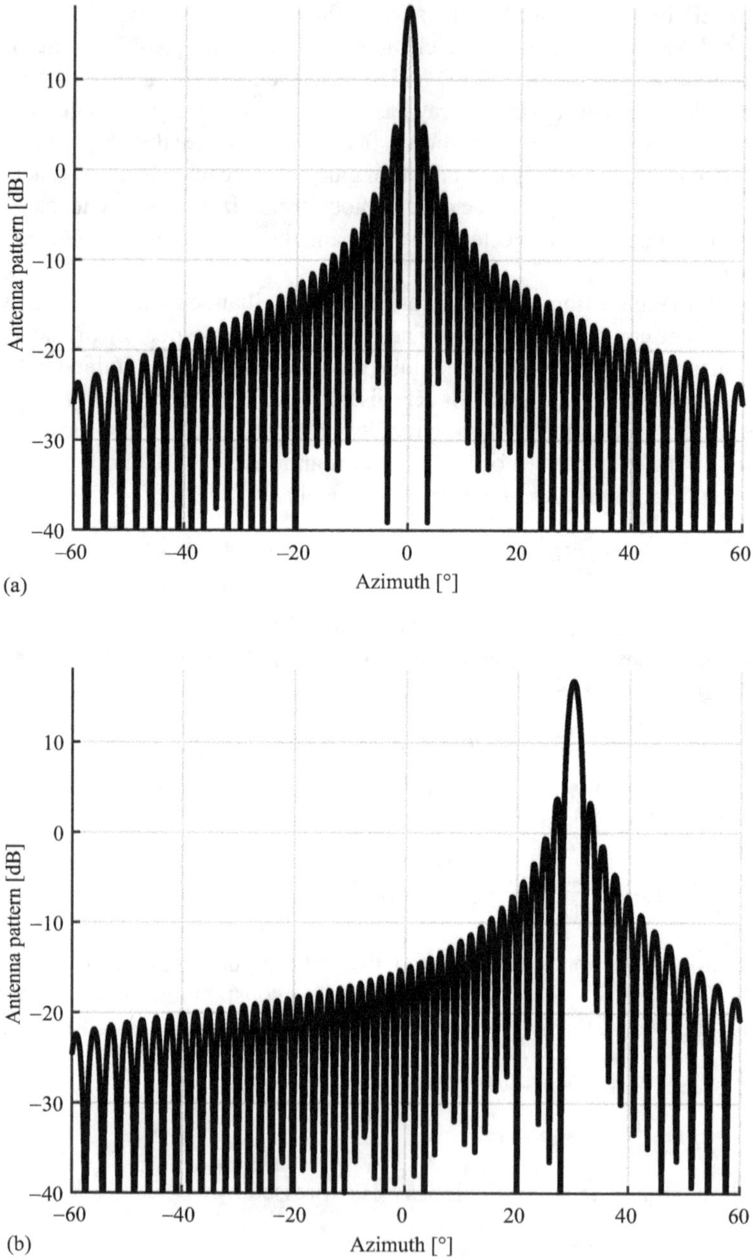

Figure 4.2 ULA pattern 64 elements. (a) 0° scan and (b) 30° scan

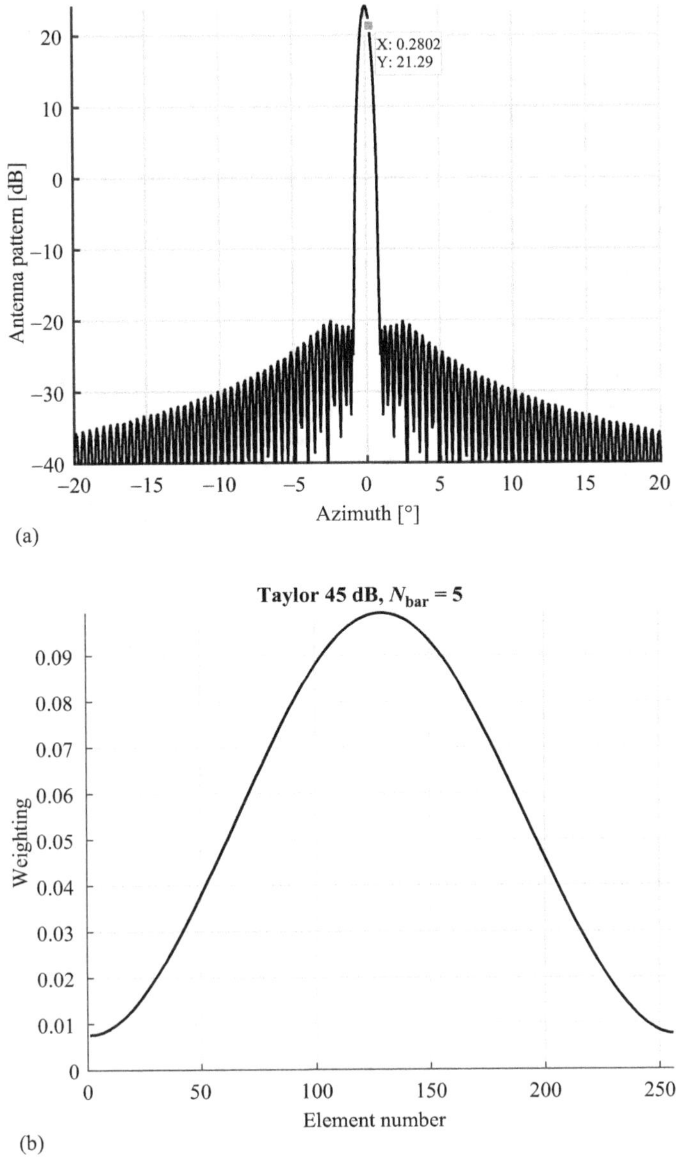

Figure 4.3 Taylor weighting −40 dB, N_{bar} = 5 for a 64-element L-band array. (a) Array pattern and (b) weighting function

Figure 4.3 illustrates the Taylor −40 dB, N_{bar} = 5 weighting along with the linear array pattern at 0° scan. The antenna pattern clearly shows the five near sidelobes at –40 dB with respect to the mainbeam peak. This antenna pattern has also been normalized to the peak gain to emphasize the reduction in sidelobes. Due to the weighting function to control the element energy distribution, the gain of the antenna has been reduced by – 1.14 dB and the beamwidth has been broadened by 1.43.

Similarly, Figure 4.4 shows the weighting function for Taylor −20 dB $N_{bar} = 2$ performance for the same array. Only the two near-in sidelobes are maintained at the −20 dB level, and the remaining sidelobes are reduced linearly with the array factor. The weighting of the sidelobes broadens the main beam by 1.13 dB and reduces the antenna gain by −0.21 dB with the sidelobe control constraints. It can be clearly seen that the weighting of the −40-dB SLR has significantly more attenuation of the signal return at the elements away from the center versus the 20-dB SLR [5].

4.1.2 Bayliss weighting

Surveillance radars often require monopulse processing to improve the detection and tracking capabilities of the system. As with the sum pattern weighting, it is important to have lower sidelobes on monopulse or difference patterns. The Bayliss weighting function is analogous to the Taylor patterns, where the individual elements are steered to control the mainlobe or sidelobes. The pattern function is written as a finite sum over N_{bar} zeros [4]:

$$F(u) = u\cos(\pi u)\sum_{n=0}^{N_{bar}} \frac{(-1)^n B_n}{\left(n+\frac{1}{2}\right)^2 - u^2} \tag{4.10}$$

The aperture distribution is given by

$$g(p) = \sum_{n=0}^{N_{bar}} B_n \sin\left(\frac{\pi p}{n+1/2}\right) \tag{4.11}$$

where the coefficients B_m are given by

$$B_m = -(-1)^m \left(m+\frac{1}{2}\right)^2 \frac{\prod_{n=1}^{N_{bar}-1}\left(1-\frac{\left(m+\frac{1}{2}\right)^2}{\sigma^2 z_n^2}\right)}{\prod_{\substack{n=0\\ n\neq m}}^{N_{bar}-1}\left(1-\frac{\left(m+\frac{1}{2}\right)^2}{(n+1/2)^2}\right)} \quad for\ m = 0, 1, 2, \ldots N_{bar}-1 \tag{4.12}$$

The weighting function of Bayliss 30 dB, $N_{bar} = 2$ is given in Figure 4.5. Note that the amplitude weighting is asymmetrical and passes through zero at the middle of the array.

It is important to recognize that the weighting on transmit is very difficult for active ESA antennas. Because of the need for maintaining the power added efficiency of solid-state transmitters, the final amplifier is run-in saturation. As a result, the taper cannot be applied with the precision illustrated in the weighting functions above. There are many approaches that will approximate the sidelobe control. These

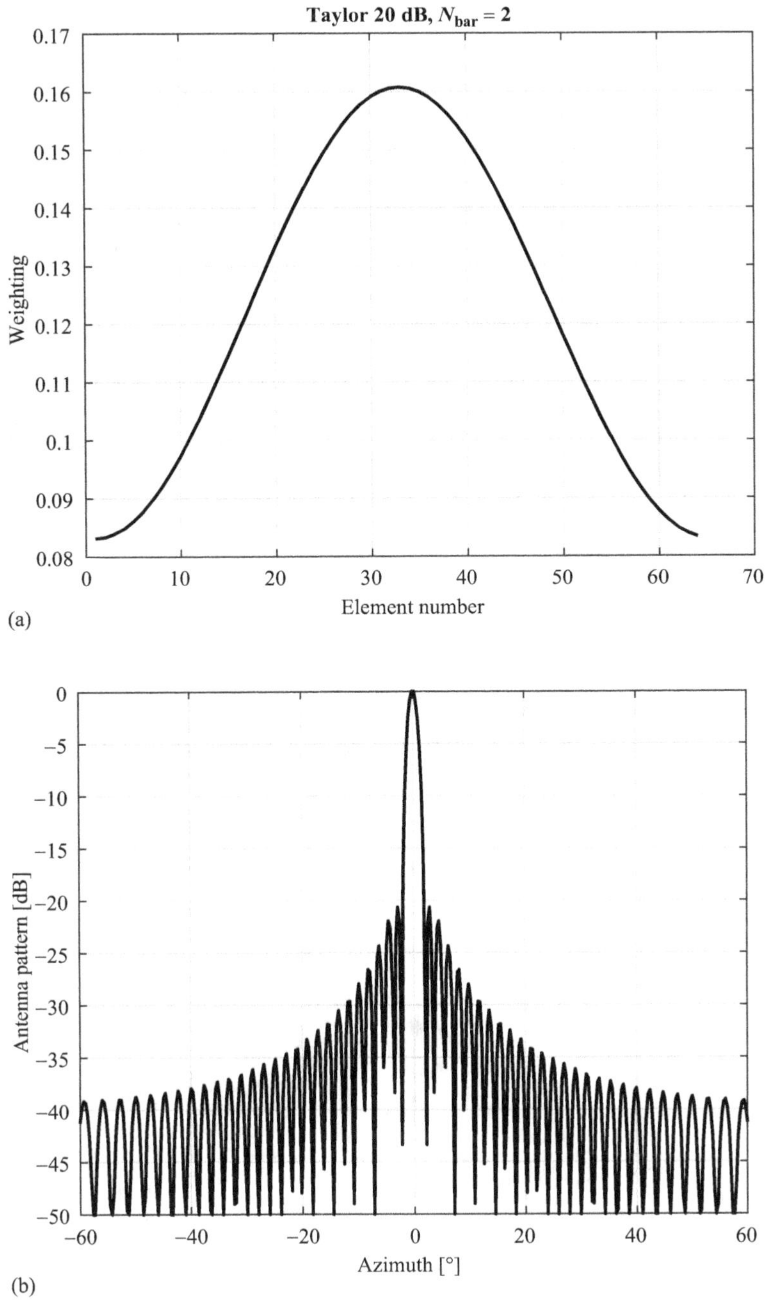

Figure 4.4 Taylor weighting −20 dB, N_{bar} = 2 for a 64-element L-band array. (a) Weighting function and (b) array pattern

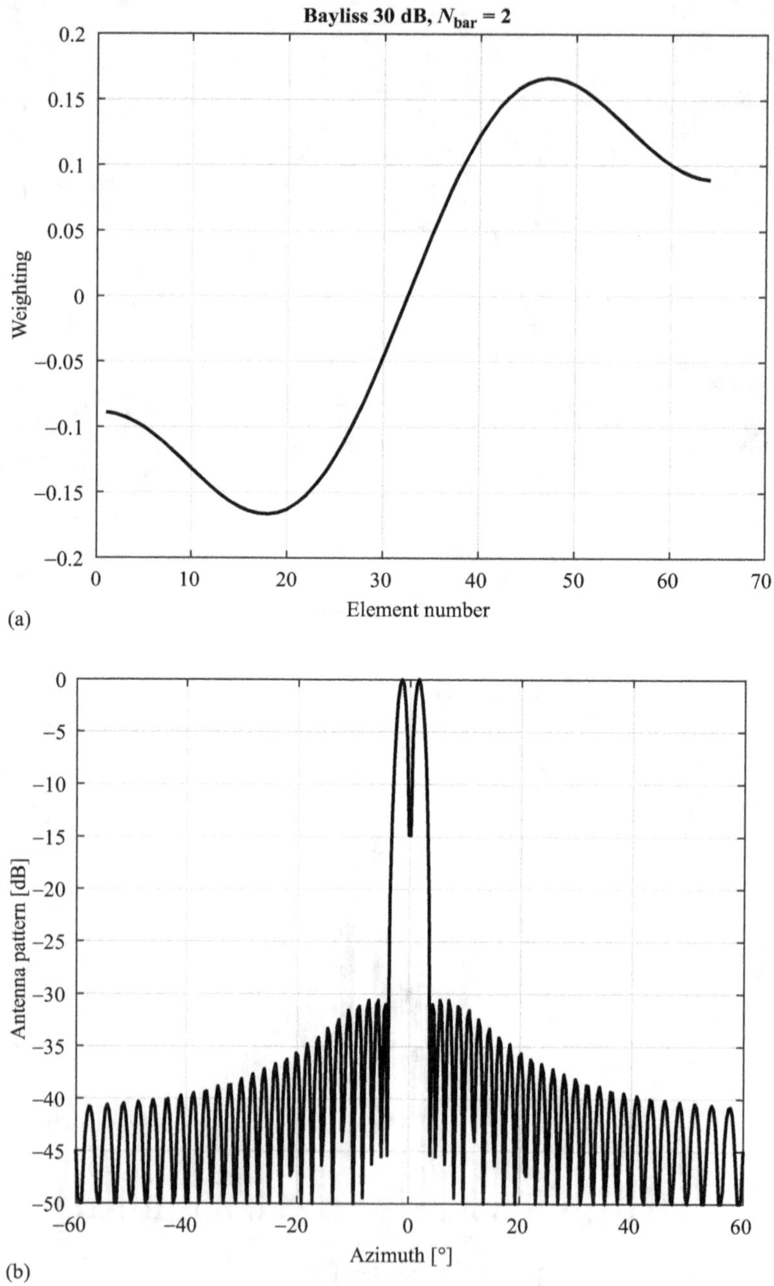

Figure 4.5 Bayliss antenna difference pattern –30 dB, N_{bar} *= 2. (a) Weighting function and (b) antenna pattern*

approaches include space density taper (SDT), where the transmit receive modules are on an irregular grid, or segmented apertures where the T/R modules have different output power levels approximating the Taylor functions. These approaches will be explored later in this chapter.

4.2 Airborne ESAs

Several experimental passive ESA systems were designed to experimentally determine the merit of electronic scan in terms of tactical applications. As each radar was tested, new applications of ESA technology would be derived and subsequently prototyped. The benefit of electronic scan was immediately recognized. However, there was a strong desire in the radar technical community to have more degrees of freedom in the antenna for detection, clutter cancellation, exploitation of fixed and moving targets and multiple phase center and beam systems. The following sections will show the antenna technologies. Later chapters will explore the benefits of new radar modes for fixed and moving targets.

4.2.1 Hostile Weapons Locator System radar antenna

One of the earliest battlefield surveillance radar phased arrays was developed by GE Aerospace in 1977 for MIT Lincoln Laboratory under The Defense Advanced Research Projects Agency (DARPA) Hostile Weapons Locator System (HOWLS) program. It was a linear array of 208 elements, as shown in Figure 4.6, consisting of dipole transitions from a waveguide feed to a radiating dipole in a sector horn. The primary objective of the passive ESA design was the potential for a lightweight, low-cost antenna for use in remotely piloted vehicles (RPVs). The phased array

Figure 4.6 HOWLS antenna with subarray [6]

could scan ±30° in azimuth to provide search access to a wide area. The antenna had a 200-MHz instantaneous bandwidth, at 16.25 center frequency and a tuned bandwidth of 500 MHz to demonstrate the modest resolution for detecting and characterizing vehicles and artillery on the battlefield [6].

The antenna had a 4.5° root mean squared (rms) phase error along with a 0.9-dB rms amplitude error yielding a signal loss and increase in sidelobe levels. In addition, there was a 0.35-dB rms phase shifter loss variation witin each element. The array phase was measured at 2° scan increments and across the 500-MHz bandwidth. The peak sidelobes were increased from the 25-dB Taylor distribution to −21 dB. Figure 4.7 shows the computed pattern versus the measured pattern, with excellent agreement to theory with the amplitude and phase errors. And the rms sidelobe level with this weighting was −18.5 dB.

In addition to using the full aperture for radar target detection, two of the subarrays could be selectively used as the Doppler beam sharpening (DBS) feed to cover a wider azimuth illumination for terrain imaging. A fast waveguide switch directly fed the two center subarrays, with no taper between them. These smaller apertures were necessary to obtain the wide area illumination for surface imaging. Both DBS and SAR image functions utilize the subarray for their coverage.

The waveguide corporate feed shown in Figure 4.8 distributed the power to each of 16 subarrays to approximate the 25-dB Taylor weighting. Each subarray of 13 elements shown in Figure 4.9 had the identical 0.8-dB linear taper across the subarray. This taper was chosen to approximate the Taylor weighting with minimal amplitude taper error. Figure 4.10 shows the waveguide distribution to each element, as well as the electrical control connections to the 4-bit PIN diode phase shifters. This segmented feed structure deviated from the ideal Taylor aperture weighting because of the desire to have identical subarray feed assemblies. This

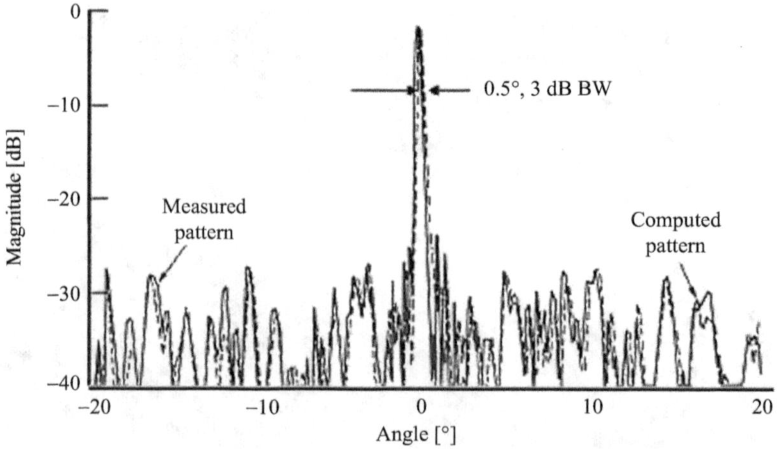

Figure 4.7 HOWLS antenna pattern (0° scan) [6]

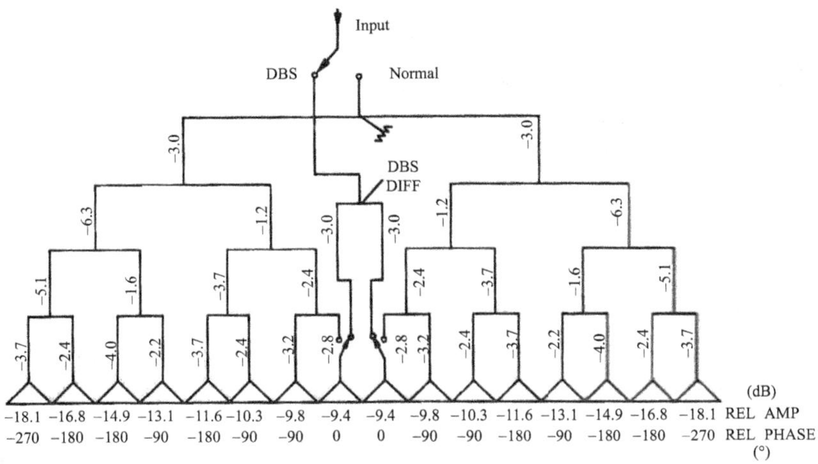

Figure 4.8 HOWL array feed distribution [6]

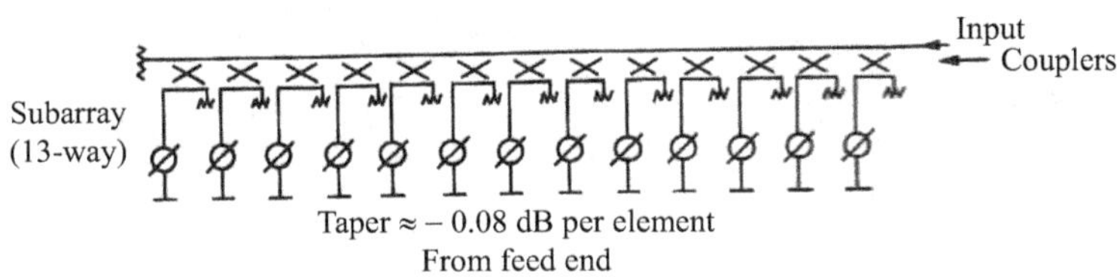

Figure 4.9 Subarray signal distribution [6]

choice of identical taper across each subarray was made in an attempt to demonstrate subassemblies that could be manufactured in large batches in order to keep the recurring cost down [6].

Fortunately, both the Taylor and Bayliss functions are available as Matlab™ functions. The computational implementation of the equations above is straightforward, while simulating many radar implementations. These analyses will be developed in later sections to illustrate the advantages of low-sidelobe antennas.

The original objective of HOWLS was to prototype an ESA that could be carried by a medium-size RPV. However, this antenna weighed a total of 24.9 kg, due to the extensive use of aluminum waveguides. This weight was higher than many small UAVs could accommodate. As a result, there was a parallel development of technology to reduce the weight by 3:1 using plastic or thin waveguide assemblies. These techniques in the 1970s were considered to be applicable to a high-volume manufacturing process. The other weight reduction innovation was the "connectorless" transition from the waveguide to the phase shifter alumina substrates. Each phase shifter only weighed 8.5 g, which was about 20% of the weight of comparable phase shifters with individual connectors [7].

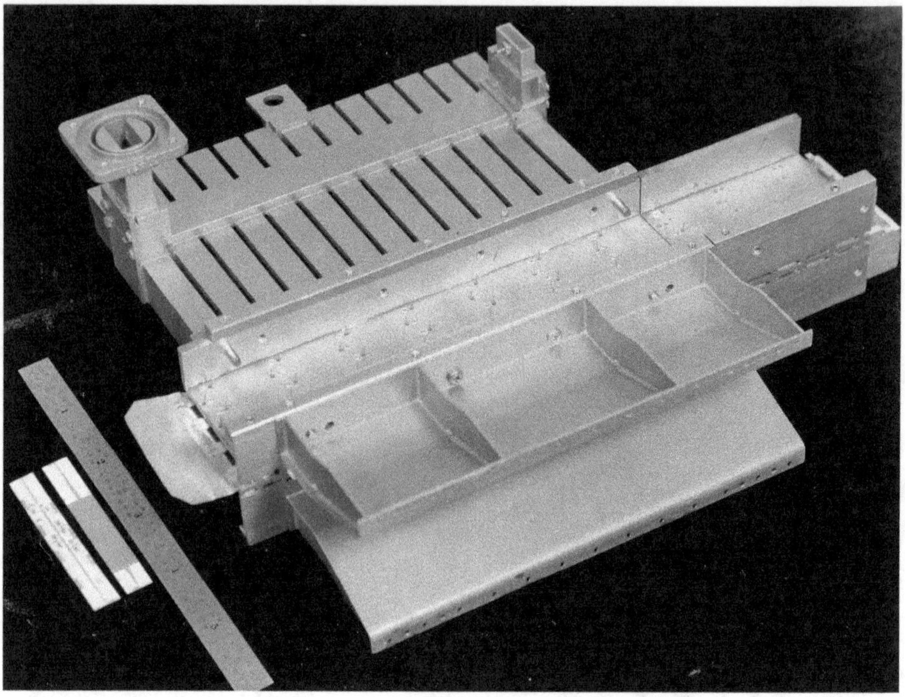

Figure 4.10 HOWLS antenna subarray [6]

4.2.2 Multibeam phase array

There was considerable interest and research in the 1980s for multiple channel radar designs. The multiple channels were intended for monopulse, polarimetric and multiple adaptive beam processing. These requirements, when coupled with the UWB system design, required considerable innovation in design and manufacturing. Not only the design of the antenna, but also the control of the waveform and spectral control quickly led to the need for digital beamforming and compensation for channel match. The research on multiple beam and channel ESAs started at the same time and was soon followed by component and assembly techniques to provide the channel match required for the various radar applications.

The desire for reciprocal operation between transmit and receive states caused a tradeoff in components. Passive solid-state phase shifters were lossy, and ferrite phase shifters required significant electromagnetic control to meet the switching time and reciprocal phase operation. Early multichannel systems would use a passive ESA along with a high-power transmitter tubes and a corporate feed analogous to HOWLS. But with the advent of the MMIC component development, the focus on AESA systems was soon becoming practical and affordable. The major drawback on the AESA is the difficulty of maintaining transmit power efficiency while achieving a low-array sidelobe design.

Multiple channels or beams are often required for improved radar system target detection performance. Those requirements also provide a major improvement in angular resolution for tracking. The multiple channels are also beneficial in providing an advantage for clutter or interference cancellation. New applications for interferometric SAR operation for terrain elevation measurement or Doppler moving target discrimination also required better channel time and frequency match. And finally, a multiple beam radar would provide the desired improvement in area coverage rate for surveillance search.

The Air Force Research Laboratory (AFRL)/DARPA-sponsored multibeam modular survivable radar (MSR) antenna had 32 subarrays of 8 elements, for 256 total columns, as shown schematically in Figure 4.11. The system was designed for very low sidelobes, by making a taper on transmit to obtain sidelobes well below isotropic gain. The transmit beam was spoiled with a quadratic phase function over the face of the array to achieve up to a 10:1 beam broadening, while maintaining the electronic scan. This wide-angle scan was demonstrated over ±60°, while maintaining the low antenna sidelobes.

Each of the 256 columns consisted of 18 radiating slots to form the elevation pattern and 8-bit ferrite phase shifters for scanning in azimuth. The ferrite phase shifters were designed to provide reciprocal phase equivalent to 8-bit phase accuracy across the 50-MHz instantaneous bandwidth, along with ±60° azimuth scan.

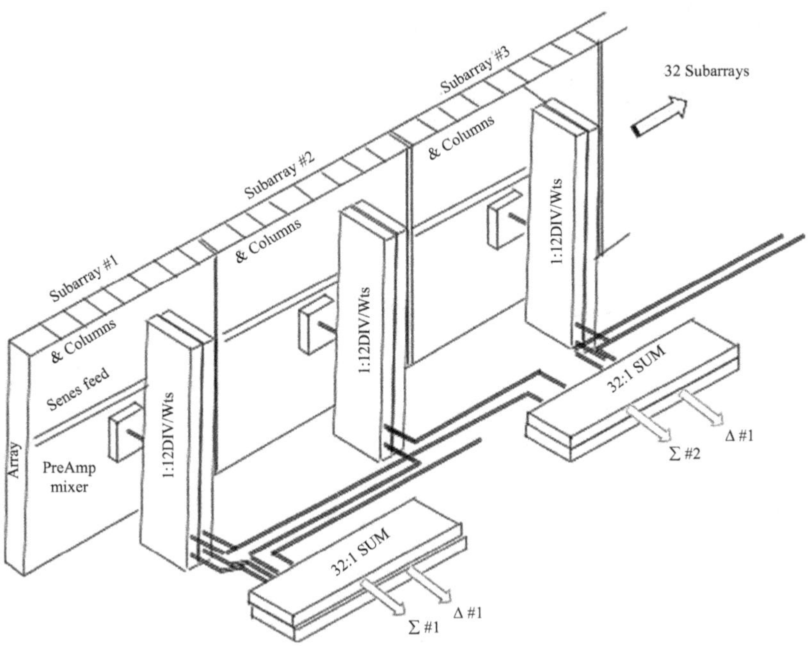

Figure 4.11 Multibeam phased array antenna [8]

In addition, the feed structure was designed for 500-MHz tunable bandwidth for frequency-agile detection of both fixed and moving targets.

The multibeam electronic scanned antenna was developed by employing emerging RF component technologies for demonstration of an advanced concept radar. The details of the Ku-band antenna was built with 32 integrated subarrays, similar to HOWLS in concept [4]. In order to maintain flatness and element match for ultra-low sidelobes, the face of the array was manufactured using a lithography technique on dielectric waveguides. Finally, the subarrays were formed with a waveguide combiner, and the RF received signals were amplified and down-converted to intermediate frequency (IF), as shown in Figure 4.12. This architecture enabled the formation of multiple independent beams, along with multiple signal processing channels.

Each subarray had a Ku-band low noise receiver and down-converter to an L-band IF frequency. The receiver gain and noise figure was sufficient to compensate for the multiple channel combining for each independent beam. Finally, the receiver beams were formed using a Butler matrix going from 32 input subarrays to 11 sum and 11 difference beams. These beams could be scanned along with the transmit beam coverage for subsequent signal processing. The 12th receive channel was implemented by a waveguide beamformer, with the same sidelobe weighting as the transmit channel. This

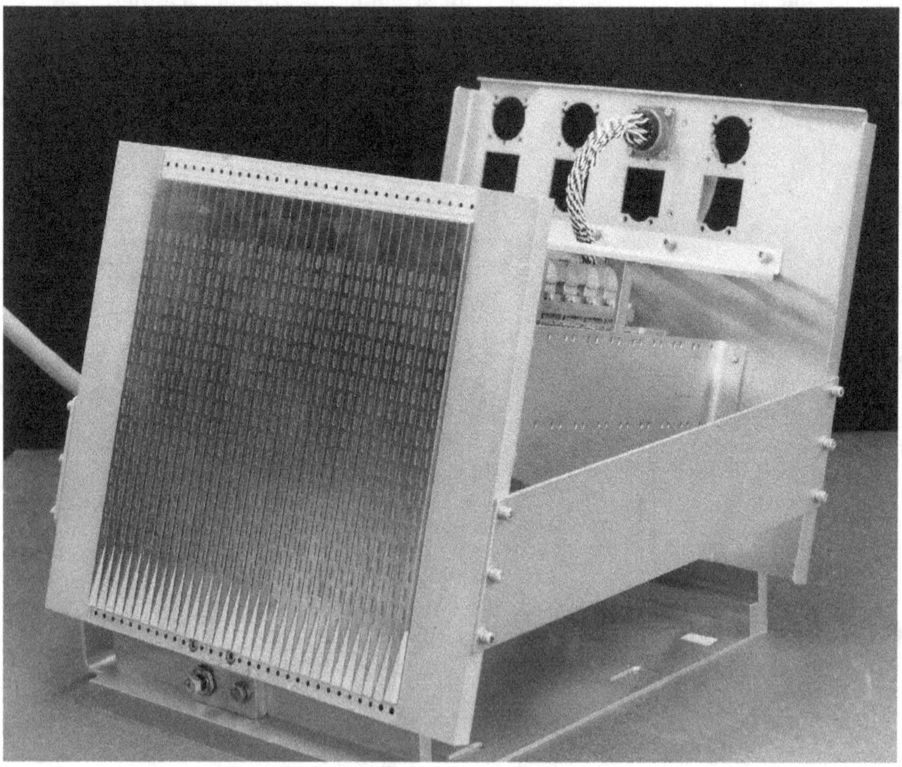

Figure 4.12 Ku-band multibeam subarray [8]

feature provided a high-quality, single-beam operation of the radar, switchable between radar modes.

The following sections will provide details of spoiling the transmit beam with only phase shifting and multiple beamforming utilizing a Butler matrix. The multibeam ESA performance will be described in Section 9.1, along with the radar performance for fixed and moving targets. The multibeam MSR system was flight tested in 1984 as a candidate for several battlefield surveillance radar concepts being pursued by the Air Force and Army. This system demonstrated the ability to electronically scan multiple RF antenna beams and operate a multimode wideband radar for ground area surveillance.

4.2.3 Array beam spoiling

The major objective for the multibeam antenna was broadening the transmit beam using phase-only weighting and obtaining a transmit beam that scanned in conjunction with multiple receive beams. It has been known from optics that quadratic phase would provide a widening of a collimated beam. The technical uncertainty to the beam broadening was whether phase spoiling of the transmit beam could maintain the desired low sidelobes with the amplitude weighting at the subarray structure [9].

The phase spoiling of the beam starts with (4.1), where there is a phase shift per element to steer to an angle and a phase progression to broadening the beam [10].

$$Y = \sum_{n=1}^{N} A_n \exp\left[-j\pi\left(\frac{2}{\lambda}nd\left(\sin\varphi - \sin\varphi_s\right) + \delta_n\right)\right] \tag{4.13}$$

where the beam spoiling phase δ_n is quadratic function [11] is given as

$$\delta_n = \frac{1}{N_{el}^2}\left(n - N_{el}/2\right)^2 \quad \text{for } n = 1, 2, \ldots, N_{el} \tag{4.14}$$

and the quadratic phase is centered on the middle element of the array. In order to reduce the antenna sidelobes, the waveguide provided a power division to each element with a Taylor −45 dB, $N_{\text{bar}} = 5$ for A_n in (4.13). For the focused transmit beam, the array pattern was given by Figure 4.13, along with the transmit feed weighting in azimuth of each element.

The quadratic phase was applied to the whole antenna's azimuthal elements, resulting in the main beam pattern given by Figure 4.14a. An amplification of the azimuth beam angle is shown in Figure 4.14b, showing the rather smooth amplitude pattern comparable with Figure 4.12. The 3-dB point in the beam is indicated, with a 10.2:1 increase in the beam width, and a loss of 10 dB in array gain. The spoiled beam was shown to easily scan within the beam space.

Figure 4.15 shows the pattern at three scan angles [–30 0 30]°. The low sidelobes of the antenna are maintained as the beam was scanned. However, the quadratic phase has significant amplitude weighting across the array face. As a result, the beam spoiling is caused by the center elements, and the sidelobe levels are maintained with the amplitude weighting. The phase weighting across the array is given

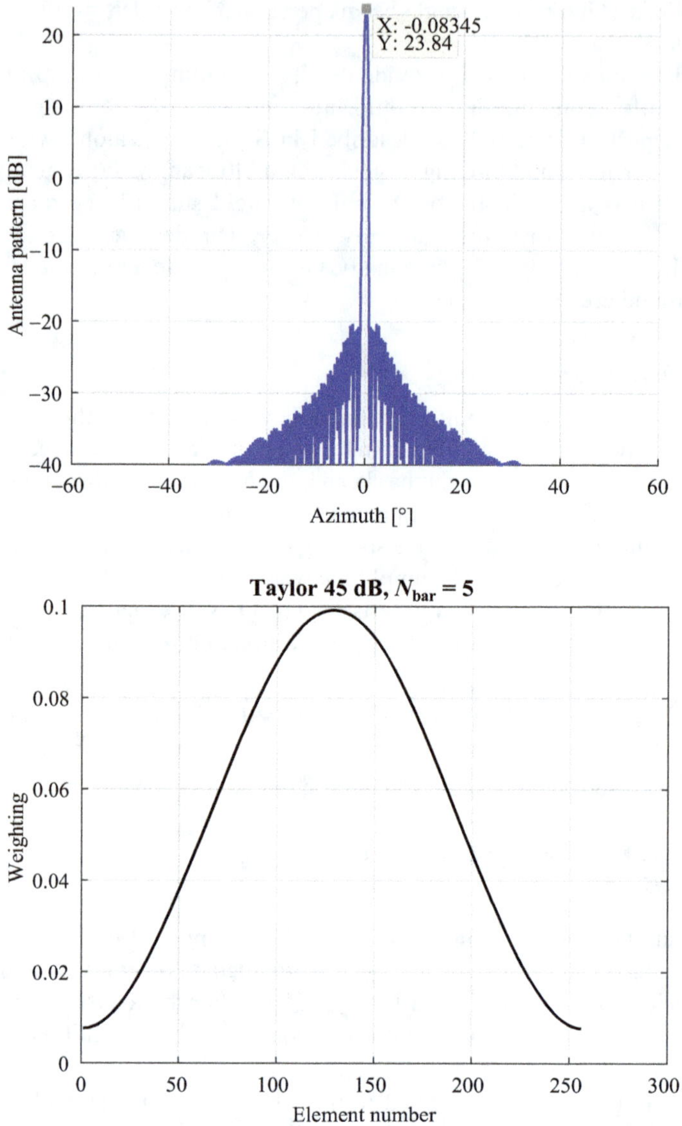

Figure 4.13 Focused 256-element antenna with Taylor −45 dB, N_{bar} *= 5. (a) Antenna AZ pattern and (b) amplitude weighting*

in Figure 4.15b. There is a significant degree of phase taper, but the phase difference between elements is small: 9 rad/128 or 70 milliradians per element.

When quadratic phase spoiling was used with active element arrays, a new phase characteristic was recognized [10]. With $A_n = 1$ in (4.13), the pattern was drastically changed as shown in Figure 4.16. For this approach to be used in the AESA design, a

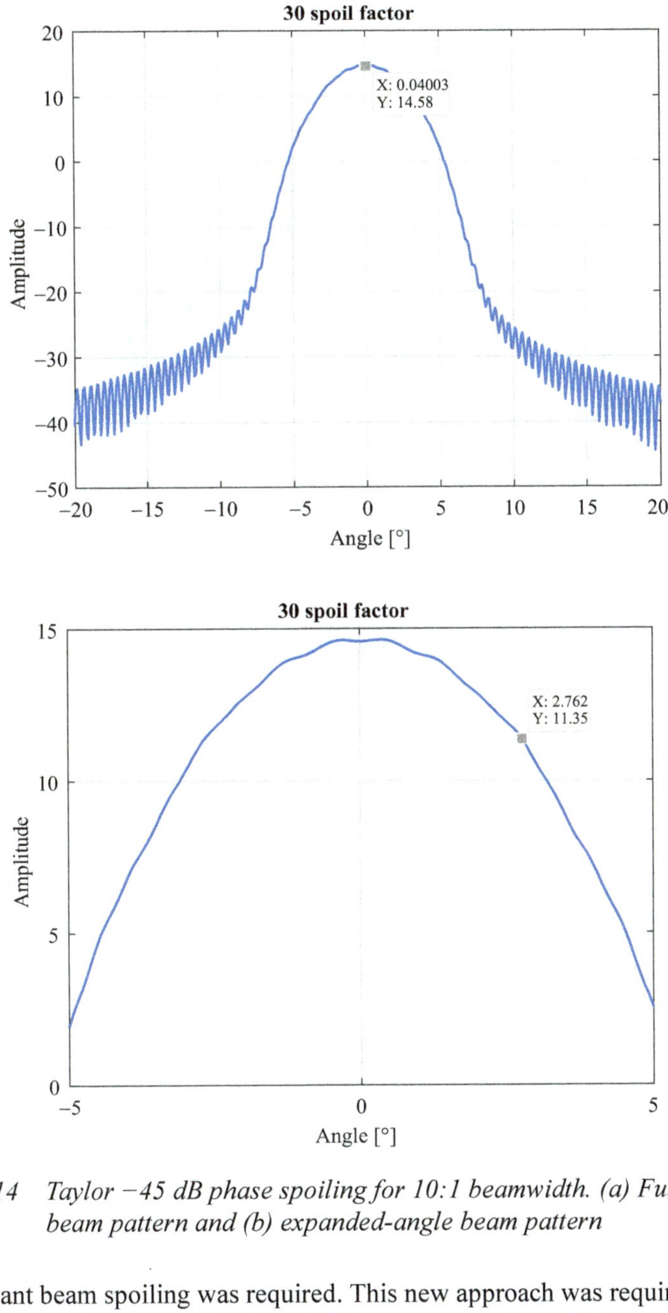

Figure 4.14 Taylor −45 dB phase spoiling for 10:1 beamwidth. (a) Full spoiled beam pattern and (b) expanded-angle beam pattern

more elegant beam spoiling was required. This new approach was required for both AESA designs without amplitude weighting on transmit, and for very wideband arrays where the beam scanning is required over many frequency components. The broadened beam is defined by a set of steered directions that span a range of angles, as illustrated in Figure 4.17. In addition, phase weights are calculated to steer power

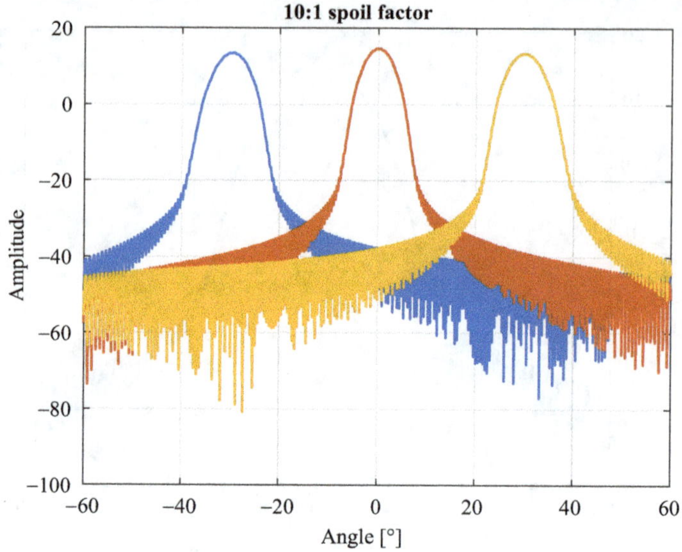

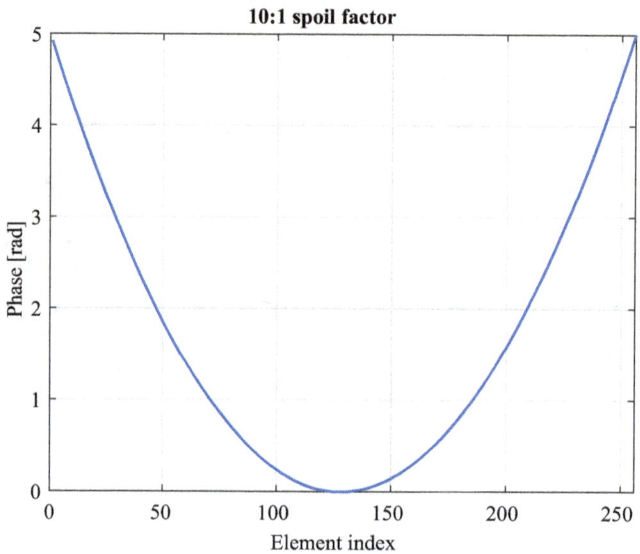

Figure 4.15 Taylor 45-dB phase 10:1 spoiled beam scanned. (a) Scan [−30 0 30]° and (b) phase taper over antenna

from each radiating element toward the corresponding desired direction. Note that the edge elements, which transmits less power due to the amplitude taper, are steered more closely together than the center elements. This produces a flat main lobe [12].

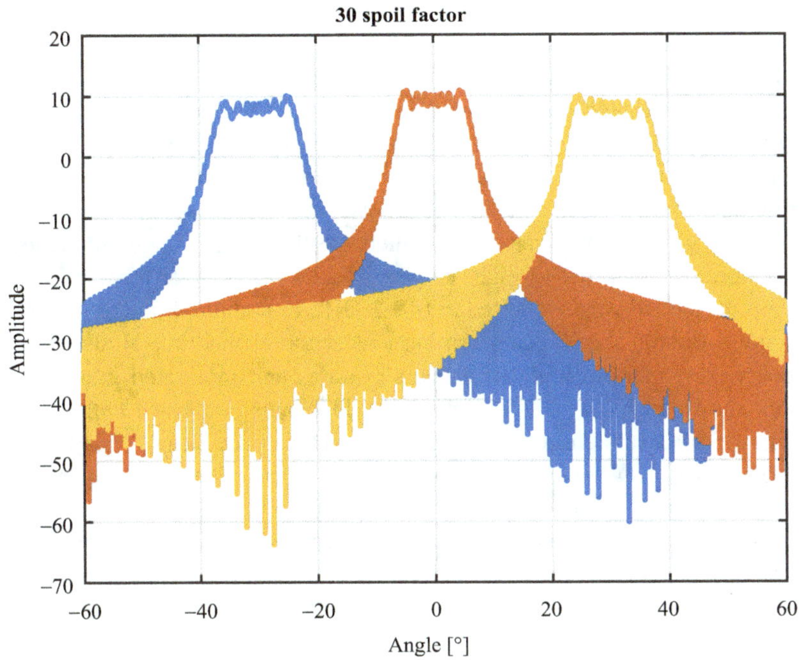

Figure 4.16 Uniform antenna with 10:1 beam spoiling

4.2.4 Multiple Beam Former

The subarray design and amplitude weighting of the multibeam antenna leveraged the experience on HOWLS. Both Taylor weight and Bayliss weight were applied at the subarray level, to obtain very low azimuth sidelobes over a wide scan angle.

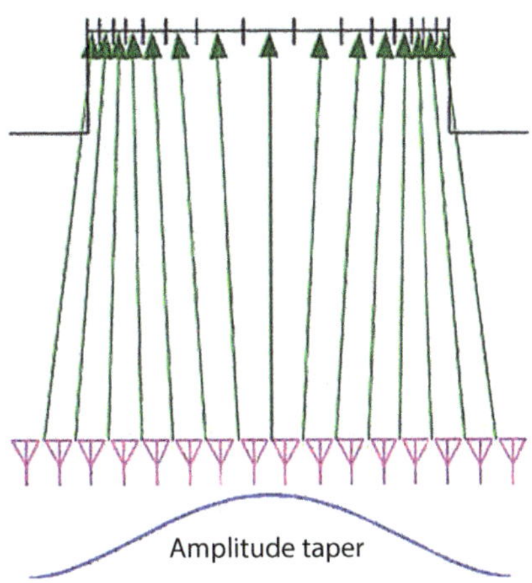

Figure 4.17 Modified phase taper for beam spoiling [12]

The Taylor weighting was also provided in the RF beamforming on transmit. The waveguide feed enabled an accurate distribution of the power split up to the subarray level. At that point, the distribution at the subarray level was a simple linear feed with identical linear taper per subarray, as shown in Figure 4.18. A Butler feed was chosen due to its simplicity and its ability to handle the bandwidths of the MSR radar [13]. The generic Butler beamformer shown in Figure 4.19 with the central subarrays and combining three adjacent beams.

On receive, the taper was also applied by the intermediate frequency (IF) Beamformer. The subarray of eight elements had a linear taper for each subarray. At that point, the RF was converted in to 1.2 GHz IF and amplified to maintain the system noise figure. The IF beamforming was accomplished in a Butler matrix for each beam fabricated on high-dielectric substrates and integrated into a compact assembly shown in Figure 4.20. Each of the 11 beams employed both the Taylor weights for Sum Beam and Bayliss weights for Difference Beam. This was implemented by a two-way power splitter out of each IF amplifiers and separate Σ and Δ combiners for each of the 11 pairs of beams. This Butler matrix power combiner implementation included the amplitude and phase required for each beam, including the amplitude weights for the low-sidelobe taper. It should be pointed that the 12th beam was provided in the beamformer subassembly for operation with the unspoiled transmit beam.

Butler matrices have been used for many fixed beamformers. They utilize the variable $u = (d/\lambda)(\sin\theta - \sin\theta_j)$ to provide a fixed-angle offset θ_j between multiple beams. The separation of each element is given by (d/λ), and θ_j is the beam pointing of the jth beam measured from broadside. The Butler beam pattern is given by the function F(u) [5]:

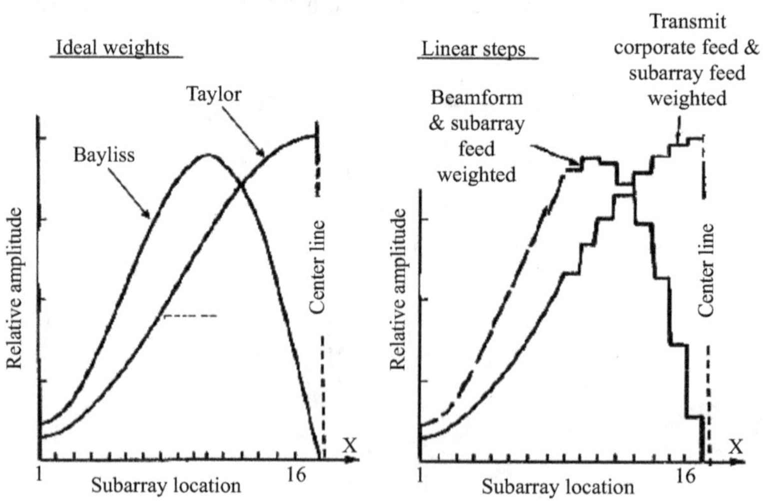

Figure 4.18 Segmented receive beam weighting [8]

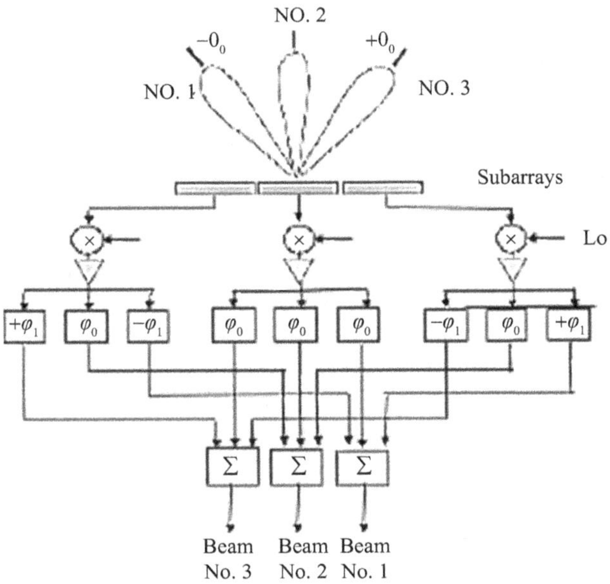

Figure 4.19 Multiple beamforming circuits [11]

Figure 4.20 Multiple beamforming assembly [8]

$$F(u) = \frac{\sin(N\pi u)}{N\sin(\pi u)} \tag{4.15}$$

All of the beams have the same shape in u-space. And they were demonstrated to scan well as a group, while the multibeam antenna scans between ±60°. By using half-wavelength spacing between elements, the beams are orthogonal while they cross at the 3-dB points. The main limitation of this approach lies in the design of quadrature hybrids in receive path splitting. For an analog beamformer, the ability to maintain this orthogonality will be limited by these components, which may be less than an octave bandwidth. These orthogonality design factors provide a very important characteristic for simultaneous detection and characterization of the earth surface or targets.

The beam position for any element spacing is given by

$$\sin\theta_j = \pm\frac{j\lambda}{2Nd}, \qquad j = 1, 3, 5, \ldots, (N-1) \tag{4.16}$$

where all of the beams are spaced by $1/N$ in u-space. When the excitation is uniformly excited, the sidelobe level is –13 dB, and the cross-over level between the beams does not change. The coverage by the N beams is given by

$$\theta_{\text{cover}} = 2\arcsin\left[\frac{(N-1)\lambda}{2Nd}\right] \tag{4.17}$$

When the spacing between the elements depart from $\lambda/2$, the beams will be closer together, and grating lobes will appear in the spatial coverage. In addition, as the waveform frequency changes, the electrical spacing between the elements will also change, i.e., increased frequency narrows the beams and decreased frequency broadens the beams.

The Butler beamformer has normally been designed with uniform split of the power. The cross-over level F_{xov} between the beam patterns obeys:

$$F_{\text{xov}} = \frac{1}{N\sin(\pi/2N)} \tag{4.18}$$

The uniform split of power is continuous and the cross-over levels approached $2/\pi$ as N increases, and the space factor approaches sinc($N\pi u$). The variation of the SLR and the cross-over factor F_{xov} is given for four Butler designs in Table 4.1.

Table 4.1 Butler array parameters [4]

N	SLR (dB)	Cross-over level (dB)
4	11.30	–3.7
8	12.80	–3.87
16	13.15	–3.91
32	13.23	–3.92

The multibeam antenna used the Butler matrix at the subarray level, with an unequal split of signal. As a result, the power split had to be adjusted for both the Taylor and Bayliss weighting to achieve the low-sidelobe levels. This was achieved by a combination of the subarray amplitude taper and the weighting in each of the 22 individual beamformer assemblies. It should be noted that this analog beamformer was a heavy and costly approach. Today's digital beamforming will greatly reduce the complexity for future system applications. Digital beamforming will be covered in Section 4.4.3.

4.3 UWB antenna performance compensation

Early UWB antennas for SAR required high percentage bandwidth and polarization channel isolation. These development antennas were fabricated with precision metal structures and analog beamforming. It was a major advance in the antenna state of the art to have over 50% bandwidth and better than 20 dB of channel isolation. This section summarizes the capabilities used for early UWB SAR testing and the processing needed to compensate for gain and phase over such a wide bandwidth. In more modern radar antennas, digital compensation has been demonstrated to obtain almost two orders of magnitude improvement in match.

Polarization was recognized early for target and terrain characterization. The earliest application was demonstrated on the DARPA/Navy P3 UWB radar system. The UWB antenna for the P3 FOPEN SAR was designed as a flared notch horn, as shown in Figure 4.21. It employed a Vivaldi notch radiator, integrated into a brass structure, in order to maintain the polarimetric characteristics over a wide bandwidth. The objective was to maintain a minimum of 20-dB cross-polarization isolation over the band. This objective was achieved over all but the lowest frequency extent.

The antenna was very large in order to provide the desired gain across the UHF band. The beam width varied from 40° at the high end to 100° at the low end of the transmit spectrum. It was installed in the "tail cone" of the aircraft, with an installation aimed at suppressing the back lobe of the radiation pattern. The high mass of the antenna made for a stable instrument. But it was clear early in the development cycle of tactical FOPEN SAR that a much lighter system needed to be developed [14].

Because of the uncompensated gain spectrum of the P3 UWB antenna, there was a concern about the use of polarization for target characterization. This was particularly true in the FOPEN applications, where the polarization would be modified by scattering in the foliage volume between the radar and the target. As such, the technique of spectral compensation was utilized to more accurately match the desired power versus frequency for UWB SAR application. Figure 4.21b shows the measured gain of the horn, with the roll off of the actual gain versus frequency, above 400 MHz. A compensation of the antenna gain as a function of frequency had to be inserted in the SAR image processing. It is clear that the resolution could be improved using several weighting functions. However, they suffered anticipated signal-to-noise ratio (SNR) loss.

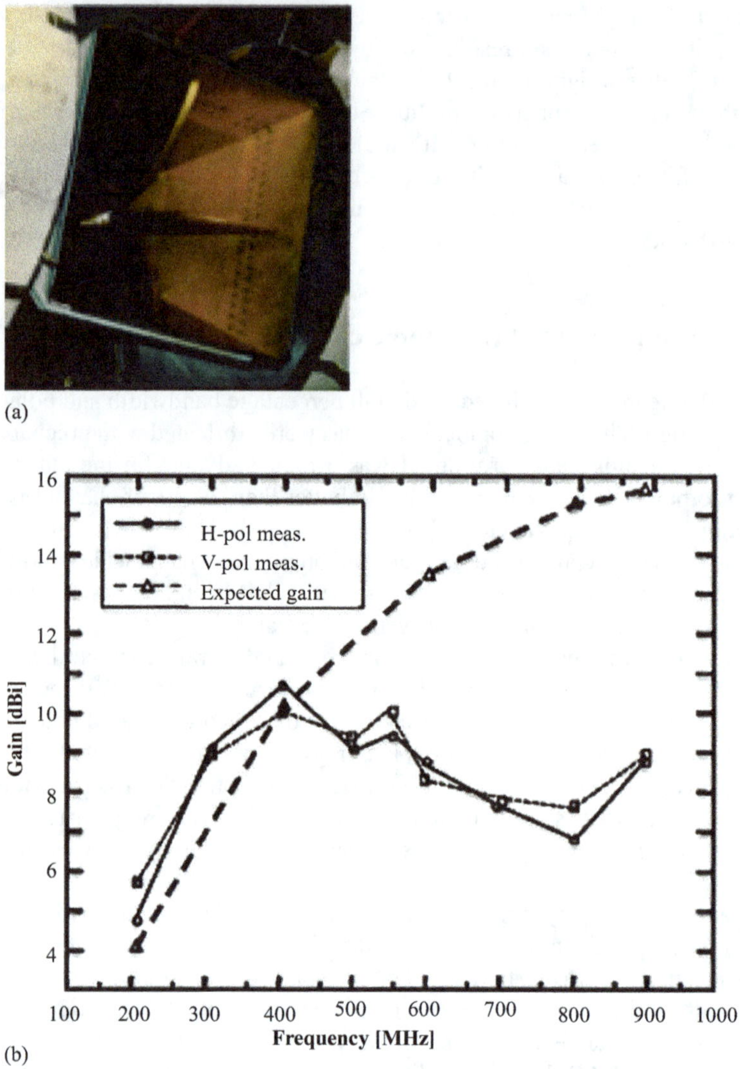

Figure 4.21 UWB polarimetric antenna [14]. (a) P3 UWB antenna and (b) antenna gain performance

The frequency response of the transmit waveform from a point scatterer is proportional to the product of the transmit power, the transmit gain and the receive effective aperture area [15]:

$$\Upsilon(f) = \bar{P}(f)\, G(f)\, A_e(f) \tag{4.19}$$

The average power $\bar{P}(f)$ will depend on the design of the specific transmitter over the band. For the initial assessment, the average power will be assumed to be uniform over the band. The effective aperture A_e, which is a function of the measured antenna gain, is given by

$$A_e(f) = \frac{G(f)\,\lambda^2}{4\pi} \tag{4.20}$$

Thus, the impulse response function will be proportional to

$$\Upsilon(f) \propto G^2(f)\,\lambda^2 = \frac{G^2(f)\,c^2}{f^2} \tag{4.21}$$

This impulse response function, assuming Hamming weighting, is given by

$$I^2(t) = \frac{f^2(2R/c)}{f^2(0)} \tag{4.22}$$

where $f(t)$ is the inverse Fourier transform of the waveform

$$f(t) = \int_{f_1}^{f_2} \exp(-j2\pi ft)\sqrt{\Upsilon(f)}H(f)df \tag{4.23}$$

and $H(f)$ is the Hamming window function.

The impulse response for the 200–715 MHz band is shown in Figure 4.22. The solid line is the desired response for uniform gain and antenna response $\Upsilon(f) = 1/f$. The short-dashed curve represents the case that uses the predicted antenna gain to compute $\Upsilon(f)$, and the long-dashed curve is the case using the measured antenna gain. An

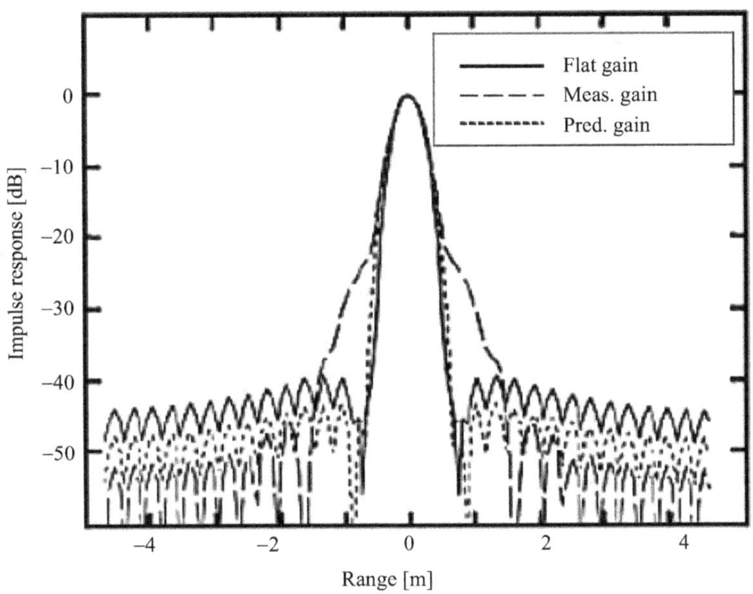

Figure 4.22 SAR impulse performance after equalization [15]

Table 4.2 Impulse response for candidate weighting techniques [15]

Gain	None	Hamming	$1/\sqrt{\Upsilon(f)}$	Hamming/$\sqrt{\Upsilon(f)}$
Uniform	0.6	2.7	2.2	3.3
Measured	0.53	1	2.6	1.8
Predicted	0.52	1.4	2.8	2.0

important factor to note is there is little difference in the 3-dB width of the impulse response mainlobe among the three curves. In addition, there is some broadening of the mainlobe pattern at the –20 dB points for the measured pattern. However, this analysis only considers the amplitude of the antenna pattern function, and not the phase. When phase errors are included the sidelobe levels will be adversely affected.

The response for gain variation can be compensated in signal processing by applying a correction factor $C(f)$, which is essentially the inverse of the errors:

$$C(f) = \frac{1}{\sqrt{\Upsilon(f)}\exp(j\phi(f))} \tag{4.24}$$

This factor takes into account the amplitude and phase of the pattern effects. In practice, the correction factor will include the variation in the amplitude and phase of both the antenna and the transmitter. Processing on receive for the gain variation errors will restore the impulse response errors, but at the expense of a mismatch loss. The mismatch loss Λ_p can be computed as

$$\Lambda_p = \left(\frac{|\sum_i S_i W_i^*|^2}{\sum_i |W_i|^2 \sum_i |S_i|^2} \right) \tag{4.25}$$

The measurements shown in Figure 4.21b was used to correct the impulse function, using a Hamming weighting. An estimate of the mismatch loss that was made for each of the antenna gain values is shown in Table 4.2.

The major development in the past 20 years has been in manufacturing lightweight polarimetric antennas, which have provided an improvement in gain versus frequency for UWB SAR. One such antenna, which reduced the UWB antenna weight by over 50:1 compared to the P-3 UWB, is shown in Figure 4.23. This multiple octave bandwidth, polarimetric antenna had very small depth ($<\lambda/10$) and demonstrates efficient radiation over the full bandwidth. As a result, the need for equalization was greatly reduced, improving the waveform performance. The antenna has been transitioned to the Army TRACER program and has been flown on Predator unmanned aerial system (UAS) platforms [16].

Figure 4.23 UWB UHF antenna [16]

4.4 Antenna technology advances

This section focuses on the significant technology advances that affected the design and employment of UWB phased array antennas. These advances are focused primarily on active elements to place the transmitter and receiver at the face of the antenna. This significant change in phased array architectures reduced the loss on transmit and receive between the transmitter and receiver and the radiating elements. This reduction in loss is a major improvement in the signal-to-noise level for detecting any target. More importantly from a system implementation standpoint, the solid-state array took most of the high weight and cost out of the radar front end.

The single most impactful development was the introduction of MIMICs that provide the opportunity to put most of the T/R module on a single chip [1]. This component design breakthrough is the equivalent of very-high speed integrated circuit (VHSIC) custom circuits in digital processing. Since the high-speed digital processing can now be integrated into these devices, the match in amplitude and phase provides improved channel match over a wider waveform bandwidth, which improves the target to clutter ratios.

At the same time, as the development of MIMIC devices, wideband digital processing has progressed into the gigahertz regime. Direct digital waveform synthesis and direct down-conversion from RF to IF, or even baseband, makes the new radars more capable. Integrating the digital matching into the MIMIC devices and onto the T/R modules increases the coherence of the time bandwidth performance of both transmit and receive chains.

Finally, the increase in digital and analog sampling enabled digital processing directly on the face of the array. Digital beamforming will remove all of the heavy RF plumbing required for earlier AESA designs. Digital beamforming with software-controlled formation of multiple channels and beams provides an immediate improvement in the area coverage rate of the systems. The impact of these technologies will be briefly summarized, along with their benefits in the critical UWB surveillance radar functions.

4.4.1 Active phased arrays

Phased array architectures have improved significantly with modern RF and digital technologies. Figure 4.24 illustrates two similar architectures that are considered and compared in alternative surveillance radar applications. The passive phased array utilizes analog phase shifters and an RF signal distribution path from the output of the transmitter and input to the receiver. The system loss involved with this circuitry occurs in both directions along the transmit and receive paths. This two-way loss directly affects the target detection capability of the overall system. As noted earlier, the taper losses through the system is also a factor in the system performance. However, the lower sidelobes on transmit are a direct benefit for both subclutter visibility as well as reduction of RF interference on the radar systems. The losses were reduced by using waveguide distribution of the signal on both transmit and receive. But these waveguide ESA assemblies are generally very heavy for airborne applications, especially on remotely piloted air vehicles. These losses were overcome by employing a high-power microwave tube amplifier. This amplifier met the coherence and average power requirements of most surveillance radar systems. In general, the early airborne, passive ESA systems were heavy and costly.

An AESA has both the transmit output stage and the receiver gain section immediately behind the radiating element. This antenna architecture provided an immediate improvement in SNR by the avoidance of the two-way loss. With today's solid-state MIMIC transceivers, the system efficiency is greatly improved over earlier active ESA designs. However, the major disadvantage to AESA is the difficulty of tapering the transmit pattern for low sidelobes, since the MIMIC amplifiers need to be operated in

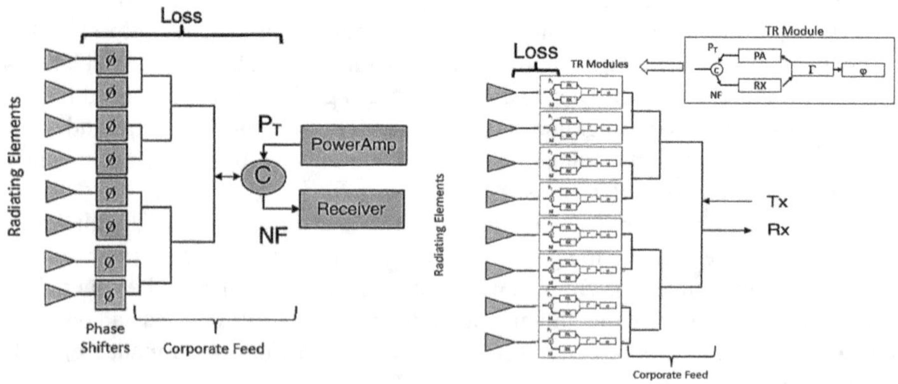

(a) (b)

Figure 4.24 ESA architectures. (a) Passive phased array and (b) active phased array saturation for maximum power added efficiency. An added system issue was that the components suffer an amplitude and phase mismatch over the UWB frequency bands, affecting the benefits of multichannel operation. It is important to factor in channel match signal processing in designing these emerging systems.

The development of AESA systems has been a major benefit for UWB surveillance and small UAV applications. The latest advances in active ESA technology employs digital preselection to improve the waveform match between channels, and channel equalization on receive to provide the necessary clutter and RF interference cancellation.

An important system measure is the power efficiency of the combination of ESAs, transmitters and receivers. Figure 4.25 illustrates a top-level consideration of the prime power delivered to these subsystems along with the critical radar parameters of transmit power, receive noise figure and losses to the RF energy between the transmitter and the receiver. It is important to compare the architecture of a passive ESA versus that of an active ESA, in terms of the overall efficiency of the the average transmitted power compared to the total input power to the ESA subsystem.

There is a significant difference between the two architectures. For a passive ESA, the transmitter is typically a tube amplifier with a high-voltage power supply. The amplifier may have a high-power added efficiency for the waveform, typically 60–70%, depending on the bandwidth. But there is loss between the final amplifier stage and the output of the array. The one-way loss is typically 6–8 dB depending on the phase shift technology. Furthermore, if there is an amplitude taper across the array, the average radiated power is lowered by the efficiency of the Taylor weighting. The losses to through the receive array is similar to transmit. So, there is a two-way loss from a reduction in the average power on transmit and an increase in the noise figure on receive. And each of the phase shifters typically require between 1 and 10 W of drive power depending on whether they are diode or ferrite phase shifters, respectively.

The active ESA has a different structure due to the favorable position of the T/R modules near the face of the aperture. The active elements combine the transmit and receive amplifiers and the amplitude and phase control of the antenna. Solid-state

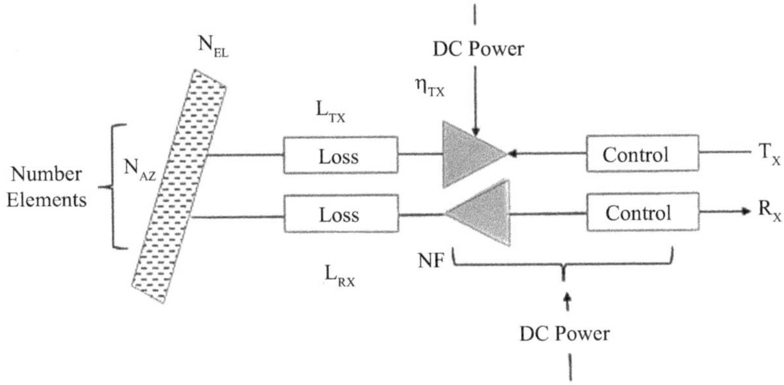

Figure 4.25 ESA system power efficiency

transmitters need to be run in saturation to get the most favorable power added efficiency. The current state of the art for an amplifier with 500-MHz bandwidth is 40–45% efficiency. The receiver noise figure in MIMIC chips can be as low as 2 dB, but with front-end protection, it is more likely to be 3 dB. However, the losses between the T/R module and the radiation element is less than 1 dB. Furthermore, the control of the amplitude and phase will typically be 20–50 mW.

The analysis of the two architectures is centered on effective radiated power and required average power into the radar system for the transmitter and antenna functions. The parameters for a typical tradeoff are given in Table 4.3. And the calculation of ESA power efficiency is given by

$$\eta_{\mathrm{ESA}} = \frac{P_{\mathrm{TX}} * (L_{\mathrm{TX}} + L_{\mathrm{RX}})}{P_{\mathrm{DC}} + P_{\mathrm{TX}} * (L_{\mathrm{TX}} + L_{\mathrm{RX}})} \tag{4.26}$$

where P_{TX} is the average transmitted power, and L_{TX} and L_{RX} are, respectively the transmit and receiver losses between the aperture and the amplifiers. The two-way loss is used for the transmit to account for increased noise figure on receive, requiring more output power. The average transmit power is calculated by the peak power times the duty factor of the waveform.

The RF power is calculated by the radiated power and the impact of losses between the transmitter and receiver into the aperture. For the passive ESA, the full transmit power passes through the feed structure and the phase shifters attenuate the effective radiated power. The losses on transmit and receiver also account for the aperture weighting for sidelobes. For the active ESA, each T/R module radiates the transmit power into the radiating element with modest losses on transmit and receive. There is no aperture weighting on transmit due to the requirement for saturated output amplifier. And the receive weighting is after the low-noise amplifier and sufficient gain that there is minimum degradation of the noise figure.

The prime power PP into the combination of the antenna, transmitter and receiver is different for passive and active ESA designs, PP_{PESA} and PP_{AESA}, respectively:

$$PP_{\mathrm{PESA}} = \frac{P_{\mathrm{pk}}\,DF}{\eta_{TX}} + \sum\nolimits_{N_{\mathrm{el}}} P_{\mathrm{ctrl}} \tag{4.27}$$

Table 4.3 Tradeoff parameters for passive versus active ESA design

ESA design factor	Passive ESA	Active ESA
Transmitter peak power (w)	4,000	10 per element
Transmitter duty factor (%)	10	10
Transmitter PA efficiency (%)	70	35
Transmitter aupply power (w)	570	2.9 per element
Tx loss (dB)	7	1
Receiver noise figure (dB)	3	3
Rx loss (dB)	7	1
Receiver control power (w)	1 per element	0.02 per element

$$PP_{\text{AESA}} = \sum\nolimits_{N_{\text{el}}} \frac{P_{\text{pk}} DF}{\eta_{PA}} + \sum\nolimits_{N_{\text{el}}} P_{\text{ctrl}} \tag{4.28}$$

It should be noted that the prime power for the passive ESA considers the centralized transmitter, and the efficiency of the power supply η_{TX} for achieving the output of the transmitter. Conversely, the AESA sums the output of all of the T/R modules based on the amplifier power added efficiency η_{PA}. Both architectures need to sum the control power into the array amplitude and phase control. Figure 4.26 shows the efficiency as the size of the power aperture increases, given a fixed limit on radiated power. The passive ESA shows a drop off of efficiency as the radiating aperture increases, primarily due to the large loss in the transmit and receive paths. For the active ESA, there is a drop off of the total efficiency due to the lower power added efficiency per module, and the addition of control logic at each T/R module. There is large area where small, lightweight AESA arrays will be more power and weight advantageous to the passive ESA designs. However, the cost of the newer technology has to be considered in the final system design.

4.4.2 Active element T/R modules

Active element scanned arrays had promise for reducing the size and weight of phased arrays for airborne radar applications. The antennas shown for HOWLS and multibeam MSR in Section 4.2 were too heavy and inefficient in the direct input average power to the radiated power for employment in tactical aircraft. The Joint Surveillance and Target Attack Radar System (JSTARS) antenna is an example of a passive ESA mounted on a large military aircraft, which illustrated the effectiveness of electronic scan and multimode radar operations. However, it has very large size, and weight could not be scaled for smaller aircraft. With the development of solid-state technology, several experimental antennas were developed to determine efficiency and lower cost (per element) for application on fighters and RPVs.

Early active element T/R modules were made possible from microwave integrated circuits, such as the L-band T/R module illustrated in Figure 4.27 [17]. The size of the circuitry could be reduced significantly by integrating solid-state components onto small high-dielectric substrates. These MIC components were used extensively in the 1980–2000 time frame for active element arrays. The module illustrated was developed in 1976 for the Naval Air Warfare Center, as a prototype for a vertical/short take-off and landing aircraft. This needed a high-power aperture assembly that could fit within the half-wavelength element spacing of the antenna. The MIC manufacturing produced 200 T/R modules, with amplitude and phase match over 200-MHz bandwidth and consistent with 45-dB channel match.

The problem with active elements is the ability to provide taper on transmit to reduce the sidelobes. Low sidelobes are needed to minimize clutter returns and avoid intercept by threat receivers. However, the solid-state amplification process requires a saturated output to maintain module efficiency. The prototype was tested and demonstrated an SDT on the array for reducing the azimuth sidelobes, as pictured in Figure 4.28. This SDT design was a technique to overcome the limitations of solid-state power amplifiers, which need to operate in Class-C saturation for maximum efficiency [18].

A novel approach was developed by the US Naval Air Development Center for a large airborne surveillance array, called electronically-agile solid-state universal surveillance (ESUS) array. The array is shown in Figure 4.29, illustrating the high density of modules in the middle of the antenna structure, and SDT at the ends. The

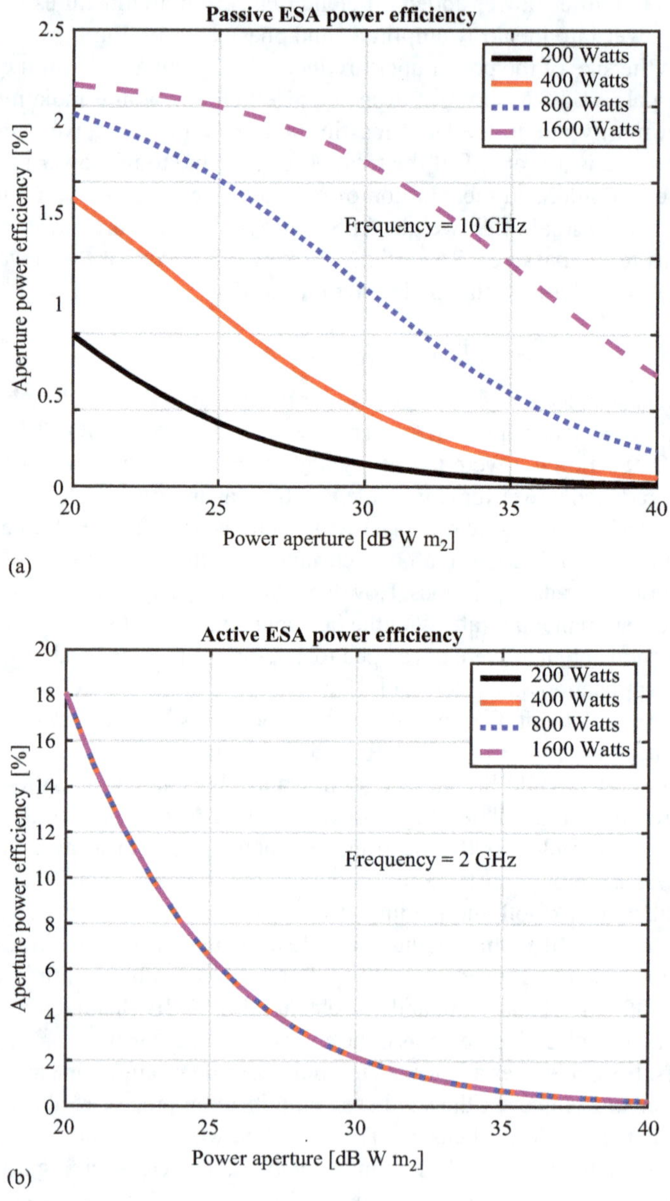

Figure 4.26 ESA power efficiency. (a) Passive ESA and (b) active ESA

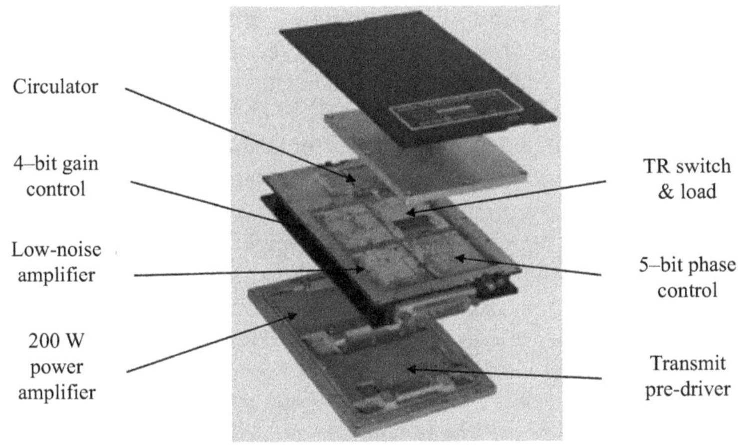

Figure 4.27 Active L-band transmit receiver module [17]

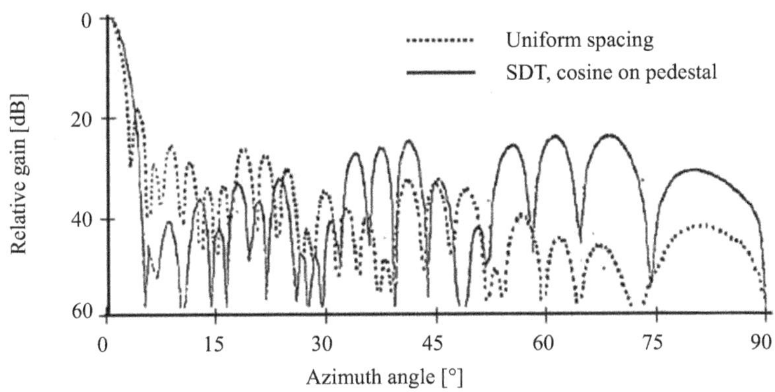

Figure 4.28 ESUS antenna SDT pattern compared to uniform spacing [18]

Figure 4.29 ESUS antenna prototype [18]

solid-state T/R modules, as shown in Figure 4.27, are housed behind each of the dipoles with amplitude and phase control on only receive [18].

The most important breakthrough of solid-state MIMIC design was the integration of many components on a single chip. Figure 4.30 illustrates two of the MIMIC devices produced by Thales in Europe for future fighters. In this case, it was a prototype for an AESA for the French Rafale fighter, demonstrated in 2008. The use of GaAs gave high-power output at X-band in 1,200 mm^2 area and a wideband core chip for low-noise figure gain along with amplitude and phase control. The X-band GaAs HBT MMIC high-power amplifier with 10-W peak power at 40% power added efficiency was integrated with a multifunction core chip, RF phase shifter, gain control, low-noise amplifiers and all T/R switches [19]. This level of RF integration of the AESA components on a very small number of chips makes an affordable active array feasible for many radar applications.

These types of devices are widely being developed globally with the proliferation of GaAs, GaN and SiGe devices using MMIC manufacturing. More importantly, this class of devices are now being manufactured for cellular radio and telephone communications, GPS receivers and AESA subsystems for direct TV reception on commercial aircraft. The manufacturing throughput and conservation of touch labor has brought the cost of AESA T/R modules down by two orders of magnitude since 2010. Research is continuing for the development of linear output amplifiers with high dynamic range and power added efficiencies over 20%. These experimental devices have yet been made for manufacturing. But future AESA T/R modules should be able to provide the needed amplitude taper on transmit for reduction of sidelobe.

4.4.3 Digital processing in wideband arrays

The awareness of channel match for radar has been emphasized for almost a half of the century. We have seen the major control of manufacturing tolerances on radars such as Multiple Aperture Surveillance Radar to get relevant clutter cancellation. This elemental control of T/R modules is critical for radar systems in a tough clutter and RF interference environment. Highly integrated components are available for high dynamic range, wideband digital processing, which have greatly improved

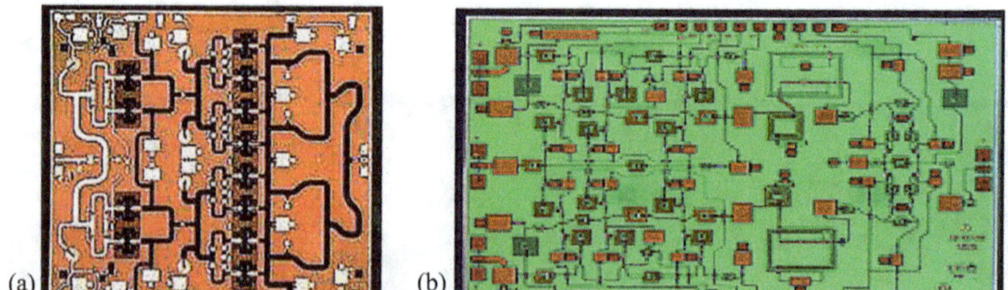

Figure 4.30 Microwave monolithic integrated circuits [19]. (a) X-band transmitter chip and (b) wideband core chip

the channel match for multichannel adaptive processing. Several improvements in UWB radar system architectures are being fielded to include compensation for antenna gain over wide band or angles. These techniques are being demonstrated in communication arrays with channel match for correlation processing between channels and digital beamforming for multichannel operation. Radar will soon be applying these component and signal processing techniques in small, UWB radar applications.

As we see in the analysis presented in Chapter 3, the radar architecture needs to control the amplitude to a fraction of a dB, and phase to a few degrees to get even 35 dB of channel match. The relationship between amplitude and phase to achieve high channel match is shown in Figure 4.31. The channel match is measured by the interference channel ratio (ICR) ρ_{ICR}. This ratio is measured between the parallel channel in a multichannel system and is compared to the main channel in terms of interference or error power levels [21]:

$$\rho_{ICR} = \frac{p_{\text{aux}}}{p_{\text{main}}} \tag{4.29}$$

When the peak transfer function between channels (amplitude or phase) across band ε_{p} is small, then ρ_{ICR} is given by

$$\rho_{ICR} = 2\,\varepsilon_P^{-2} \tag{4.30}$$

An example of this measurement is when the mismatch is either $\varepsilon_{\text{amp}} = 0.02$ or $\varepsilon_{\text{phase}} = 1.0°$, the peak inter-channel ratio will be $\rho_{\text{ICR}} = -35$ dB minimum. For many applications, the ICR needs to be 50–60 dB to meet system requirements. Hence, the channel match needs to be improved by 20–25 dB.

There will be a mismatch between channels caused by variation in the spatial input to the channel and the temporal variation due to spectral mismatch. These

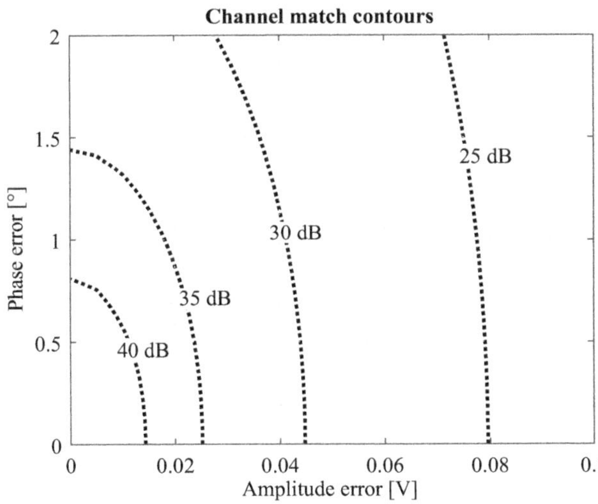

Figure 4.31 Interference channel ratio [20]

errors can be caused by variations in the RF and IF components, as well as mismatch in the path through the AESA from the radiation element to the ADC. The use of wideband digital processing can provide significant improvement in these channels for use in clutter and RF interference masking of targets. Because most of the AESA designs are using Class-C amplifiers, there will be peak sidelobes between –13 and –20 dB on transmit. This level of match needs to be reduced on receive by at least 20 dB for AESA to properly operate in a strong clutter and RF interference environment.

Components that provide high dynamic range signal sampling, along with real-time signal processing, are readily available. Several examples have been reported that provide match over wide angle scanning, UWB waveform processing and digital beamforming. The mismatch between channels is caused by RF and IF errors across the bandwidth of the channel in both angle and frequency. The adaptive weights are calculated by injecting wide band or impulse signals into each channel. The ideal equalizer is approximated by a digital finite impulse response (FIR) filter having L taps and T seconds apart.

In order to compensate for mismatch within the passband of a channel, a transfer function of receiver needs to be obtained. If there is a sinusoidal ripple within the frequency response, this can be analyzed by the transfer function $H_M(f)$ and $H_A(f)$ of the main and auxiliary channels as

$$H_M(f) = 1 + \varepsilon_P \sin\left(\theta_M + \frac{2\pi f}{f_M}\right), H_A(f) = 1 + \varepsilon_P \sin\left(\theta_A + \frac{2\pi f}{f_M}\right) \tag{4.31}$$

where ε_P represents the amplitude of the sinusoidal ripples, θ_M and θ_A are the unknown phases with random phase over $[0, 2\pi]°$, and the respective ripple frequencies are f_M and f_A. Figure 4.31 illustrates the ICR for these channels at various combinations of amplitude and phase distributions.

There are two steps in channel equalization. First, the primary channel is used as a basis for aligning the parallel channels. These weights are estimated offline, as shown in Figure 4.32 by injecting wideband or pseudo-random noise into the channel. The output of each channel is measured as a covariance between the physical channels and the temporal delays for the inputs. These weights are stored for real-time channel equalization. And next, the equalizer is implementing in high-speed digital processing, as shown in Figure 4.33 using an FIR filter. The number of time taps are determined by the bandwidth of the signal and an estimate of the variance between the channels.

The objective is to have the calibration signal a flat spectrum to characterize the channel. The equalization filter will have L taps and L weights to be applied in the auxiliary channel, and have the following transfer function:

$$H_{ID}(f) = \frac{H_M(f)}{H_A(f)} \tag{4.32}$$

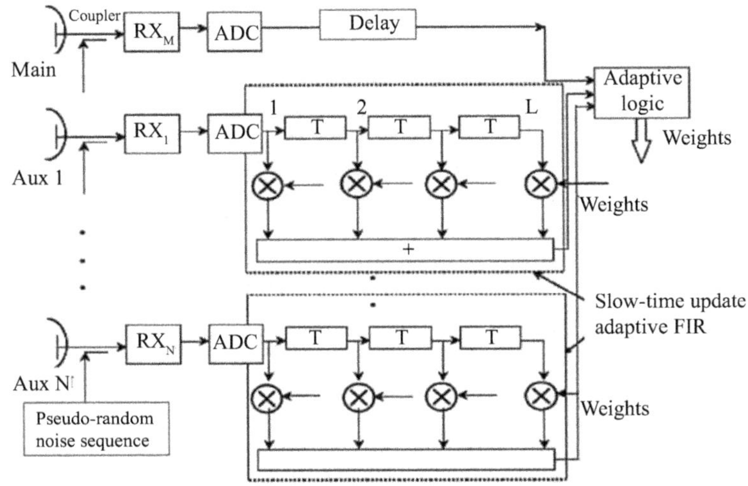

Figure 4.32 Training for channel match [20]

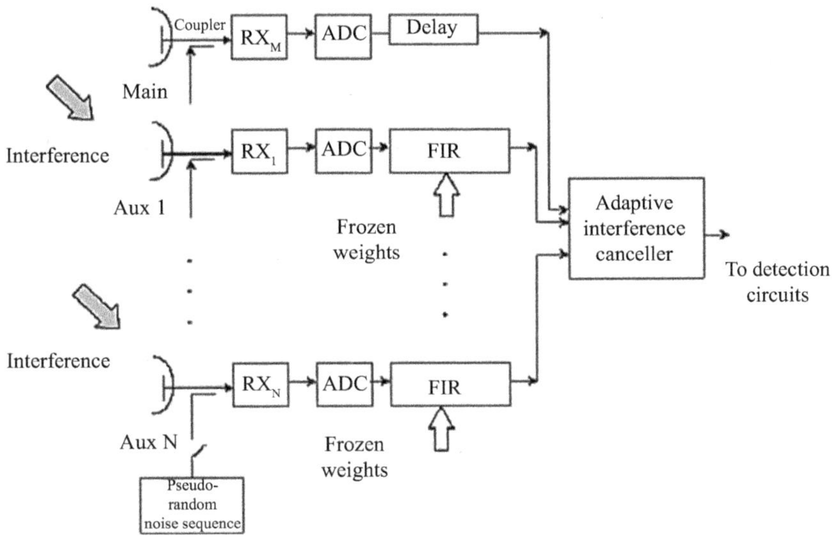

Figure 4.33 Adaptive channel equalization [20]

The ideal equalizer can be approximated by a digital FIR filter having L taps, T seconds apart. The L-dimensional weight vector $w = [w_1\ w_2\ \ldots.\ w_L]^T$ is calculated using the classic least-squares process, as shown in Figure 4.34:

$$w = R^{-1}r \tag{4.33}$$

This requires the estimation and inversion of the covariance matrix of the calibration signal into the auxiliary channel:

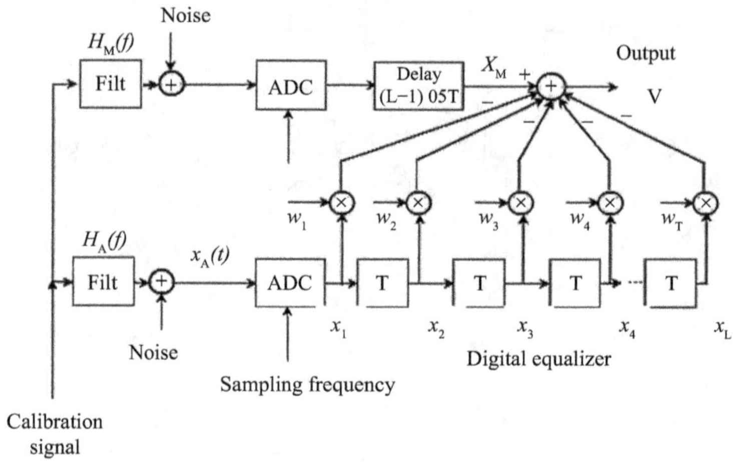

Figure 4.34 Calculation of equalization filter weights [20]

$$R = E\{x^* x^T\} \tag{4.34}$$

And the sequence $x^T = [x_1\, x_2 \,....\, x_L]$ to multiply to the estimate cross-correlation:

$$r = E\{x^*\, x_M\} \tag{4.35}$$

For a multichannel system, this process needs to be carried out for each channel, based on its cross-correlation with the main channel.

4.5 Wideband ESAs

Transitioning wideband AESAs to UWB requires a critical change in the normal ULA array steering vector. For the narrow band implementation, the center frequency can be used in calculating the phase progression across the face of the array. However, for wide band, it is important to use the actual frequency within the signal bandwidth. Because the wave front is delayed for each increment of the array, a phase correction cannot be used. The actual distance divided by the speed of light needs to be employed.

The most accurate approach is to use a true time delay steering at each element, as shown in Figure 4.35 employing the distance from a planar front to each element. This will require long time delays for wide scan angles or UWB waveforms. For lower frequencies where the separation is a large physical distance, true time delay can be used at each element. However, for microwave frequencies, it is not feasible due to the large increase in the number of time delay units.

Hence, for microwave frequencies, subarrays have been used with phase shifters on the subarray, and time delay units at the subarray output as shown in Figure 4.36. These subarrays should be sized such that the time delay across the subarray at the maximum scan angle is minimized. The objective is for the subarray to not scan

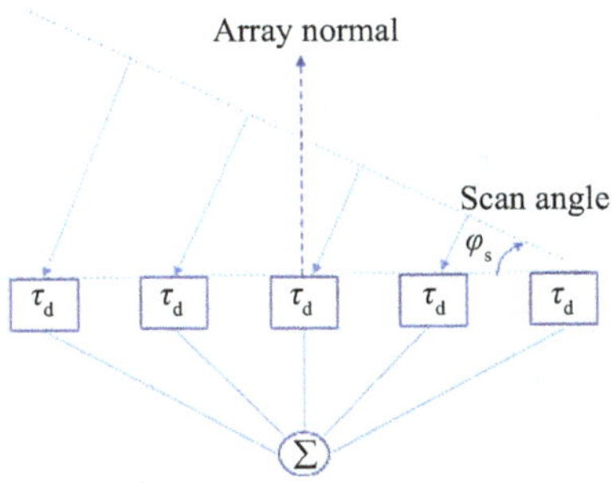

Figure 4.35 Wideband array scanning [22]

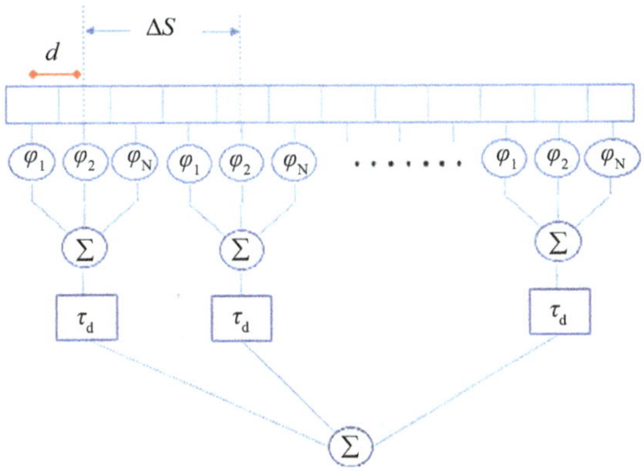

Figure 4.36 Linear wideband array [21]

more than a specified angle or fraction of the main beam width. Each of the subarrays need to receive the signal within the coherence limits of the radar system.

There are two control factors in the subarray steering of an UWB ULA. First, the subarray pattern, where the pattern is "allowed" to steer with frequency. The approach is to pick a size of the subarray so that there is not sufficient time delay between the edges of the subarray to affect the waveform over the total bandwidth. Second, the array factor needs to be evaluated, where the digital time delays are applied to provide collimated addition of the signal with appropriate time delay for the scan angle. The time delay effectively takes the beam steering effects out of the aperture. With today's high-speed digital processing, the UWB antenna can be used for a multibeam architecture, employing both time delay and adaptive weighting of each of the subarrays (for each of the M independent beams).

The array steering approach is shown in Figure 4.37 with the following analytic expressions (4.36). The array factor is steering of each element in the array, and the subarray pattern is shown in the right portion of the equation. The use of λ and λ_0 indicates that the actual wavelength λ has a separation in steering from the center frequency wavelength λ_0.

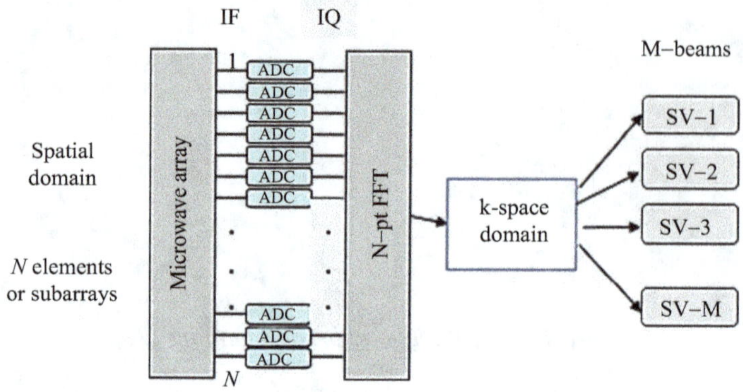

Figure 4.37 Multiple beam digital processing

$$E(\varphi,\lambda) = \overbrace{\sum_{m=1}^{M} \exp\left[-j\frac{2\pi}{\lambda} m\Delta S(\sin\varphi - \sin\varphi_s)\right]}^{\text{Array factor}} \overbrace{\sum_{n=1}^{N}\left[\exp(-j2\pi\, n\, d)\left(\frac{\sin\varphi}{\lambda} - \frac{\sin\varphi_s}{\lambda_0}\right)\right]}^{\text{Sub array pattern}} \tag{4.36}$$

where

$$\tau_d = \frac{md\sin\varphi_s}{c} \quad d \le \frac{\lambda}{(1+|\sin\varphi_{s|})} \quad \theta_s = \frac{2\pi}{\lambda} nd\sin\varphi_s \tag{4.37}$$

Normally, a combination of phase and time delay steering required to maintain constant phase front over wide bandwidth. The linear phase across the array will steer the beams. And the time delay at the subarray level will alleviate any beam steering over the bandwidth. A thorough design analysis should be carried out to make sure the time delay will compensate for any dispersion in the channels over the full azimuth or elevation steering.

The advantage of digital beamforming is that multiple beams can be synthesized based on the digital output of the ADC at each element or subarray. Once the signal has been digitized at the Nyquist rate, it can be converted from the time domain to the frequency domain using fast Fourier transform (FFT) processing. In the Doppler or k-space, multiple beams are then synthesized and steered digitally to cover the spatial extent of the search area. This is illustrated in Figure 4.38.

More details will be given in Chapter 9 in the development of multimode processing with multiple beams and UWB waveforms.

4.5.1 Digital beamforming computation

Multiple beams are very beneficial for multimode surveillance radars to provide wide area search. Early versions of multiple beam radars used analog beamformers, due to the limited technology for digitizing and processing coherent signals. However, with the rapid advances in ADCs, it is now feasible to process data from

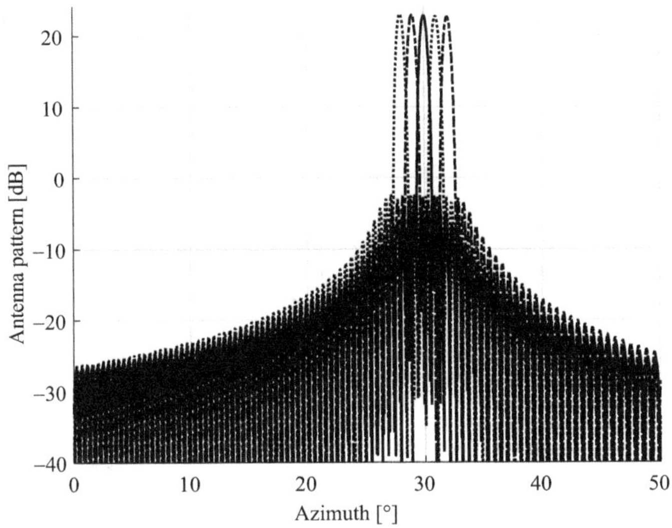

Figure 4.38 Digital beam array pattern

multiple elements or channels. As part of the systems engineering approach, the sizing of a coherent processor to form and control these multiple beams is essential. Figure 4.39 illustrates the approach for analyzing a digital multiple beamformer.

The array architecture can take many forms. The most common is a rectangular array with N_{az} and N_{el} elements, uniformly distributed in two dimensions. Other architectures include cylindrical, conformal and nonuniform arrays. The front-end components include the wideband receivers (one per element) and ADCs with bandwidth B and dynamic range N_{bit}. The number of ADCs will either be two per channel at B for an analog down-converter to I and Q or one per channel at $2B$ for a digital down-converter. It is also possible to have multiple analog transmit and receive circuits per radiating element and analog combination of the received signal into

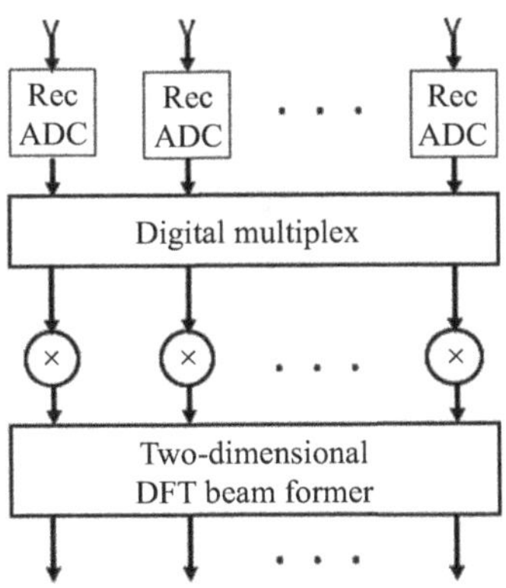

Figure 4.39 Digital processing for multiple beams [23]

subarrays. For this estimation of the operations count, a uniform two-dimensional array will be considered with one receiver per element.

A ULA of N elements (and N is divisible by 4) would require no more that $N^2/4$ complex multiplications and N^2 complex additions to form N beams using direct Fourier transform (DFT) processing. To apply a DFT algorithm to form multiple beams with uniform spacing in both azimuth and elevation, the elevation DFT is applied to the columns to preweight and scan the beam in elevation. Next, the elevation beams are preweighted and steered in the azimuth direction to form N beams.

In a square array with N elements in each dimension, and all divisible by 4, the required complex weights is N^2 multiplication for preweighting and steering. In addition, there are $N^2/4$ or column beamforming and $N/4$ for azimuth beam steering. The total processing count for a single beam is [23]

$$P_{1\text{beam}} = \frac{5\,N^2 + N}{4} \tag{4.38}$$

The number of complex multiplies for K beams is given by

$$P_{\text{Kbeam}} = N^2 + K\frac{N}{4}\,(N+1) \tag{4.39}$$

It should be pointed out that the number of complex multiplies will be simplified if each of the azimuth beams are at the same elevation angle. Table 4.4 summarizes the number of complex additions and multiplications for a 16 × 16 array and a 64 × 64 array. These numbers are for a 5-MHz sampling signal. As the bandwidth ADC sampling rater increased, the number of complex multiplies and additions increase proportionately.

If the number of elements in both dimensions are binary numbers, then the more efficient FFT algorithm can be used. In this case, the FFT of order $N = 2^n$ will need $N \log_2 N$ complex additions and $\mu N/2 \log_2 N$ complex multiplications, where μ is a factor less than 1, typically 1/3. This will reduce the number of operations compared to the DFT. However, it will require the same number of multiplications for preweighting and balance as DFT. An important point to remember is that the FFT algorithm provides phase shifting in beam steering. For wide band, multiple beamforming, the following algorithm must be employed to produce the time delay steering of the beams.

Table 4.4 Example of two-dimensional rectangular arrays

Subsystem	Array 1	Array 2	Comments
Elements	16 × 16	64 × 64	
Receivers	256	4,096	5 MHz signal
ADCs	512	8,192	8-bit I&Q
Complex weighting multipliers	32	512	Variable weights, calibration
Complex multipliers/s	2.56×10^9	269.1×10^9	
Complex adds/s	10.24×10^9	245.8×10^9	

Preweighting of the beams can be obtained using the amplitude and phase control on the multiplier prior to FFT for several sidelobe processing algorithms. This only requires one complex multiply per element. However, each beam will possess the same sidelobe weighting. If the desire for independent weighting of each beam is required, postprocessing on each beam is required to shape the beam. When an N-point FFT is used to form N unshaped beams, the output of the beamformer is expressed by [22]

$$u_k(p) = \sum_{n=0}^{N-1} v_n(p)\ \exp\left[j\frac{2\pi}{N}kn\right],\quad k = 1, 2, \ldots, N. \tag{4.40}$$

The uniformly weighted beams can be combined after formation. For example, if the three adjacent beams $[u_{m-1}\ u_m\ u_{m+1}]$ are combined to form a beam f_m

$$f_m = -\frac{1}{2}u_{m-1} + u_m - \frac{1}{2}u_{m+1} \tag{4.41}$$

The post-combination of beams is created by

$$f_m = \sum_{n=0}^{N-1}\left(2\sin\frac{\pi n}{N}\right)^2 v_n\ \exp\left[j\frac{2\pi}{N}mn\right] \tag{4.42}$$

This shows that preweighting with weights v_n is equivalent to postweighting the weights of

$$\left(2\sin\frac{\pi n}{N}\right)^2 v_n \tag{4.43}$$

Algorithms for wideband arrays need to compensate for progressive time delay of the wave front across the array, in addition to the amplitude weighting and phase shifting for narrow band-array operation. For wideband arrays, the most common compensation is time delay steering at the element or time delay steering at the subarray. Figure 4.40 shows the basic processing steps for a digital beamformer. The complex input signal has the I and Q components coming from the low-noise receiver, synchronous demodulation and ADC. The time (range) resolution δ_R of the signal is based on the signal bandwidth B

$$\delta_R = \frac{c}{2B} \tag{4.44}$$

And the dynamic range of the signal depends on the number of bits in the ADC. These two parameters established the input parameters for the arithmetic operations. The complex number samples in the nth channel are established by the complex waveform

$$s\left(t + t_n\right)\ e^{j\,n\,(\omega_0\tau+\alpha)} \tag{4.45}$$

where τ is the incremental time advance of the signal across between elements of the array, at an incident angle of θ:

$$\tau = \frac{d\cos\theta}{c} \tag{4.46}$$

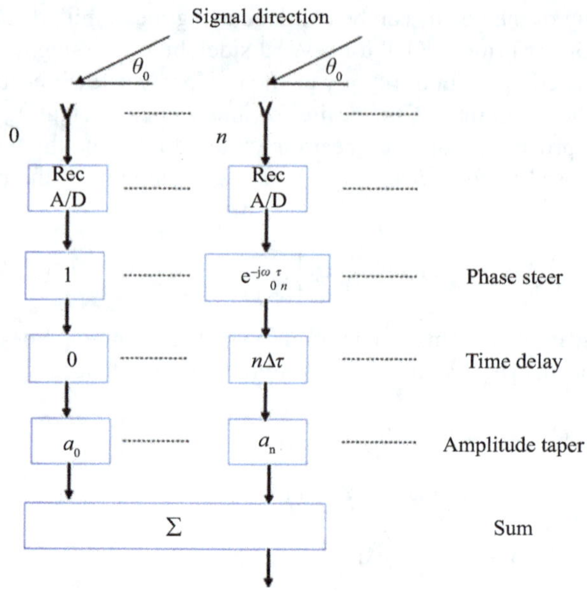

Figure 4.40 Basic operations of digital beamforming [22]

The first stage of the processing is the phase steering where the successive phase across the array, given at the center wavelength ($\lambda_0 = 2\pi/\omega_0$) is

$$\alpha = -\frac{2\pi\, d}{\lambda_0} \cos\theta_0 \tag{4.47}$$

However, the phase correction has modulo 2π degrees available. Because of the frequency contribution to wave front phase in ω_0, it is necessary to adjust the time advance of the wave by steering with $\Delta\tau$. After the time delay, the amplitude taper is applied across the array to control the sidelobes. For narrow-band phased arrays, the time delay is omitted in the beam steering. For wideband arrays, an incremental time delay is given by

$$\Delta\tau = -\frac{d}{c} \cos\theta_0 \tag{4.48}$$

For analog beamforming, an incremental time delay can be obtained in discrete steps. However, for digital beamforming, the time increment can be obtained by digital increments of time in the processing. More importantly, the multiple beamforming can be obtained using FFT steps to steer the beam to incremental sectors. The fine steering can then be implemented with the phase shift, as indicated in (1.11). Typically, the array will be made up of subarrays with phase shifters at the individual elements and time delay steering at the subarray level. The processing is greatly simplified with FFT processing, where FFT(*N*) is a *N*-point FFT, as summarized in Table 4.5. If the number of elements in the array is *N*, and the number of elements in a subarrays is *S*, then the number of subarrays is *N*/*S*. The number of

Table 4.5 Digital beamforming operations [23]

Beamforming	Parameter	Computation count
Elevation	N_{el}	$N_{el}/2 \log_2 (N_{el})$
Azimuth	N_{az}	$N_{az}/2 \log_2 (Naz)$
Subarray	N_s	$N_{az}/N_s \log_2 (N_s)$

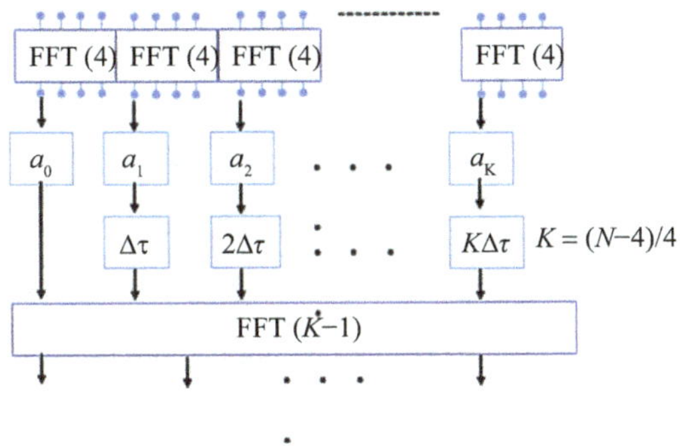

Figure 4.41 Wideband digital beamforming [23]

time increments is given by N independent of S. The subarray phase steering matrix requires N/S FFT(S) processing to form the subarray beams and the fine beam steering within the subarray requires S FFT(N/S). As a result, the total data processing requirement for the wideband array will be FFT(N). Figure 4.41 illustrates the use of FFT(4) for the subarray and beamforming for N element linear array.

An example of the complex beamforming processing is illustrated below: there are $N_{el} = 4$ elevation elements, $N_{az} = 64$ azimuth elements and $N_s = [1, 5, 11]$ subarrays.The parameters for the analysis are given in Table 4.5. The plot in Figure 4.42 shows the number of DBF processing complex operations needed for the case of three subarray sizes. When the subarray has only one element, there is no reduction in the processing of the full azimuth array. For the case of $N_s = 4$ and $N_s = 8$, there is a reduction in the processing for time delay steering. The total processing including the sample bandwidth varying from 16 to 512 MHz, where the subarray elements are only providing element phase shifting. When the signal bandwidth increases to UWB (e.g., greater than 50 MHz), the amount of digital processing grows proportionately. This is clearly the case for use of VLSI circuits and multiple parallel processing cores to achieve the real-time throughput.

4.6 UWB antenna summary

In summary, there has been extensive radar development to provide a technical background on many modes for air-to-ground surveillance. Passive ESA systems

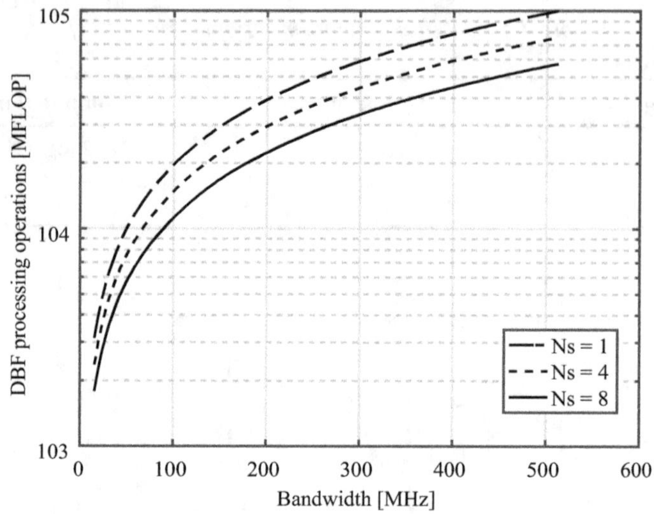

Figure 4.42 Wideband digital beamforming with subarrays

have been implemented for several military battlefield surveillance radars. However, multiple channel systems for small UAS and aircraft platforms were not feasible with earlier RF/IF design approaches.

Because of widespread development of MIMIC technology for both radar and communications systems, active ESA systems began rapidly developed by 2000 for military and geoscience applications. The enabling technologies were single-chip T/R modules and digital waveform generators and equalization techniques integrated into compact AESA designs.

UWB AESA designs are reported in development by integrating high percentage bandwidth components and maturing channel matching with IF FIR filters. This high-speed computational approach will solve producibility issues with lower cost device technology. One of the most important development has been multichannel and multibeam radars, employing RF or IF beamforming. However, these technologies are not applicable for future small manned and unmanned platforms.

Fortunately, the progression of MMIC and VHSIC technologies have progressed rapidly – following the Moore's Law of doubling the speed every two or three years. Because of the great progress in commercial communications technologies, there is a wide area of commercial off-the-shelf components to enable low-cost, lightweight surveillance systems. Active ESA technologies make it feasible to design and fly small, low-cost radars for both military and civilian applications. The radar literature has many examples of UWB AESA designs. This tutorial will attempt to summarize the current state of the radar art for future development applications.

References

[1] Cohen E. 'The mimic program—A retrospective'. *IEEE Microwave.* 2012;**13**(4):77–88.

[2] Hansen R.C., Antennas P.A. 'Linear array pattern synthesis, Chapter 3'. *Phased Array Antennas.* New York, NY: Wiley and Sons; 1998.

[3] Taylor T.T. 'Design of line-source antennas for narrow beamwidth and low side lobes'. *Transactions of the IRE Professional Group on Antennas and Propagation.* 1955;**3**(1):16–28.

[4] Bayliss E.T. 'Design of monopulse antenna difference patterns with low sidelobes*'. *Bell System Technical Journal.* 1968;**47**(5):623–50.

[5] Mailloux R.J. *Phased Array Handbook.* 2nd edn. Norwood, MA: Artech House; 2005.

[6] Barker J., Davis M.E. 'Ku-band linear phased array'. *Microwave Journal.* 1977;**20**:44–57.

[7] Aerospace G.E. *"General Electric HOWLS Electronic Scanned Array".* Utica, NY: General Electric Marketing Brochure; 1977.

[8] Banister W., Grove C.E., Song J. *Multibeam Antenna Development and Test Program – Final Technical Report.* AFWAL-T/R-83-1049, AFWAL/AARM, Wright Patterson Air Force Base, Contract F33615-79-C-1907; 1983.

[9] Kinsey R. 'Phased array beam spoiling technique'. *Transactions Antenna and Propagation Society Symposium.* 2. Montreal Quebec Canada; 1997. pp. 698–701.

[10] Leifer M.C. Revisiting a method of beam shaping using phase weights. *Proceedings 2016 IEEE International Symposium on Phased Array Systems and Technology*; Waltham, MA; 2016.

[11] Hofmeister E.F., Szczepanski W.E., Oot R.F., Dalpe D.C., Davis M.E. Modular surveillance radar for battlefield surveillance applications. *Proceedings RADAR-82*; London England; 1982.

[12] Leifer M.C. An airborne radar phased array with low sidelobes and a spoiled beam. *Proceedings 2018 IEEE Radar Conference*; Boston, MA; 2018.

[13] Butler J., Lowe R. 'Beam forming matrix simplifies design of electronically scanned antennas'. *Electronic Design.* 1961;**9**:170–3.

[14] Tomlinson P.G. 'Implications of the frequency dependence of the P-3 UWB antenna gain'. *Unpublished Memorandum.* Arlington, VA: DSA Inc; 1994 21 July.

[15] Ayers E.L., Ralston J.M., Maloney R.P., Tomlinson P.G., McCorkle J. Antenna measures of merit for ultra-wide synthetic aperture RADAR. *Proceedings IEEE 1998 National RADAR Conference*; Dallas, TX; 1998 May. pp. 331–6.

[16] Lalezari F. *"Compact, Lightweight Antenna Technology Enables VHF/UHF SAR Operation on Small UAV Platforms".* Boulder CO: FirstRF Company Marketing Brochure (Public Release by DARPA); 2009.

[17] Davis M.E., Smith J.K., Grove C.E. 'L- band T/R module for airborne phase array'. *Microwave Journal.* 1977;**20**:54–60.

[18] Smith J.K. Space density tapered arrays for airborne surveillance radar. *Proceedings IEEE International Radar Conference*; Washington, DC; 1980. pp. 263–70.
[19] Mancuso Y. Thales components and technologies for T/R modules. *Proceedings 2008 IEEE Radar Conference*; Rome Italy; 2008.
[20] Farina A. 'Digital equalisation in adaptive spatial filtering for radar systems: a survey'. *Signal Processing*. 2003;**83**(1):11–29.
[21] Nitzberg R. 'Sidelobe canceller, Chapter 4'. *Adaptive Signal Processing for Radar*. Norwood, MA: Artech House; 1992.
[22] Aalfs D. 'Adaptive Digital Beamforming, Chapter 9'. *Principals of Modern Radar - Advanced Techniques*. Scitech Publications; 2013.
[23] Oehling G., Applebaum S. *Data Processing Requirements for Digital Multiple Beamforming*. EHM-12-299, Army Contract DASG60-76-C-0027, GE Aerospace. Syracuse, NY; 1976.

Chapter 5
Ultra-wideband synthetic aperture radar processing

This chapter covers the image formation processing for ultra-wideband (UWB) synthetic aperture radar (SAR) applications. As the desire for more spatial resolution is considered, the signal bandwidth and SAR integration angles needed to be pushed to the limit of current component capabilities. However, if Moore's Law continues, these radio frequency (RF) and digital bandwidths will continue to be available from the device development communities.

The initial approach is to look at the geometries for SAR image collections. This will include aspects of standoff range and available frequency allocation to assess feasible resolutions of the image. The longer wavelength of UHF and L-band requires significantly wider synthetic apertures for a given cross-range resolution. Range migration is a major factor in the image formation process. The finer the range resolution, the more range walk that will occur during the SAR collection path, directly affecting the coherent focus of each pixel. This migration will also be important in terms of the amount of data to be collected, including the secondary data of platform motion and image plane orientation to the flight path.

There are two prevalent SAR image formation algorithms. One approach is a temporal collection of RF signal returns, called the back projection algorithm (BPA) [1]. The other is a frequency domain collection of the data, called the range migration algorithm (RMA) [2]. There are competing approaches for both algorithms based on the need for real-time signal processing with available signal processing computational architectures. Both techniques have been used for UWB SAR, and their techniques are widely covered in the literature.

Due to errors in navigation measurements or incomplete data on the digital terrain elevation data (DTED), there will be incomplete focus of the images for UWB SAR. These factors have led to several autofocus techniques for SAR image formation. The autofocus relevant for UWB SAR will be summarized in Section 5.3.

Finally, the requirements for motion compensation (MOCOMP) will be covered. An accurate measurement of the position and speed of the radar platform is essential. By using an onboard inertial measurement unit (IMU) and the global positioning system (GPS), the measurement of platform distance to the imaging plane can be accurately measured. For UWB SAR, this factor will be important in terms of processing efficiency, depending on the speed and type of airborne platform.

With longer standoff ranges and wider image collection geometries, the accurate MOCOMP designs are greatly needed, as summarized in Section 5.4.

5.1 SAR data collection

Long wavelength (ultra-high frequency–UHF or L-band) SAR processing has many of the same characteristics of high-resolution microwave SAR. However, to obtain the maximum image resolution, very wideband waveforms and large integration angles are required. Figure 5.1 shows the integration angles versus center frequency for SAR imaging between UHF and X-band for three different cross-range resolutions, based on (3.16) [2].

$$\delta_{CR} = \frac{k_{CR}\,\lambda_c}{4\sin\left(\theta_I/2\right)} \tag{5.1}$$

When the cross-range resolution δ_{CR} is finer than 1-m, the SAR integration angle requires more frequency and angle diversity to support the data collection requirements. These requirements on waveform and integration angle make demands on both the amount of processing to form the image and the motion measurement and compensation to ensure geospatial accuracy and focus over the full image.

UWB SAR is complicated because of the wide angle and dispersion of the signal in angle (Doppler) and time (bandwidth). With these wide angles, the collection must include not only the range extent of data for the swath width, but also the projected range extent for the wide-angle support data to provide the along-track image resolution. This section will go into the impact of integration angle on

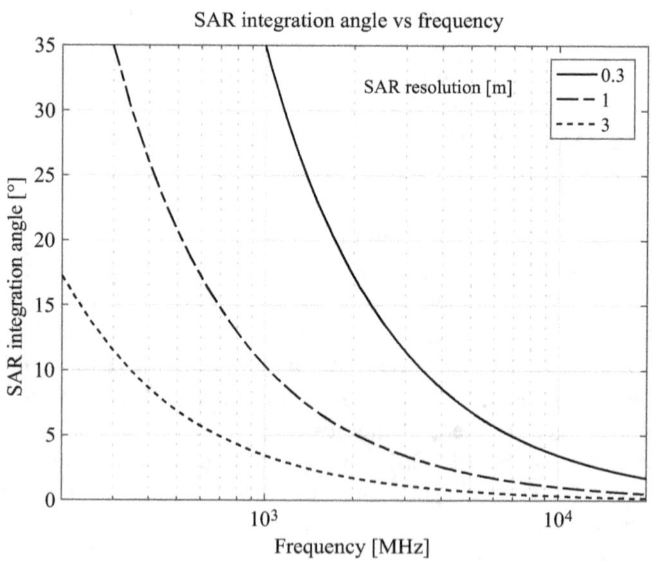

Figure 5.1 SAR integration angle versus frequency

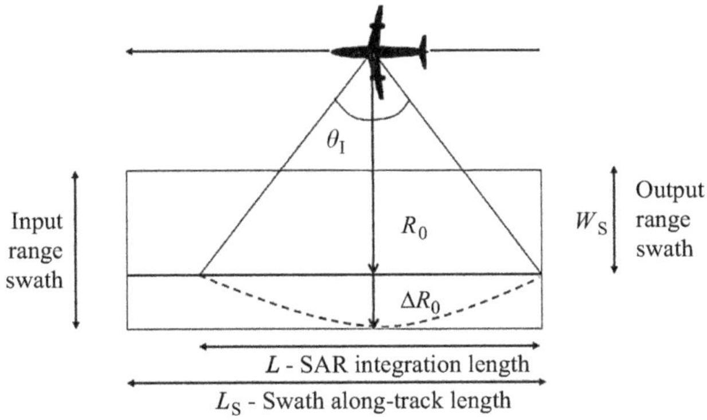

Figure 5.2 Impact of range curvature on SAR data support [3]

cross-range resolution and the impact of range curvature on both data collection and image formation.

The two most common algorithms for UWB SAR outlined with their impact on the image formation processing are the BPA (temporal) and the (RMA) (spectral). It is important to have an onboard, motion measurement system that determines the position, velocity and orientation of the radar antenna phase center at each pulse. MOCOMP then coherently aligns the pulses in fast time (range) and slow time (Doppler) to provide a focused image.

The first aspect of UWB SAR to be considered is the range swath W_S needed for meeting the SAR range extent, as illustrated in Figure 5.2. The slant range R_0 to the farthest edge of the desired image strip is shown at θ_I=0. However, as the platform moves along the collection geometry, the slant range to the far edge of the image needs to be longer than R_0. This extra length is needed to obtain range samples within the SAR integration angle required for the desired cross-range resolution. As a result, this extra range ΔR_0 is given by [3]

$$\Delta R_0 = \left[\frac{1}{\cos(\theta_I/2)} - 1\right] R_0 \tag{5.2}$$

A sufficiently long integration aperture must be collected to form the desired cross-range resolution. The number of pulses to be integrated can be approximated based on the range to the center of the image R_C and the desired cross-range resolution δ_{CR}. The number of pulses required to satisfy the cross-range resolution can be expressed by

$$N_P = \frac{k_S\, R_C\, \lambda_C}{2\delta_{CR}^2} \tag{5.3}$$

where k_S is the oversample ratio for data collection, R_C the range to swath center and λ_c the wavelength at the center of the spectral bandwidth.

For UHF and L-band SAR, the number of pulses can be very large. These are determined by the number of range bins in the desired swath and the additional data samples required due to range curvature. The other aspect of UWB SAR is the effect of range curvature on the image. As illustrated in Figure 5.2, data to support the resolution in the middle of the swath length data must be collected from the beginning of the swath through to the end. So, the range extent of the data must be larger than the desired swath width, such that the data at the ends contributes to the pixel resolution in the middle of the swath. This requirement on the total range bins N_R to satisfy the collection requirements can be expressed by

$$N_R = \frac{(W_s + \Delta R_0)}{\delta_R} \tag{5.4}$$

where δ_R is the range resolution. These considerations will now be quantified.

The range curvature during a long SAR collection establishes a requirement on the data collection. The excess range extent ΔR_0 is illustrated in Figure 5.3 for frequencies from UHF to microwave. Because of the large integration angles, the range curvature for longer wavelength SAR is significant. As the standoff range is increased, this extra range collection becomes several kilometers of data. At UHF, the image requires approximately 0.5 km–2.0 km of additional sample support for the 1-m cross-range resolution. And at L-band, the collection requires almost 300 m of extra range extent collection. With range resolutions below a meter, this quickly becomes several thousand range bins. Consequently, for L-band and UHF SAR images, this geometric factor will impact the data collection, MOCOMP and signal processing operations. Range curvature is not as significant in microwave SAR (C-band, X-band and Ku-bands), but some extra data collection is still required due to the range curvature.

The along-track data collection length is similarly determined by the cross-range resolution and the speed of the platform. The synthetic aperture length is related to

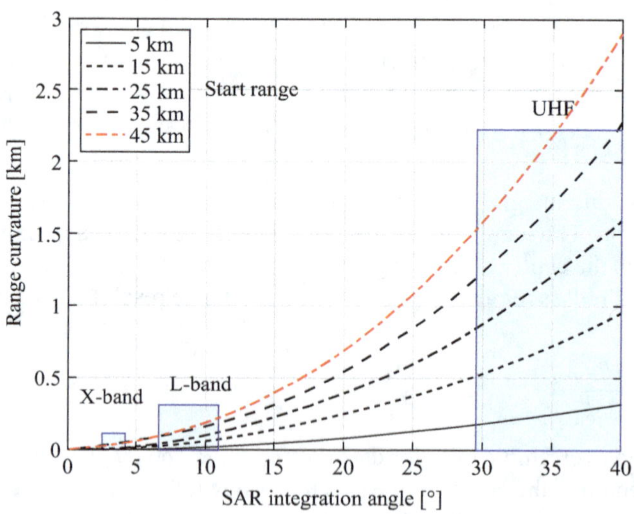

Figure 5.3   Additional swath width versus integration angle

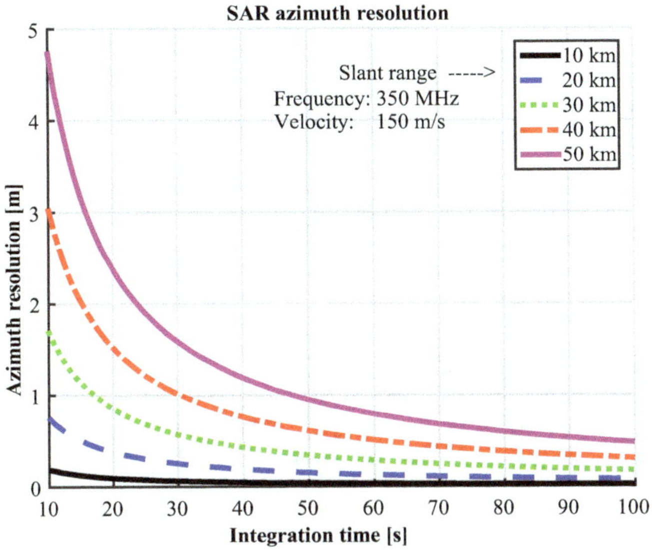

Figure 5.4 *UHF SAR cross-range resolution*

the integration angle and the standoff range. Figure 5.4 illustrates the integration times required for UHF SAR at 150-m/s platform velocity and at a variety of standoff ranges. Similarly, Figure 5.5 shows the integration times for L-band SAR at the same collection geometries. The integration times are proportionately lower at L-band due to the smaller integration angles for the same resolutions.

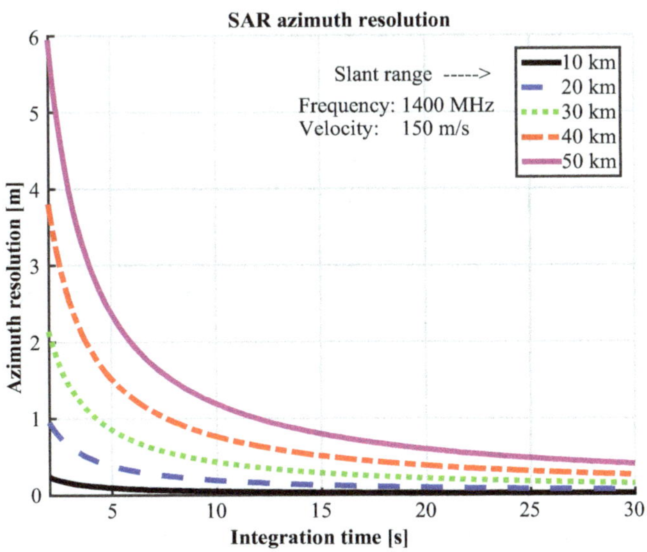

Figure 5.5 *L-band SAR cross-range resolution*

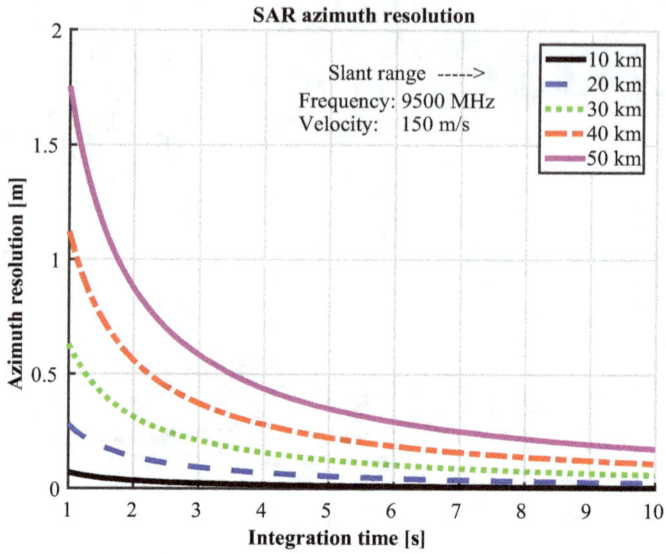

Figure 5.6 X-band SAR cross-range resolution

At microwave frequencies, a cross-range resolution finer than 1-m can be easily obtained in under 2-s out to 25 km, as shown in Figure 5.6. This is in contrast with 48-s with UHF FOPEN SAR and 12-s collection times for L-band. It should be noted that the UHF and L-band collections are characterized as UWB due to their frequency bandwidth percentage. And the X-band collections are only considered to be UWB when the bandwidth exceeds 500 MHz.

For long-wavelength SAR, the wide-angle data collection impact on the system is major. First, an entire pass is consumed in collecting a single swath of data. In addition, the variation of the terrain can be large over this geographic area. The demands on the motion measurement system and MOCOMP are amplified by the long collections and variation of the terrain in calculating the slant range to each pixel. As will be seen with the image formation processing, there are limitations on choice of algorithms for efficient and effective focus of the image. For microwave SAR operation, the collection angle is significantly smaller than long wavelength. However, the UWB frequency operation must be taken into consideration as being UWB for RF operation and frequency allocation. The latter factor is a major system design consideration and is covered in Chapter 8.

5.2 SAR waveform

The SAR system needs to have sufficient bandwidth for range resolutions and integration angle for cross-range resolution, as covered in Section 3.2.3. The former is provided by a combination of the transmitter waveform, and the latter is due to the integration angle and antenna gain versus frequency. UWB systems present a requirement on the spectral coverage in both the time and Doppler dimensions. If the RF hardware does not have the amplitude and phase needed for supporting the waveform coherence, then an additional processing step is required to correct for any amplitude and phase errors required for reconstructing the signal in matched filtering.

The ground range resolution δ_R in the ground plane is given by

$$\delta_R = \frac{k_R\, c}{2\, B \cos(\gamma_g)} \tag{5.5}$$

where c is the speed of propagation, B the signal bandwidth, k_R the range resolution degradation due to weighting for sidelobes and γ_g is the grazing angle.

In order to achieve a cross-range resolution δ_{CR}, the scene must be illuminated within the SAR collection over an integration angle θ_I. If the antenna has a physical width D, the illumination angle will ideally follow

$$\theta_I \geq \arcsin\left(\frac{k_{CR}\, \lambda}{D}\right) \tag{5.6}$$

If the SAR antenna is fixed in azimuth angle, the minimum cross-range resolution will be determined by the azimuth antenna pattern. However, if the antenna is gimbaled or has electronic scan to maintain illumination on the scene, or the platform can fly a circle around the scene (e.g., circle SAR), the cross-range resolution will not be limited by the antenna pattern. In the following performance derivation, it is assumed that the antenna is fixed, looking normal to the velocity vector.

Note that the convention for cross-range resolution is to specify integration angle in (5.6) at the center wavelength. Figure 5.7 illustrates the ideal spectral support required for obtaining both range and cross-range resolution. It is clear that for UWB operation, the wavelength λ varies significantly during the waveform transmit (and receive) time. One can also expect that the beamwidth of the aperture will also vary with wavelength, due to practical details of the antenna design over these wide-frequency excursions. As illustrated by the shaded region in Figure 5.7, the ideal integration angle is approximately linear in wavelength. So, the convention of using the center wavelength λ_c should not introduce significant error in the imaging processes. However, the system design for the transmitter and antenna characteristics over the UWB SAR collection bandwidth does not typically follow these ideal conditions. An efficient radar system needs to characterize the frequency variations

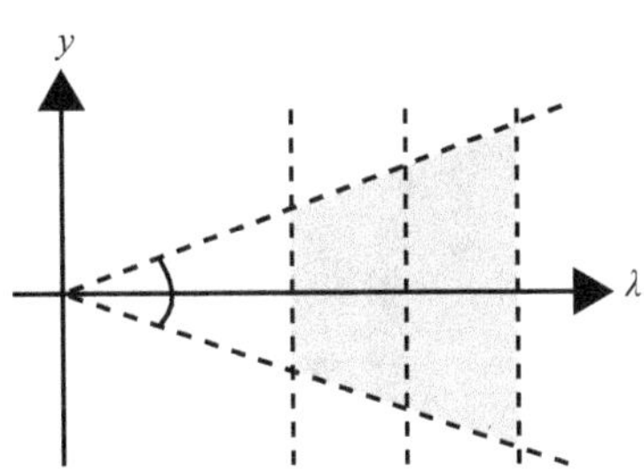

Figure 5.7 Waveform support for SAR image quality [2]

in the components and to compensate for them in image formation processing. The consideration of these factors was covered in Chapter 4.

The development of the SAR imaging requirements is normally conducted at an arbitrary squint angle α_{DC} with respect to the aircraft velocity vector. However, since UWB SAR involves very wide integration angles, the imaging is generally accomplished with a squint angle of 90° with respect to the velocity. In those cases where a different squint angle is desired, the integration time $t_{I,sq}$ and resolution $\delta_{CR,sq}$ should be obtained by projection onto the broadside dimension:

$$t_{I,sq} = \frac{t_I}{\sin(\alpha_{DC})} \tag{5.7}$$

and

$$\delta_{CR,sq} = \frac{\delta_{CR}}{\sin(\alpha_{DC})} \tag{5.8}$$

The imaging problem can be viewed in terms of frequency versus the along-track set of coordinates as shown in Figure 5.7. At each pulse in the SAR collection, the time and cross-range units will vary along a different vector, as shown in the figure. More importantly, the image support will vary in the along-track and cross-track geometry, due to the projection of the waveform onto the surface of the ground. It is important to take range curvature into consideration during both the waveform design and MOCOMP processing.

The SAR system transmits a series of pulses as it creates the synthetic aperture, as illustrated in Figure 5.8. The position of the SAR at each pulse is illustrated by a coordinate u_n along the y-axis. In this simplified coordinate system, the illumination

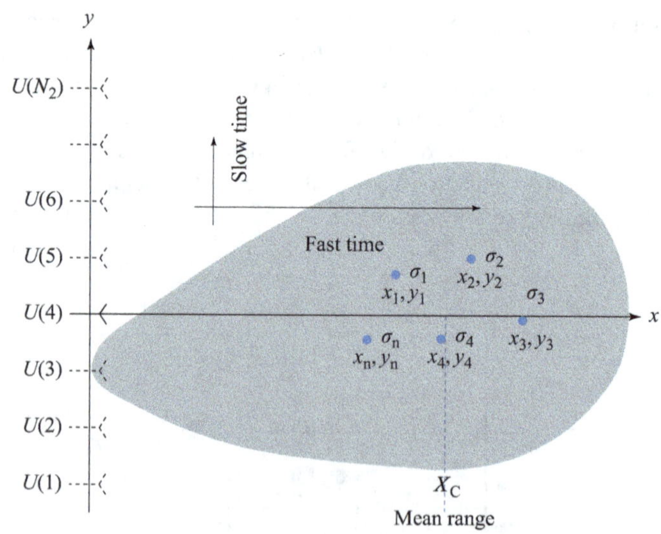

Figure 5.8 SAR image formation coordinate system [1]

is in a two-dimensional coordinate system, and the flight path is along a straight line on the y-axis. In a real system, the flight path will not be perfectly linear; hence, the vertical projection of the beam on a nonplanar surface must be considered. Each scatterer in the scene is illuminated by the radar transmit beam, as shown by the scattering variable σ_n, located at positions (x_n, y_n). In the SAR image, there is a field of scatterers that are physically separated. The ability to characterize these objects is determined by the range and cross-range resolution of the radar and the interaction of the electromagnetic scattering in the UWB SAR imaging process.

In Figure 5.8, there are two "time coordinates": *fast time and slow time.* Fast time t is the time for a single radar pulse to travel to the scene and return. Slow time $\hat{t}$ is the pulse-to-pulse variation of the nth pulse, within the pulse repetition interval T, expressed by the relation [1]

$$\hat{t} = t - nT \tag{5.9}$$

Each radar pulse will illuminate this field with signal intensity set by the peak power and the transmit antenna gain. And the received signal will depend on the reflectivity of the imaged surface and from any objects in the field.

It should be noted that the transmit coordinate system $u(n)$ moves with the velocity of the aircraft. The image formation challenge is to coherently add up these signals during the SAR integration time. If the signal transmission and reception geometry conform to a uniform linear array, the processing is simplified. However, due to aircraft accelerations typically experienced during the very long UWB SAR collections, uniform aperture spacing cannot be achieved. It is the function of the MOCOMP processing to correct the signal phase history.

The average power of most recent UWB SAR systems has received the benefit of using a long pulse, or a series of pulses, combined with UWB frequency modulation and the development of UWB pulse compression processing.

5.2.1 Linear frequency modulation waveform

The linear frequency modulation (LFM) technique has been successfully demonstrated on many UWB SAR systems. Each pulse has a modulation that changes the frequency of the signal as a function of time, as illustrated in Figure 5.9.

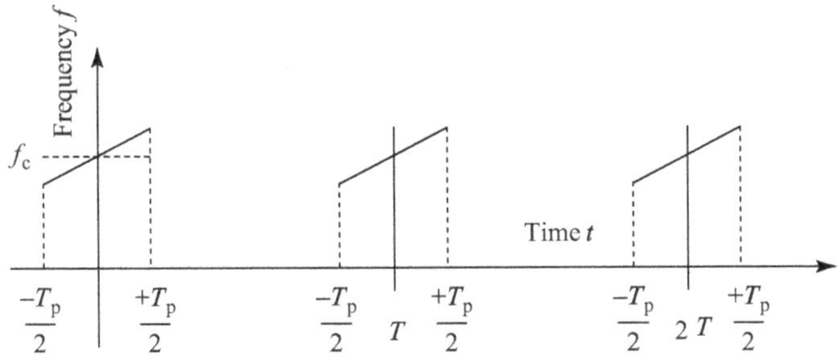

Figure 5.9 LFM waveform [2]

The waveform considerations for LFM determine the pulse-compressed slant range resolution, with sidelobe weighting factor k_R satisfying the equation

$$\delta_{R,\ PC} = \frac{k_R c}{2B} \tag{5.10}$$

The waveform is very simplistic, as long as the waveform generation and transmission can be properly provided by the hardware. In the early days of SAR, this was difficult because of lack of microwave components and fast digital processing. The most important part of the waveform is to have a pulse that can transmit the full bandwidth to give the required range resolution. An LFM pulse has a finite duration, which will limit the near range coverage. And the pulse repetition frequency (PRF) needs to be sufficient that the Doppler spread across the antenna provides the cross-range resolution [2].

$$\begin{aligned} \omega &= 2\pi f = 2\pi f_c + \pi \gamma_p \left(t - nT\right) \\ &= \omega_c + \pi \gamma_p \hat{t} \end{aligned} \tag{5.11}$$

where ω is the instantaneous radian frequency, f_c the center frequency of LFM modulation and γ_p the LFM modulation index (B/T_p) in radians per second.

Range resolution is obtained with the matched filter of the received signal provided by a replication of the transmit waveform. If the received signal is not weighted (i.e., $a_0 = 1$), then the sidelobes will be –13 dB down from the peak.

$$s_T\left(n, t\right) = a_0 \text{rect}\left[\frac{\hat{t}}{T_P}\right] \exp\left[j2\pi f_c t + \pi \gamma \hat{t}^2\right] \tag{5.12}$$

where T_p is the modulated pulse width, rect() the Functional equal to 1 for $\hat{t} \in [-T_p/2, T_p/2]$ and 0 otherwise, and a_0 the signal amplitude.

One should also note that the transmit amplitude a_0 is assumed to be constant. This constant amplitude is for an ideal system where the transmit amplitude does not fluctuate from pulse to pulse or across the frequency band. For more exact image processing, the waveform model must take into account any changes or modulation of the transmit function caused by either intended amplitude changes or unintended changes due to transmit and antenna power-spectral characteristics.

A representative UWB SAR LFM pulse having a center frequency of 400 MHz and a bandwidth of 200 MHz, and that is modulated over a pulse width of 10 μs, exhibits the characteristics in frequency and fast time domains, as shown in Figure 5.10. The slant range pulse resolution, with a uniform weighting fast time aperture, is 0.75 m. There is a problem with a wide-area SAR aperture, where strong discrete scatterers will obscure small nearby returns. The sidelobe control for the pulse can be applied on receive, yielding an improvement in the impulse response (IPR) characteristics. It is typical to use Taylor weights on LFM pulses, with the attendant reduction in sidelobes along with broadening of the mainlobe.

An alternative to the LFM pulse is a frequency jump burst (FJB), where a sub-band of the full bandwidth is transmitted on each pulse. And the additional sub-pulses are transmitted at a different time, often at a fraction of the PRI for each sub-pulse. This implies a higher PRF for the sub-pulses, combined with a narrower

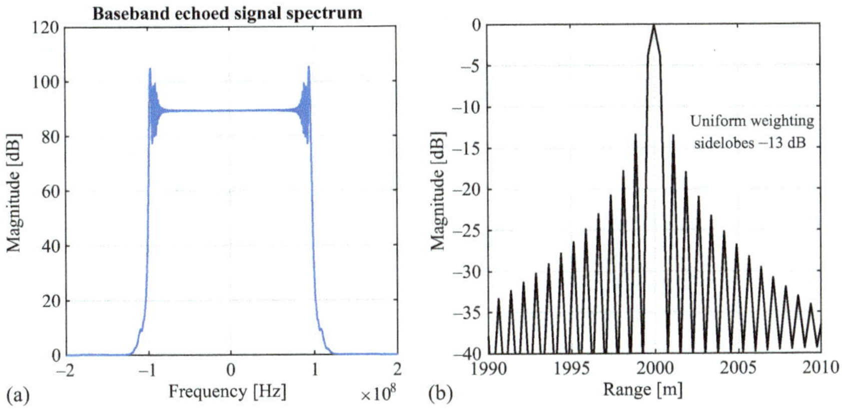

Figure 5.10 LFM range compression uniform weighting. (a) Pulse spectrum and (b) matched filter range resolution

intra-pulse receiver window. This FJB waveform will be presented in Chapter 9, along with its advantages and disadvantages for SAR mode operation.

5.2.2 SAR phase history

The SAR operates by collecting a series of coherent pulses shown above in Figure 5.8, yielding a phase history to be processed for image formation. This phase history changes continuously as a function of time along the SAR integration path. It is normal to define a *reference point* at a range R_c to calibrate the coherent integration process. The reference point is different in competitive algorithms depending on the assumption of a constant reference point. The radar transmits and remembers a reference pulse s_{ref} in order to match filter the signal with the time and frequency modulation [1]:

$$s_{\text{ref}} = \exp\left[j\left((\omega_c\left(t - \frac{2R_c}{C}\right) + \pi\gamma_\rho\left(\hat{t} - \frac{2R_c}{c}\right)^2\right)\right] \tag{5.13}$$

The received signal is obtained from the return from each point $R_t\,(x_t, y_t, z_t)$ in the scene with a signal amplitude of $a_n = \sqrt{\sigma_n}$, based on the radar backscatter area of each point. Each point has a round trip time delay from the transmit to receive t_d. Two factors make up the time delay: motion between successive transmit pulses motion during transmission and reception of a pulse, and motion between transmission and reception of a pulse. Motion between and during pulses does occur and contributes to a space invariant defocusing of the image. This effect can be removed with autofocus up to a limit on the change in spatial frequency during the pulse. In the extreme, the effects need to be compensated using inertial sensors on the radar. For the current discussions, the motion during a pulse is ignored in the image formation process.

The two-way time delay to each target point in the SAR image is given by

$$t_d = \frac{2\,R_t}{c} \tag{5.14}$$

The distance R_t is the two-way distance from the transmitter to each of the clutter cells and back to the receiver and varies during each point u in the SAR collection path. This distance is shown in Figure 5.8 and given by the relation

$$R_t(u) = \sqrt{x_n^2 + (y_n - u)^2} \tag{5.15}$$

The coordinates $R_t(x_n, y_n)$ are the true two-dimensional location of the scattering points on the surface, and the points $R_a(x_a, y_a)$ are the measured locations of the antenna phase center on each received pulse. Note that R_a will have errors due to MOCOMP inaccuracies as well as geolocation errors in the exact position with respect to the image surface. Given these coordinate systems, the received signal phase history is expressed by

$$s_R(n,t) = a_T\text{rect}\left[\frac{\hat{t} - 2R_t/c}{T_p}\right]\exp\left[j\omega_c\left(t - \frac{2R_t}{c}\right) + j\pi\gamma_p\left(t - \frac{2R_t}{c}\right)^2\right] \tag{5.16}$$

More importantly, the location of the radar $\boldsymbol{u}$ is a three-dimensional location, and (5.16) needs to be generalized into three dimensions.

The final step in defining the SAR phase history is to consider the demodulation of the signal. This is accomplished by digitally mixing the reference signal in (5.13) with the received signal in (5.16). The process can be done either before or after pulse compression, depending on the SAR image formation algorithm and the method of waveform modulation used. If the waveform varies in the fast time dimension such that compensation to a fixed range will distort the signal spectral characteristics, this factor needs to be either removed prior to SAR demodulation or incorporated into the image formation algorithm. Given these coordinate definitions, the received signal phase history is expressed by

$$S_R(n,t) = a_t\text{rect}\left[\frac{\hat{t} - \frac{2R_t}{c}}{T_p}\right]\exp\left[j\omega_c\left(t - \frac{2R_t}{c}\right)\right]\exp\left[j\pi\gamma_p\left(\hat{t} - \frac{2R_t}{c}\right)^2\right] \tag{5.17}$$

Figure 5.11 depicts several targets that were used to assess image formation steps and effects of waveform and MOCOMP on the FOPEN SAR design. The characteristics of the notional system are summarized in Table 5.1. There are three targets on each primary range and cross-range axes, and additional targets distributed within the 100 m × 100 m image area. In addition, σ_{14} is placed two-times δ_R separation from σ_5, and σ_{15} is placed two-times δ_{CR} from σ_2. These scattering targets will provide an assessment of the resolution characteristics of the waveform and the image processing algorithm and MOCOMP capability to maintain image quality.

5.2.3 Back projection algorithm

The BPA was the first successful image formation technique for UWB SAR [4]. BPA is a time domain convolution of each point on the image surface with the radar waveform in both slow and fast time. It was well known that the convolution

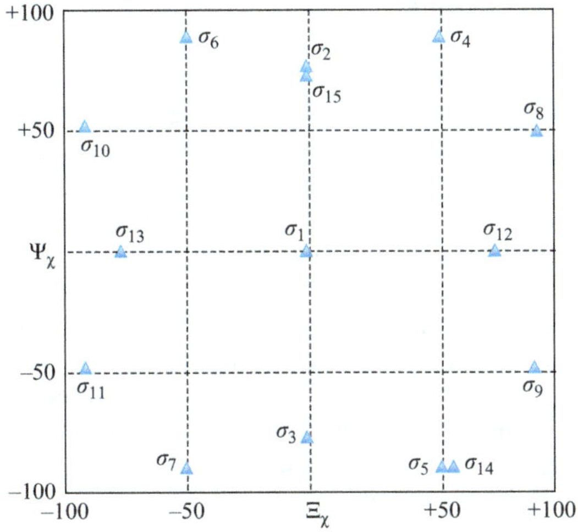

Figure 5.11 Location of example targets for image quality assessment [3]

operation can be accomplished with Fast Fourier transform (FFT) operations. The convolution $\rho_{sh}(t)$ between two signals is given by [5]:

$$\rho_{sh}(t) = x(t) \otimes h(t) = \int_{-\infty}^{\infty} x(\tau)\ h(t-\tau)\, d\tau \tag{5.18}$$

This convolution between the two functions can be computed by the Fourier transform:

$$\rho_{sh}(t) = F^{-1}\{X(\omega)\ H(\omega)\} \tag{5.19}$$

where $X(\omega)$ and $H(\omega)$ are the Fourier transforms of $x(t)$ and $h(t)$, respectively. This formulation is important with radar signal processing and especially for SAR image formation.

To form the SAR image, a time domain correlation is performed with each individual scatterer. This correlation step for an image of N_A pixels in the along-track dimension and N_R pixels in the range dimension using N_P transmitted pulses requires an order $O(N^3) = N_P N_A N_R$ of complex operations. The advantage of this SAR image formation approach is there are no approximations to the range and Doppler frequency contributions to image formation. When the transmitter location

Table 5.1 Summary of exemplary FOPEN SAR parameters for analysis

f_c	[400, 1,400] MHz	δ_R	0.75 m
B	200 MHz	δ_{CR}	0.77 m
L	[1300, 150] m	θ_I	[43, 5.9]°
x_c	2,000 m	X_0	±100 m
y_c	0 m	Y_0	±100 m

is accurately known at each pulse, and a geographically stabilized image plane can be maintained on the Earth's surface, the image will be well focused. The most accurate georegistration and focus of the image will be obtained when the image pixels are distributed on a digital elevation model (DEM) corresponding to the imaged region. If these image formation criteria can be maintained over the wide geometry, there will theoretically be no degradation of the image focus.

This return pulse is then multiplied by the reference signal, to provide the matched filter at each pixel in the image. The challenge then is digital signal processing and stable time reference to provide the coherent integration, as well as projecting this return on the image plane. If the time and the platform motion measurements are accurately known, the image will be correctly registered. Most importantly, the waveform and the antenna need to provide the image spectrum support, as shown above in Figure 5.7. This illumination must cover the image area during the synthetic aperture, with the angle and wavelength coverage requirement to meet both the range and cross-range resolution.

However, the image formation process is computationally expensive and was determined to not be suitable for real-time implementation in early FOPEN SAR systems. As the speed of computers and the combined accuracy of GPS/inertial navigation system (INS) systems have improved, the BPA has become a serious contender for a real-time airborne collection system. Furthermore, research in *fast* BPA processing has evolved an approximation based on sub-apertures, which reduces the computation by a factor of $\sqrt{N_P}$ [6].

Figure 5.8 provides the coordinate system used for the understanding of the BPA waveform transmission. The processing flow for the BPA is shown in Figure 5.12, along with an estimate of the processing requirements for each stage. The factor C_P is a multiplier for interpolation processing, as described in (5.26) below. The first

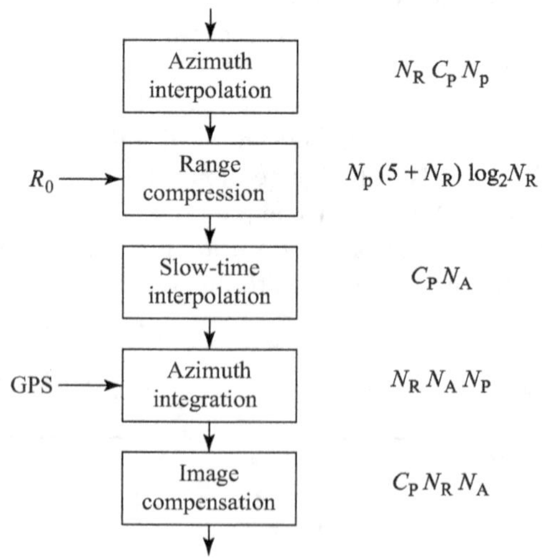

Figure 5.12 Algorithm flow for back projection image formation [3]

block is needed to compensate for the nonuniform location of the transmitter during radar transmission. Each pulse is transmitted at a start time determined by $(n-1)T$, the nth pulse repetition interval of the radar. These points, which are nominally established to be on half-wavelength centers, are determined from the velocity of the aircraft and the center wavelength λ_c of the waveform. Based on the inertial measurement system, the position and velocity of the aircraft are measured and compared with the desired transmit position.

The SAR image is formed by correlating the transmit waveform with each pixel on the image surfaces [1]:

$$s_M = s(t, u) \otimes s_T(-t) \tag{5.20}$$

where ⊗ is the convolution symbol. The time domain correlation image is given by

$$f(x_i, y_j) = \int_u s_M[t_{ij}(u), u]\, du \tag{5.21}$$

where the time delay t_{ij} from the SAR platform position u to each pixel in the image is given by

$$t_{ij} = \sqrt{\frac{x_i^2 + (y_j - u)^2}{c}} \tag{5.22}$$

Therefore, the time delay to each point on the surface must be calculated based on each position along the synthetic aperture flight path. It can be appreciated that knowledge of the position, velocity and orientation of the SAR platform is needed at each point in the SAR map. Moreover, having a DEM of the terrain is highly desirable to correctly calculate the time delay to the surface. Otherwise, if the image is formed only in a plane, there can be issues in the focus and position for each voxcel of the SAR image. The term voxcel is used here because of the three-dimensional propagation of the waveform and the real possibility that the scattering near any object will be affected by layover. In addition, when there is significant volumetric clutter near the objects (such as foliage), the clutter will be between the radar and the surface.

Since the distance from the transmitter to each point on the ground varies with slant range and angle, it is important to interpolate the pulse-compressed data to each point on the ground, on a pulse-by-pulse basis and along the slant range dimension. Figure 5.13 shows the basic interpolation used for resampling data in a generic slant range and angle (i_{SR}, θ) coordinate system into the image coordinate system (x_i, y_j). The interpolation is accomplished by upsampling the phase history function, from the non-uniform sample spacing to a uniformly spaced coordinate system [2]:

$$y(m) = \textstyle\sum_k \hat{x}_m(k)\, h(m-k) \tag{5.23}$$

where

$$\hat{x}_m = \begin{cases} x\left(\dfrac{m}{M}\right) & m = 0, \pm M, \pm 2M, \ldots \\ 0 & \text{otherwise} \end{cases} \tag{5.24}$$

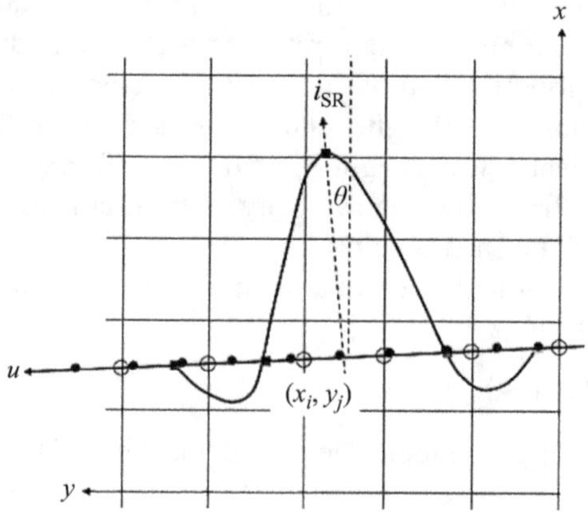

Figure 5.13 Interpolation function for along-track samples [2]

$$h(m) = \text{sinc}\left(\frac{\text{m}}{\text{M}}\right) \tag{5.25}$$

The solid circles along the u-axis give the along-track samples. The phase history is given by the unit vector i_{SR} at an angle θ with respect to the image coordinate system (x, y). The open circles are the resampled values of the function $y(m)$ on the image y-axis.

The signal processing requirement C_p, or operations complexity in number of real multiplies and additions, required for data interpolation is given by

$$C_P = 0.5N_I\left[\left(L_f - 1\right)\ \text{DSR} + 1\right] \tag{5.26}$$

where N_I is the number of output points, L_f the length of the resample filter and DSR is the down sample ratio of the output to the input samples.

Next, the radar waveform in the fast time dimension needs to be compressed so that the required range resolution is obtained. This requirement for early pulse compression differs between the alternative strip map and polar algorithm SAR processing algorithms. In order to get the necessary spatial correlation of the signal at each image point in the BPA, it is important to perform range compression prior to the time domain correlation of the transmit signal with the signal received from each point on the image surface. The pulse compression will be referenced to a range delay $2R_c/c$ that corresponds to the central reference point in the image.

The processing in back projection requires detailed and accurate knowledge of the system characteristics. The range delay to each point in the antenna beam and for each pulse in the synthetic aperture depends on several variables. First, the amplitude and phase in both the k_x and k_y dimensions affect the integration at each pixel in the scene. Then, the focus will depend on accurate knowledge of the orientation and velocity of the aircraft and antenna, as well as the image plane on the surface of the Earth.

When the waveform is modified or notched to handle the radio frequency interference, the effects on the spatial and spectral waveform characteristics must be considered. If holes remain in spectral domain, they should be compensated for in the time domain correlation, or high sidelobes will result. Because of the wide antenna angles, especially for UHF SAR, Doppler ambiguities need to be characterized and compensated. If this is not done, the returns at wider angles or Doppler ambiguities will fold into the image as artifacts. If these factors can be properly accomplished, the BPA will provide the largest depth of focus. This is due to the fact that each pixel is focused to the position, velocity and propagation delay to the radar.

It was well known that the convolution operation could be accomplished with FFT operations. Because of the wideband waveform, it is usual to first transform the fast time domain representation of the signal phase history to the spatial frequency domain, where k is the wavenumber in the fast time domain. Note that the transmit waveform is also characterized by its power spectrum $P(\omega)$. What results is a linear superposition of spherical phase modulated signals with amplitude characterized by the scattering cross-section. The next operation is to perform an FFT in the slow time u domain into the Doppler domain k_u. Again, the transformed signal phase history is a linear superposition of the scattering centers. More importantly, the phase modulation term is a linear function of the geographic locations (x_n, y_n).

The image then can be reconstructed by a two-dimensional FFT match filter using the wavenumber variables k_x and k_y [2]:

$$s(\omega, u) = P(\omega) \sum_n \sigma_n \exp\left[-j2k\sqrt{x_n^2 + (y_n - u)^2}\right] \tag{5.27}$$

where $k = \omega/c$ is the wavenumber of the radar return. After an FFT in the spatial domain, the signal is transformed to $S(\omega, k_u)$ as

$$S(\omega, k_u) = P(\omega) \sum_n \sigma_n \exp\left[-j\sqrt{4k^2 - k_u^2}\, x_n - j\, k_u\, y_n\right] \tag{5.28}$$

This is followed by the SAR spatial frequency mapping or transformation

$$k_x(\omega, k_u) = \sqrt{4k^2 - k_u^2}, \quad k_y(\omega, k_u) = k \tag{5.29}$$

where the practical reconstruction is obtained via fast time matched filtering

$$\begin{aligned} F\left[k_x(\omega, k_u),\, k_y(\omega, k_u)\right] &= P^*(\omega)\; S(\omega, k_u) \\ &= |P(\omega)|^2 \sum_n \sigma_n \exp\left(-j\,k_x\,x_n - jk_y\,y_n\right) \end{aligned} \tag{5.30}$$

Figure 5.14 illustrates the impact of mapping the signal into each pixel of the image. The time delay is quadratic in terms of the spatial location of each scatterer. As the platform translates its position, the mapping of each range slice must be interpolated on the grid. If not correctly interpolated, there will be a loss of coherence and hence a blurring of the image. The second practical consideration is measuring the position of the platform and providing MOCOMP to the pixel in the image.

Figure 5.11 showed the BPA applied to an image with several point scatterers. The scatterers are distributed in the range and along-track positions. This

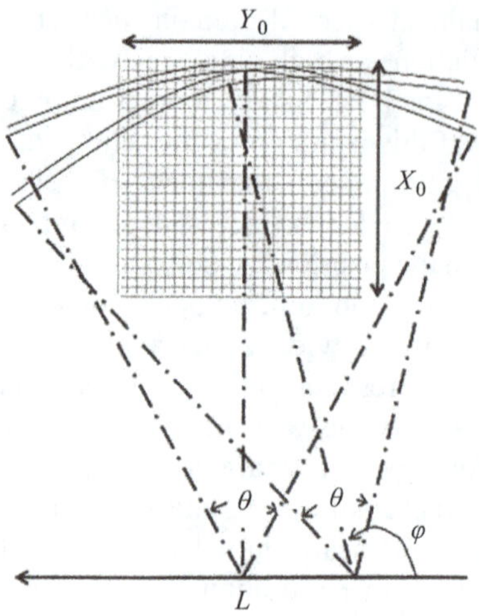

Figure 5.14 Back projection of each pulse return on image plane [1]

will provide an assessment of the impact of position on the resolution. There are also pairs of scatterers separated by a few range or cross-range resolution to test the impact on point spread function. The received signal for the UHF simulation is shown in Figure 5.15a where there is an extensive overlap in range and cross-range due to the time returns over wide integration angles. Figure 5.15b shows the fast time return after fast time matched filter (range compression). The individual targets are more defined, albeit with significant range curvature.

The received signal for the L-band simulation is shown in Figure 5.16 where there still is an extensive overlap in range and cross-range due to the time returns from the 16 targets over wide integration angles. Figure 5.16 shows the fast time return after fast time match filter (range compression). The individual targets are

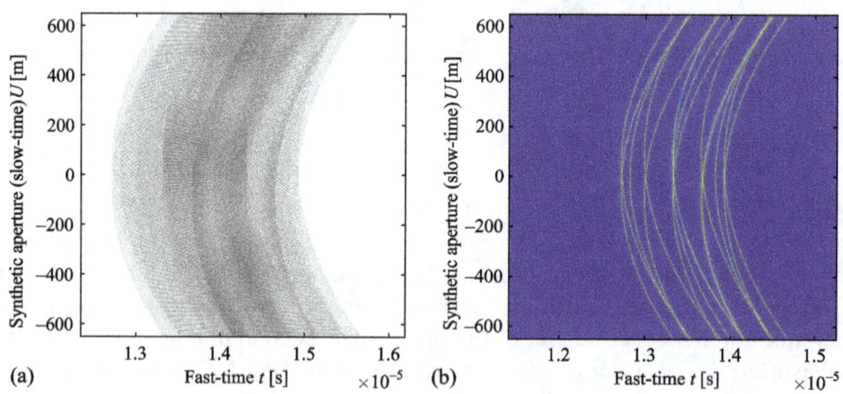

Figure 5.15 UHF SAR fast time matched filter. (a) Measured SAR signal and (b) SAR signal after fast time matched filter

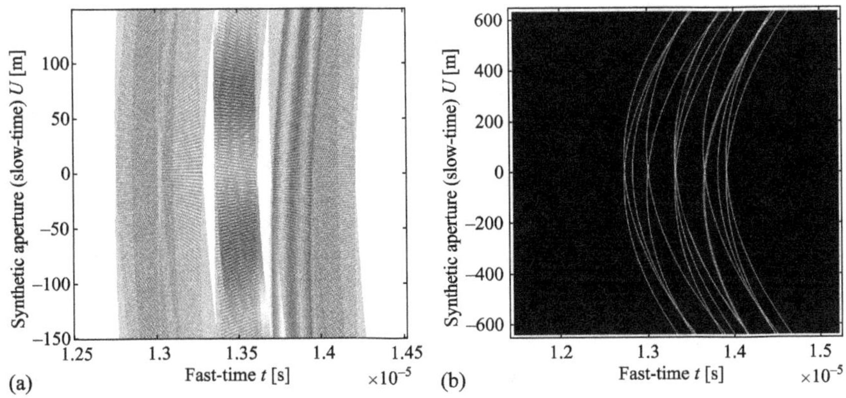

Figure 5.16 *L-band SAR fast time matched filter. (a) Measured SAR sigma and (b) SAR signal after fast time matched filter*

more defined, and the range curvature is significantly reduced by the smaller SAR integration angle at L-band frequency.

The range compression processing is carried out next with an FFT in range dimension, followed by MOCOMP to center point R_C and then an inverse FFT(IFFT) in Doppler is given by [2]

$$\begin{aligned} s_{PC}\left(u,t\right) &= \sum_{N_S} s\left(u,t\right) * s_0\left(u,t\right) \\ &= \sum_{N_S}\left(u,t\right) * \exp\left[j\omega_c\left(t-\frac{2R_c}{c}\right)-j\gamma\rho\left(t-\frac{2R_c}{c}\right)^2\right] \end{aligned} \tag{5.31}$$

For back projection processing of LFM pulses, the individual scatterers are more clearly separated. The pulse compression uses a double convolution algorithm with FFT in the range dimension, match filter with the waveform and converting back to the time domain with an IFFT. The MOCOMP for BPA is to the center point of the image. The range curvature is still clearly visible.

Figure 5.17 illustrates the along-track match filter for the UHF BPA. The point spread function is uniform for all pixels in the image. For the two areas where there are two scatterers, the pixels show good separation. However, it should be noted that this is an ideal flight path with perfect MOCOMP. The along-track match filter for the L-band image is identical to that of the UHF SAR simulation.

5.2.4 Range migration algorithm

The RMA, also called ω-k, was developed for SAR based on signal processing in the oil exploration seismic processing to handle wide-angle data collections within a dispersive media [2]. The top-level algorithmic flow for the RMA is shown in Figure 5.18. The algorithm compensates for the range curvature of the SAR signal transmission through a series of transformations. More importantly, it utilizes FFT operations to limit the number of operations for image processing.

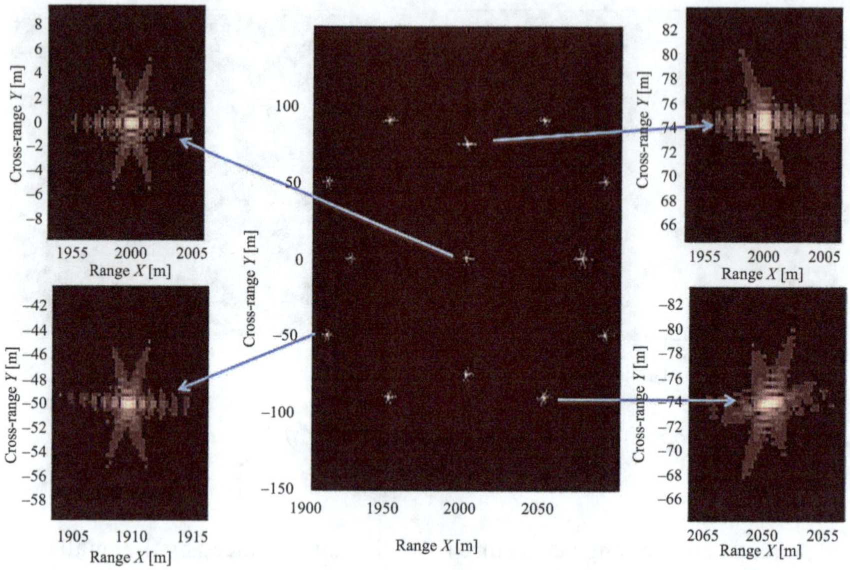

Figure 5.17 Final UHF image reconstruction [1]

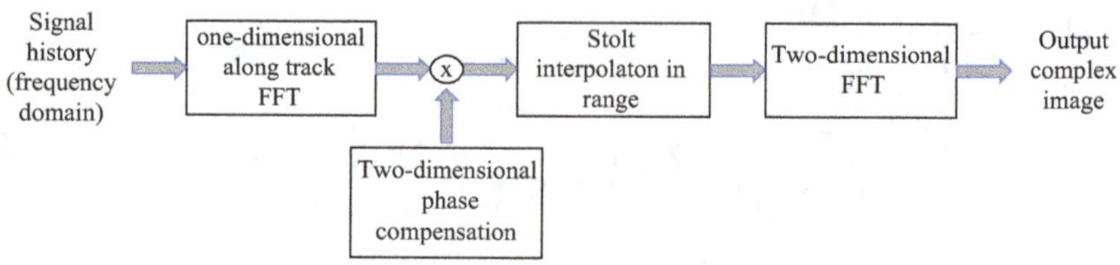

Figure 5.18 RMA functional flow [3]

The first step is to transform the phase history via an along-track FFT. This is done prior to range compression and azimuth dechirp, and it first removes the along-track dependency of the range curvature. The second step is to apply a two-dimensional phase compensation to correct the range curvature of all scatterers at the same range as scene center. The detailed algorithm steps are summarized below. Figure 5.19 shows the progression of the range curvature improvement in each stop of the RMA algorithm [3].

After the along-track Fourier transformation, the signal phase history still has a range varying curvature. The image is first focused along a straight line at the middle of the image. Along this line, there is no range curvature in the direction of the K_X mapping. However, there is a curvature at closer range and longer range that needs to be compensated. A mapping in phase function that will straighten the range curvature along the center of the image is given by [2]

$$\Phi\left(K_X, K_R\right) = -K_X K_t - R_8\sqrt{K_R^2 - K_X^2} + K_R R_S \tag{5.32}$$

where the parameters are K_X is the azimuthal spatial frequency, K_R the range spatial frequency, X_t the along-track position of platform versus slow time, R_B the minimum range from platform to the scatterer and R_s the broadside slant range from platform to scene center.

The second step is to apply a two-dimensional phase compensation to correct the range curvature of all scatterers at the same range as scene center. Matched filtering of the SAR phase is carried out to remove the last two terms in (5.32). The phase of the matched filter, which removes the linear phase with slant range R_s, is given by

$$\Phi_{mf} = -K_R R_S + R_S\sqrt{K_R^2 - K_X^2} \tag{5.33}$$

The signal phase is now given by

$$\Phi_2\left(K_X, K_R\right) = K_X K_t + \left(R_B - R_S\right)\sqrt{K_R^2 - K_X^2} \tag{5.34}$$

Along the mid-point of the image, the phase is constant on the $R_B = R_t$ line. However, at shorter and longer ranges, there is a residual phase curvature as shown in Figure 5.19 [3].

The major difference between the RMA and the polar algorithm is that the RMA motion compensates to a straight line and the polar algorithm compensates to a point at the center of the image. Furthermore, the Stolt interpolation in the RMA compensates for the differences in range curvature as a function of range. The Stolt interpolation straightens the lines of constant phase to form a linear phase in azimuth direction. The resampling provides the transformation:

$$\sqrt{K_R^2 - K_X^2} \implies K_Y \tag{5.35}$$

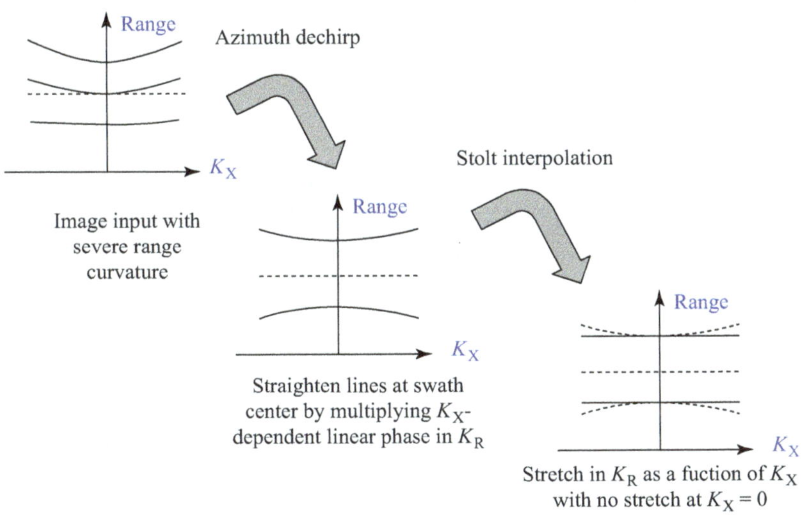

Figure 5.19 RMA range curvature correction [2]

Following the Stolt interpolation, the phase is now parallel to the K_t line for all ranges.

$$\Phi_{RMA}\left(K_X, K_Y\right) = K_X\, K_t + K_Y\left(R_B - R_S\right) \tag{5.36}$$

The signal from a point scatterer anywhere in the image has now been transformed to be linear in phase in K_X and K_Y space. As a result of this linearity, a two-dimensional IFFT can be performed to compress the signal simultaneously in range and cross-range. By using a two-dimensional FFT, the number of operations is reduced from $O(N^3)$ to $O(N^2 \log_2 N)$.

With this resampling of the data, the image phase is orthogonal in both range and cross-range dimensions. This enables image matched filter using FFT pixel processing instead of in the time domain. Hence, the computational burden is reduced from order N^3 to $N^2 \log N$. For large images, this can be a significant reduction in the operations count.

Figure 5.20 details the algorithm computational steps in the RMA algorithm to compensate for range curvature, along with the operations count for each stage of the processing. After along-track Fourier transformation, the signal phase history has a range varying curvature. The image is first focused along a straight line at the middle of the image. Along this line, there is no range curvature along the image K_X mapping. However, there is a curvature at closer range and longer range that needs to be compensated.

Next, a transformation is made to stretch K_R as a function of the slow time variable K_X. The Stolt interpolation straightens the lines of constant phase to form a

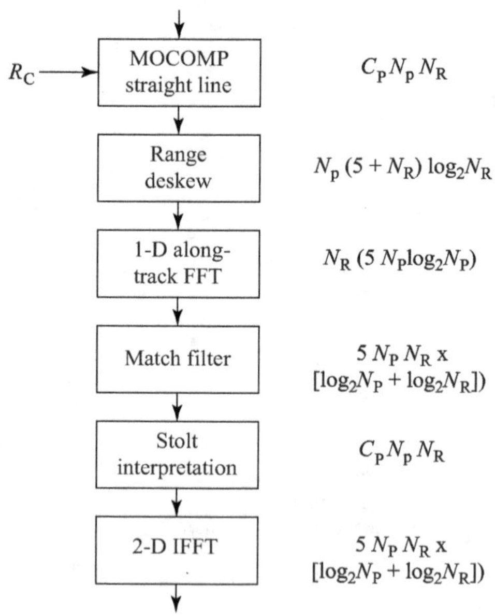

Figure 5.20 Image processing flow – RMA [3]

linear phase in azimuth direction. The advantage of this approach is that range and azimuth compression can be obtained by a two-dimensional FFT.

Range migration reduces the operation count by using FFT in the along-track dimension. The major difference from the BPA is that MOCOMP is carried out along a line through the center of the image instead of the point in the middle of the scene. This allows the image processing to approximate the range curvature to a linear phase function in the range dimension. Once this transform has been carried out, the RMA uses a Stolt interpolation to straighten out the phase in both the range and cross-range dimensions.

5.3 SAR autofocus techniques

An SAR has the function and capability of imaging an area with high resolution in both range and cross-range dimensions. Depending on the wavelength and standoff range, the length of the SAR image collection path can be quite large. There are two effects on the radar data collection that can cause a defocus of the features on any targets in the area. First, the MOCOMP between the radar and the target needs to be a small fraction of a wavelength. A combination of GPS and an IMU has been satisfactory for coarse resolution or long-wavelength SAR collections. However, if the wavelength is small or if there is a variation of the position, orientation or velocity of the platform, there will be an error in the scattering from the target in the platform coordinate system.

The second effect on the propagation can be due to anisotropic return through the path length between the radar and the scene. This is common in SAR imaging of dense forest or through severe weather conditions. These perturbations are manifested in the form of amplitude and phase fluctuations from pulse to pulse. However, the amplitude contribution will generally be small whereas the phase can vary significantly during the imaging process. Both effects cause a phase error in the returns from all surfaces in the scene. This variation of the phase in the SAR formation will defocus the image, and the image point spread function will be affected.

For motion errors beyond the measurement capabilities of the IMU or for propagation-induced phase errors, it is necessary to use data-driven autofocus techniques. The phase gradient autofocus (PGA) approach has been shown to be very robust to both motion-and propagation-induced phase effects. This claim is generally valid based on the illumination coherence within a pulse across the scene. Since the illumination source has a deterministic amplitude and phase across the scene, the measurement and propagation errors can be corrected by a data-driven autofocus. The principal assumption is that the scene has been consistently illuminated during the SAR illumination with a coherent illumination pattern, and the illumination waveform is redundant across several areas of the scene.

The PGA approach has been developed to exploit this coherency across the scene or a portion of the scene in order to capture the variation of phase error during the SAR data collection. The technique uses the phase gradient across the scene, in order to estimate the defocus mechanisms due to motion or propagation errors. Once the SAR complex image has been formed, the processing of the phase within the image is used to estimate and correct for the phase errors.

The original PGA technique starts with the range-compressed phase history

$$F_n(u) = |F_n(u)| \exp\{j[\phi_n(u) + \phi_e(u)]\} \tag{5.37}$$

where $F_n(u)$ is the return from the *n*th range bin, with amplitude $|F_n|$ and phase ϕ_n. The relative position of the imaging platform along the SAR collection is given by *u*. The second phase term is the phase due to measurement errors given by $\phi_e(u)$. Because of the coherent illumination of the scene, the uncompensated phase error along the synthetic aperture is common to all range bins of interest and is expected to be independent of range index *n*.

The azimuth compression of the phase history is obtained by a Fourier transform of $F_n(u)$:

$$\mathfrak{F}\{F_n(u)\} = \sum_m h(x) \otimes a_{m,n}\, s(x - x_{m,n}) \tag{5.38}$$

Here, $h(x)$ is the Fourier transform of the complex aperture phase error function, and the scatterer-induced IPR function is $a_{m,n}\, s(x-x_{m,n})$. The convolution between the two scattering mechanism functions is given by ⊗. The functional flow of the PGA algorithm is shown in Figure 5.21. Each part of the algorithm plays a major part in the success of MOCOMP and restoration of the image pulse response. The following paragraphs will go over these individual processing steps.

- Circular shifting

The first step in PGA algorithms is to localize the strongest scattering point a_n and shift it to the center of the image. Any resolution cell at a range ring *n* will potentially have a Doppler content due to the range rate from the radar motion. This

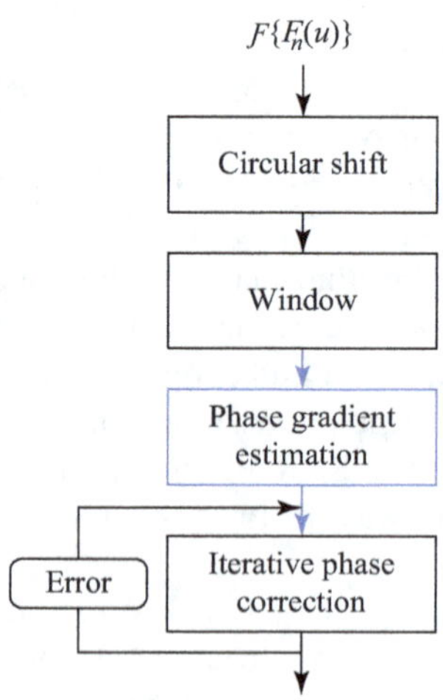

Figure 5.21 PGA algorithm flow

shifting to the center will provide a zero for any Doppler frequency offset of the scatterer.

- Windowing

The next step is to window the circularly shifted range data. This has the effect of preserving the cross-range blur for each range bin, while discarding data that does not contribute to the phase error estimation. This selection of data enables the phase error estimation to proceed with data of highest signal-to-noise ratio.

The process to select the window size W so that the dominant blur can be distinguished by surrounding distributed clutter, when there are dominant strong clutter or man-made scatterers facilitate this process. However, the PGA also works with data without a lot of strong scatterer, so long as there is structure in the clutter return. The estimate of the blur is obtained by averaging over each range bin due to the fact that the phase error is redundant in the image. W is then estimated by averaging the dominant blur over each range bin. Mathematically, the one-dimensional function of windowing is computed as

$$s(x) = \sum_n |f_n(x)|^2 \tag{5.39}$$

where $f_n(x)$ is the circularly shifted image data.

Because of the initial MOCOMP, the image will be registered and $s(x)$ will be the maximum over x. And because of the redundancy of the blur function, $s(x)$ will typically exhibit a plateau approximately W in width. Outside of this plateau, there will be a significant reduction in the clutter-to-noise ratio. With this in mind, W can be estimated by thresholding $s(x)$ at a point 10 dB down from the peak. The estimated width should then be increased by 50% to include any lower level blurring in the process. The window will be reduced in size in subsequent iteration of the PGA process, as the amount of blur is decreased, and the signal-to-clutter ratio of strong scatterers increases.

If the scene has low dynamic range, the windowing process will break down. This is because the low point and edge contrast features do not sufficiently exceed the surrounding clutter. However, by gradually decreasing the window width, even low-contrast scenes can be improved. The selection criterion is to select the maximum scan blur width and successively reduce the estimate of W by 20% for each iteration.

- Phase gradient estimation

The phase gradient is estimated after circular shifting and estimation of blur window width W. The shifted and windows image data is $g_n(x)$. The inverse Fourier transform of $g_n(x)$ is given by [7]

$$G_n(u) = |G_n(u)| \exp \{j |\phi_e(u) + \vartheta_n(u)|\} \tag{5.40}$$

where the scattering-dependent function for each range bin is $\vartheta_n(u)$. The linear unbiased minimum variance estimates the gradient of the phase error $\phi_e(u)$ is given by

$$\hat{\dot{\phi}}_{\text{est}}(u) = \dot{\phi}_c(u) + \frac{\sum_n |G_n(u)|^2 \, \dot{\vartheta}_n(u)}{\sum_n |G_n(u)|^2} \tag{5.41}$$

This is the minimum variance estimate of the gradient of the true phase error based on the circular shift and windowing process. The PGA phase gradient estimation kernel is defined as the second term inequation (5.34). This value is examined in each iteration to determine when the iteration has reached a sufficient convergence on the image focus.

- Iterative phase correction

The estimated phase gradient in the image $\hat{\dot{\phi}}_{\text{est}}(u)$ is integrated to obtain $\dot{\phi}_c(u)$, and any bias and linear trend are removed prior to correction. Phase correction is carried out by a complex multiplication of the range-compressed phase history data by $\exp\left[-j\dot{\phi}_c(u)\right]$. The estimation and correction process is repeated iteratively. On each interation of the phase gradient, the image becomes more focused, and individual scatterers become more compact. As a result, the signal to clutter of each scatterer improves, and the circular shifting more precisely removes the Doppler offsets. The phase gradient error is evaluated on each interation, and the algorithm is drive toward convergence. Once the phase gradient has reduced the error, the image have reached a satisfactory level of focus. This is qualitative condition and is an objective of the particular application of SAR imaging.

An example of PGA is illustrated using a representative image with obvious focus problems, as shown in Figure 5.22a. At this point in the processing, the error is much larger than desirable, e.g., 1.0. The phase gradient across the image and the value of the error are shown in Figure 5.22b. The PGA processing is then allowed to iterate with the objective of an error less than 1E-3. Figure 5.23a shows the focused image after 14 iteratons, and Figure 5.23b shows the progression of the phase across the image and the reduction in the error value for each of the 14 iterations.

5.4 Motion compensation

Platform navigation and terrain height variations have direct effects on the focus and geolocation accuracy of the SAR images [8]. The UWB SAR image formation is carried out either to an image plane or to a DTED height reference. As a result, the position and velocity of the radar platform need to be accurately known at the time of each pulse transmission, along with the relative location on the image plane in the DTED reference of each resolution cell. An onboard GPS and IMU are used to provide accurate position, velocity and orientation of the SAR antenna phase center during the SAR data collection. It is assumed here that the GPS-IMU has

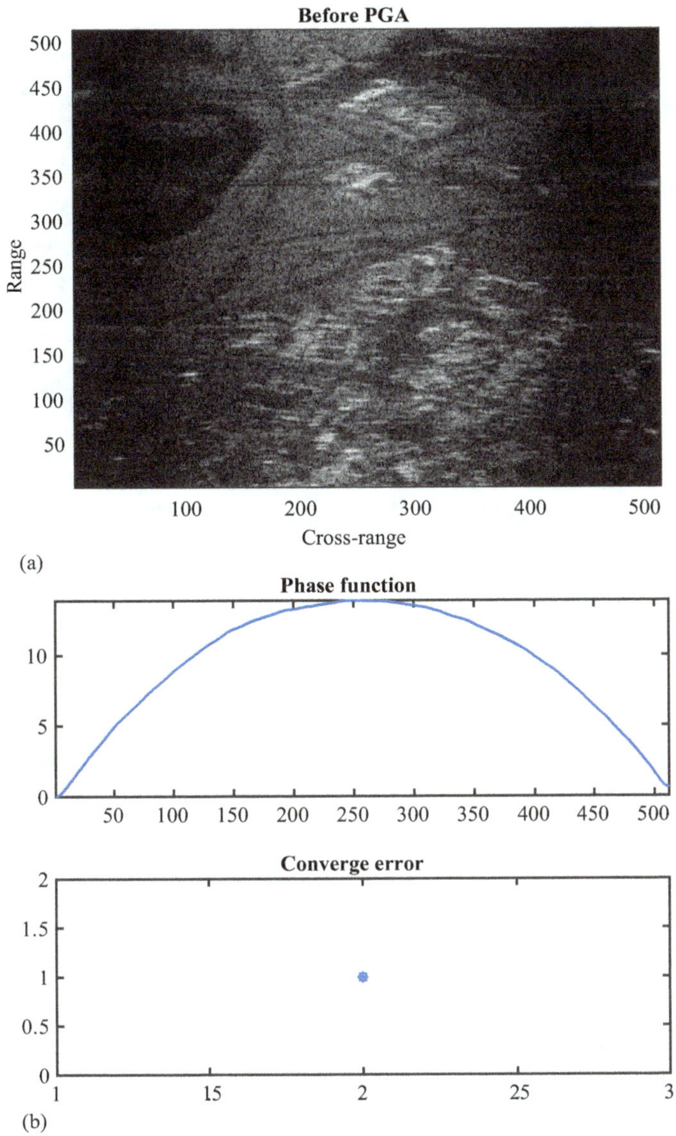

Figure 5.22 *Example of unfocused SAR Image. (a) Before PGA and (b) phase function and error*

been specified to meet the short- and long-term accuracies of the SAR image quality. The significant unknowns then are due to the terrain variation and DTED accuracy, and the departure of the sensor platform's flight path from a straight line.

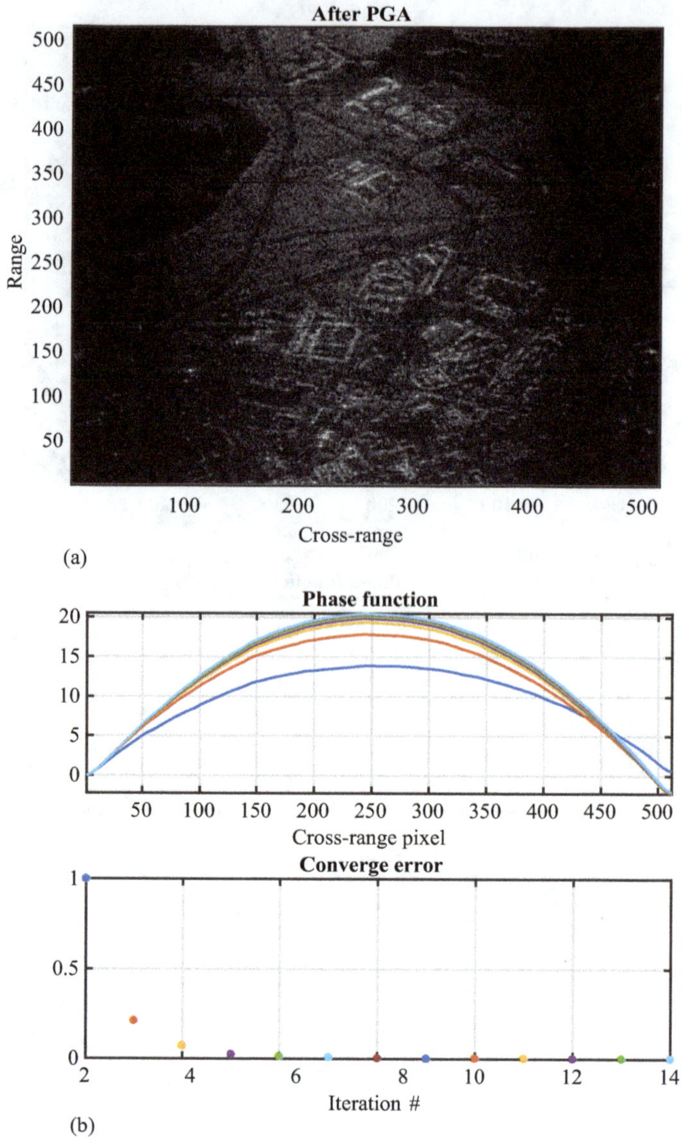

Figure 5.23 Example of focused SAR image. (a) After PGA and (b) phase function and error history

5.4.1 Motion measurement

MOCOMP is a very important aspect of UWB SAR. There are two aspects of MOCOMP required for UWB image formation [9]. First is the measurement of platform position, attitude and velocity during the long SAR collections. This will enable the image processing to calculate the time delay from the transmit pulse to each scatterer on the image plane. The second is an estimation of errors in the propagation delay to the terrain or the image plane. The use of DTED bases is instrumental in refining these errors overestimating the delay to an estimated surface.

The introduction of satellite-based navigation systems such as GPS have made UWB SAR operations more productive, since there is not a requirement of a heavy and costly navigation system. The small size of GPS receivers, and advances in IMU components, have provided a significant reduction in the size and cost of the measurement systems. The blending of the GPS and IMU measurements with frequent updates provides the necessary motion measurements for real-time processing.

It is important to define the coordinate system to measure the radar platform, as well as the locations of object on or near the Earth's surface. The GPS uses an Earth-centered, Earth-fixed (ECEF) coordinate system, which is Earth-centered, as shown in Figure 5.24. The *X*-axis is positive in the direction of intersection with the Greenwich Meridian. The *Z*-axis is parallel to the Earth's rotation axis and is positive toward the North Pole. And the *Y*-axis is in the equatorial plane and completes a right-handed coordinate system. Therefore, any point on the reference surface may be described by (*X*,*Y*,*Z*) coordinates, or alternatively in the equivalent longitude, latitude and height above an ellipsoid coordinate [10].

A WGS-8 reference ellipsoid is defined for the ECEF coordinate system, primarily because it is also used by the GPS measurement system. A reference ellipsoid is defined by the semi-major axis a and flattening factor f. For WGS-84, those are defined as a = 6378137.0 m and f = 1/298.257223.563. The relationship between geodetic (latitude, longitude and height) coordinates and ECEF is a function of a and f. The transformation is given by the following:

$$X = (N + h)\cos(\varphi)\cos(\lambda) \tag{5.42}$$

$$Y = (N + h)\cos(\varphi)\sin(\lambda) \tag{5.43}$$

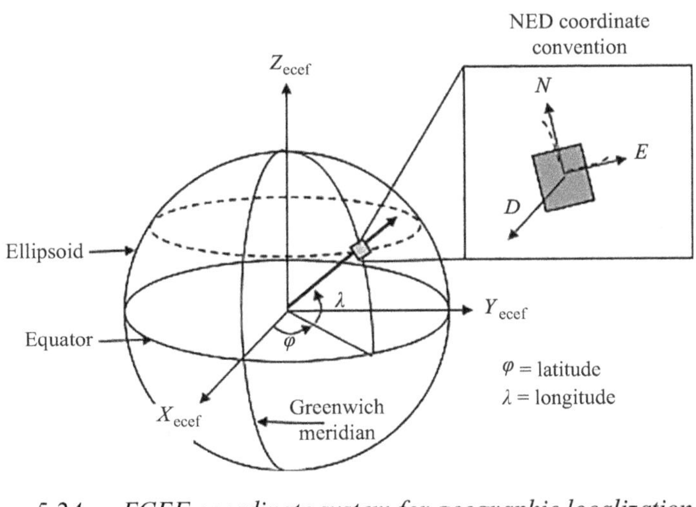

Figure 5.24 ECEF coordinate system for geographic localization [10]

$$Z = [N(1 - e^2) + h]\sin(\varphi) \tag{5.44}$$

where

$$e^2 = f(2 - f) \tag{5.45}$$

$$N = \frac{a}{\sqrt{1 - e^2 \sin^2(\varphi)}} \tag{5.46}$$

Given an accurate GPS location of the aircraft, the distance to any point on the Earth's ellipsoid is a simple vector subtraction between the (X,Y,Z) vectors. The position of the platform $P(u)$ can be expressed by [10]

$$P(u) = [x_e(u),\ u + y_e(u),\ z_e(u)] \tag{5.47}$$

where x_e, y_e and z_e are the motion errors in range, cross-range and elevation, respectively. The distance of any position on the image plane can be expressed by

$$R_{SR}(u) = \sqrt{[x_n - x_e(u)]^2 + [y_n - y_e(u)]^2 + [z_n - z_e(u)]^2} \tag{5.48}$$

Hence the radial motion error δR_{en} function is expressed by

$$\delta R_{en}(u) = -\sqrt{[x_n - x_e(u)]^2 + [y_n - y_e(u)]^2 + [z_n - z_e(u)]^2} + \sqrt{[x_n]^2 + [y_n - u]^2 + [z_n]^2} \tag{5.49}$$

For the general squint narrow beam width SAR, the radial motion error function is approximated by:

$$\delta R_{en}(u) \approx x_e(u)\cos(\theta_c) + y_e(u)\sin(\theta_c) \tag{5.50}$$

where the average squint angle of the imaged area is given by

$$\theta_c = \tan^{-1}\left(\frac{Y_c}{X_c}\right) \tag{5.51}$$

The narrow band MOCOMP of the image $s(\omega,u)$ is obtained by multiplication of the error in phase given by

$$s_{MC}(\omega,u) = s(\omega,u)\exp\left[-j\frac{2\pi}{\lambda_c}\left(x_e(u)\cos(\theta_c) + y_e(u)\sin(\theta_c)\right)\right] \tag{5.52}$$

This MOCOMP relies on the approximation of the motion in two dimensions and a narrow beam width imaging of the scene. The extension to three dimensions and a wide beam illumination is given below, including an extension to wideband illumination.

MOCOMP needs to be handled in three dimensions due to both the aircraft motion in roll pitch and altitude and to the effects of the terrain elevation on the focus of SAR images. GPS-based MOCOMP can be used for MOCOM along these three dimensions. Furthermore, the formation of the SAR image requires a coherent phase history be maintained during the time it takes the aircraft to fly the synthetic aperture. For FOPEN SAR, these apertures are very long, and the area for MOCOMP is typically much larger than for microwave SAR. It has already been shown that UWB SAR requires significant processing to compensate for range curvature. If the motion measurement is not accurate, this effect will be seriously degraded.

For back projection MOCOMP, the individual scatterers are tracked back to the antenna phase center in the fast time domain after range compression (or matched filter), as given in (5.52). The image is formed by coherently adding the contributions to each pixel in the image plane.

5.4.2 Platform motion impact on image quality

Figure 5.25 illustrates the types of errors that will result if there is a three-dimensional error in the location of the pixels in an image plane, either due to terrain height variation or due to platform navigation errors. The estimated slant range to a scatterer location on the ground R_s (in latitude and longitude coordinates) is used to calculate the phase of a path length. Over a synthetic aperture collection time, this phase is given by [9]

$$\psi(t) = \frac{4\pi}{\lambda_c} R(t) = \frac{4\pi}{\lambda_c}\left[R_0 - \int V_L \, dt\right] \tag{5.53}$$

where V_L is the line-of-sight velocity between the antenna and the ground point, and R_0 is the initial range. The line-of-sight velocity is found from

$$V_L = \hat{l}_{SR} \bar{V} \tag{5.54}$$

where $\hat{l}_{SR}$ is the unit vector along the slant range to the ground point and $\bar{V}$ is the aircraft velocity vector. The accuracy of calculating $\hat{l}_{SR}$ and $\bar{V}$ is readily determined from the accuracy of position, velocity and orientation specification on the GPS-IMU subsystem.

The actual slant range R_S will have an error δR in determining the MOCOMP processing correction, providing the estimated slant range $\tilde{R}_s$ as

$$\tilde{R}_s = R_s + \hat{l}_{SR} \, \delta R \tag{5.55}$$

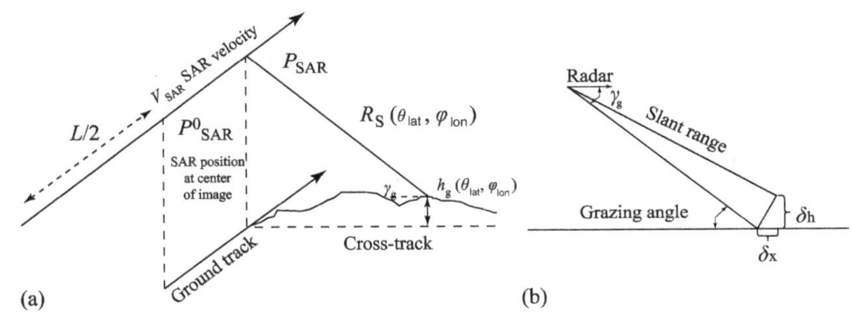

Figure 5.25 Motion measurement for SAR image formation. (a) MCOMP geometry and (b) terrain height variation

If there is a velocity error, either or both of two effects can occur. If the velocity vector is incorrect, the azimuthal angle to the image point will be in error. Thus, the image point will either suffer error in focus or be rotated in geospatial coordinates. If the magnitude of the velocity is in error, the sample distance will be in error and the Doppler at each point in the scene will be scaled incorrectly. Both these effects can give rise to focus and geospatial accuracy degradation [11].

For platform height errors, the phase to each point in the image will also be estimated incorrectly in the MOCOMP process. This contributes to a layover effect in the SAR image as shown in Figure 5.25. For a height error δ_h and grazing angle γ_g, the position error in ground range δX will be

$$\delta X = \delta h \tan \gamma_g \tag{5.56}$$

For a given point in the synthetic aperture, the MOCOMP error δR_{MC} for the SAR image focusing is given by:

$$\delta R_{MC} = \left|P_{SAR} - I_P\right| - \left|P^0_{SAR} - I_P\right| - \left|P^{'}_{SAR} - F_P\right| - \left|P^0_{SAR} - F_P\right| \tag{5.57}$$

where I_p is the location of the scatterer, $P_{SAR}{}^1$ the nominal position of the SAR platform on the ground track, $P_{SAR}{}^0$ actual position of the SAR platform, and F_p the location of the focus point.

To examine the impact of motion effects on the RMA algorithm, the results of several data collection flights were examined in terms of the image quality. The P-3 UWB SAR was instrumented to collect data to analyze the impact of navigation errors on the image. Since the MOCOMP for RMA is normally applied at broadside, the flight path was examined during an SAR integration period. This is illustrated as a departure from the straight-line flight path.

These were major impediments to real-time operation in early surveillance systems. The effects of motion measurement on image formation have been greatly reduced. The next step in the evolution is determining the impact of aircraft flight operations on providing high-quality images. It should be noted that RMA employs an approximation based on the aircraft flying a straight line in order to utilize the Stolt interpolation. There were many SAR image collections in the early UWB SAR testing; so, these data were used to understand the difference between BPA and RMA algorithms [12].

Figure 5.26 illustrates the deviation of a platform from a straight line, as required for accurate RMA image processing. When the actual SAR motion has a cyclic variation, the impact on the image IPR has different manifestations. A linear motion will cause an angular error in the geolocation accuracy. And a quadratic error will cause a broadening of the azimuth (or elevation) IPR. Two points are illustrated: the focus point is aimed at the center of the SAR collection and the image point I_p is offset in cross-range. These geometries will be explored next.

Figure 5.27 illustrates typical motion errors that occur during the UHF SAR image collection. Two types of deviations are illustrated: horizontal and vertical. The horizontal deviation for this platform is the larger and will have the greater impact on the image focus. These image degradation effects are due to the inability

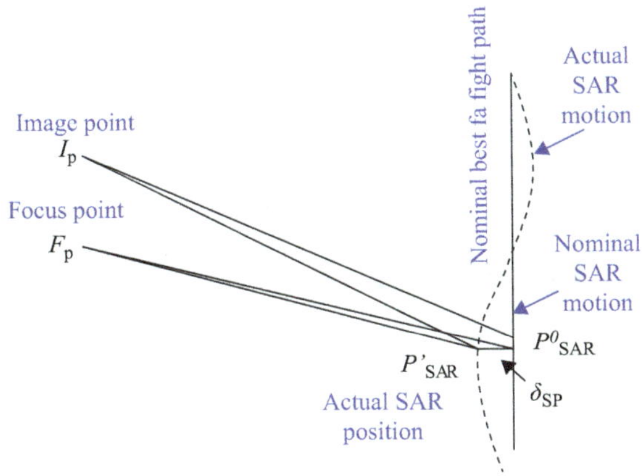

Figure 5.26 Platform motion – deviation from straight line [12]

of the RMA algorithm to remove the cross-track variation of range curvature except at the MOCOMP line at the center of the image. These data, as shown in Figure 5.28, were characterized by three classes of image collection runs: *small*, where the cross-range spatial errors were below 1.4 m, *medium* with deviations of the order of 2.6 m and *very large* in a single instance of a 10-m deviation. All of these errors are greater than a wavelength at UHF and need to be understood in terms of the image focus.

As discussed above, the focus of an image at a point distant from the image point is determined by the ability to compensate the returned phase history to a straight line. If there is an error in the GPS/INS navigation, the variation in the phase is straight-forward. However, even if the MOCOMP is perfectly accurate, there will

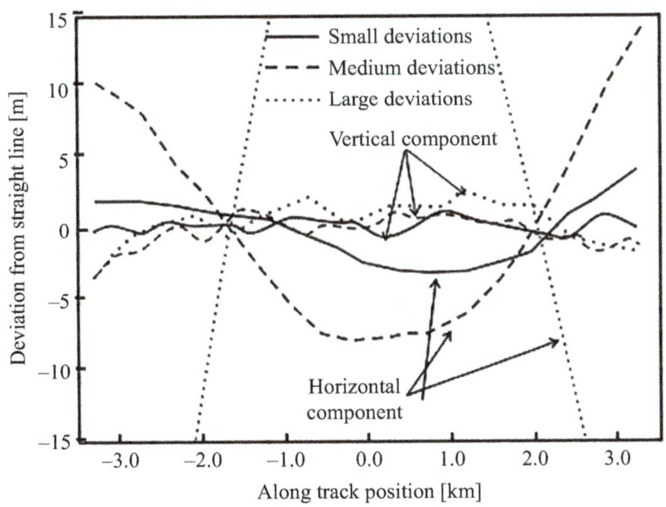

Figure 5.27 Motion measurement from UWB SAR image collection [12]

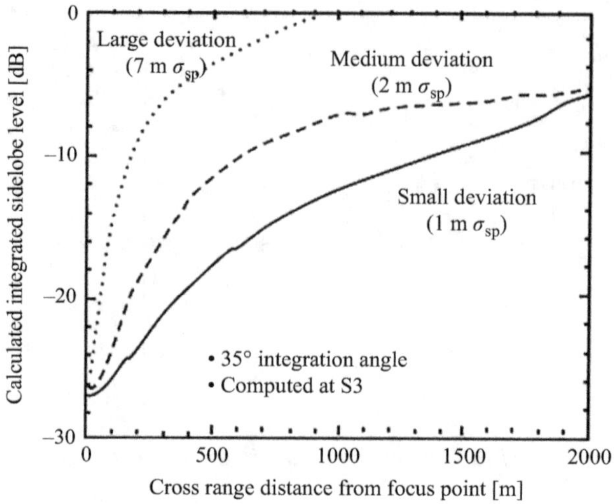

Figure 5.28 Image quality versus focus point distance [12]

be an error in the Stolt interpolation compensating the range curvature in the cross-range dimension. The deviation can be random or periodic. If it is periodic, the degradation will be equivalent to a large sidelobe level in the IPR, as illustrated in the upper chart. If the deviation is purely random, the rms sidelobes will be raised. Uncompensated deviation will directly affect the ISLR in the image. But even if the MOCOMP is perfect, the approximations in the Stolt interpolation will give rise to a defocus of the IPR away from the center of the image.

There will be an error in the RMA Stolt interpolation in compensating for the range curvature in the cross-range dimension. The MOCOMP difficulty arises due to the wide integration angles in FOPEN SAR image collection. The relationship between uncompensated motion δR and the phase error Φ_{err} is given by

$$\Phi_{err} = \frac{4\pi\,\delta R}{\lambda} \tag{5.58}$$

When the motion is cross-track, there will be a residual phase error at the scene center due to a displacement of δy. At the edge of the beam β_a, a residual phase error for targets exists, given by

$$\Phi_{err}(y) = \frac{4\pi\,\delta y}{\lambda}\left[1 - \cos\beta_a\right] \tag{5.59}$$

This residual phase error due to cross-track deviation from a straight line will be comparable for all pixels within the scene. The effect of this platform motion was examined for UHF UWB SAR images, where two image formation processes were used. The first process employed a MOCOMP that was broadside to the image formation, which is analogous to a strip map image formation. The second process applied the MOCOMP to the center of the image, which is analogous to a spotlight image formation, but using the RMA algorithm.

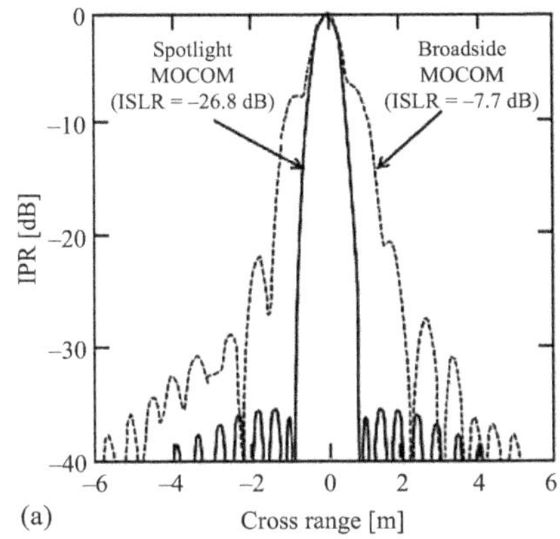

Figure 5.29 MOCOMP of RMA with platform deviations from straight line [12]. (a) IPR at focus point and (b) IPR at 500-m cross-range.

The IPR function for these two-image formation and MOCOMP techniques is compared in Figure 5.29. The dotted line corresponds to MOCOMP at broadside to the image formation, and the solid line corresponds to MOCOMP to a point in the center of the image collection. These two curves illustrate the IPR at the focal point. The strip map MOCOMP processing exhibits a defocusing of the image during collection, resulting in an integrated sidelobe ratio (ISLR) of –7.7 dB. The spotlight MOCOMP processing had greatly improved IPR at the focal point of the image, achieving an ISLR of –26.8 dB [13].

Figure 5.29 illustrates the impact of image focus for points away from the center of the image. Again, the dotted line is for the broadside compensation, but with the image point 500 m in cross-range from the focus point. There is comparable ISLR of – 7.7 dB when compared to the same condition in Figure 5.29. For the RMA image formation with MOCOMP performed to a central focus point, the displacement causes image quality degradation of the ISLR to –11.6 dB.

A modification of the Stolt interpolation to correct for small order cross-range deviation has been developed. For an UWB SAR system design, the anticipated platform motions need to be evaluated. For small UAV operation, the navigation will be determined by weather and winds aloft. As a result, there is increasing support for employing the BPA for these collections, despite the increased real-time operations count.

With the RMA spotlight approach, good image quality can always be achieved by limiting the output scene length L_{out} for a given ISLR performance objective. Figure 5.30 depicts the ISLR versus L_{out} for several cases of slant range deviation from the UHF UWB SAR. Small deviations can obtain –20 dB ISLR for scene lengths of 500 m. However, for higher cross-track motion, the scene length is significantly reduced. These conditions require more processing since L_{out} is significantly less than the SAR integration length L. It is important to understand the maximum

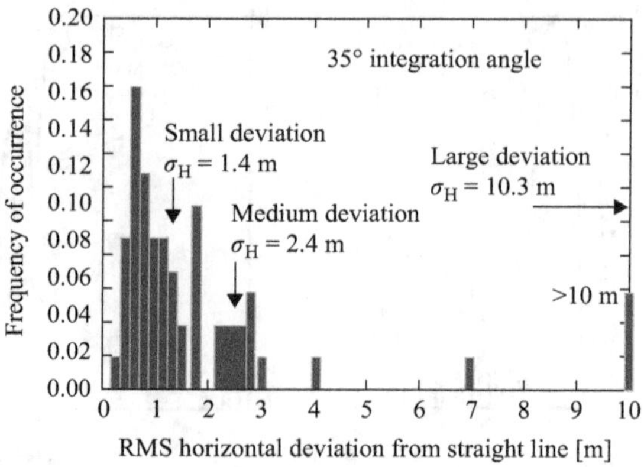

Figure 5.30 Motion deviation effects on integrated sidelobe level [12]

integration length that can be used as a function of the image quality and to plan the onboard processing to improve the overall efficiency.

The image quality, i.e., the amount of allowable defocus, can be related to the deviation from the correct path δY by using (5.59). The deviation δY will be different in each point of the SAR image collection. As such, it can be considered as a random phase error in the uniform linear array formation. The maximum output image length can then be determined by specifying a maximum allowable phase error Φ_{max} as a function of the integration angle $\Delta\theta$ by using the relation

$$L_{out} = R_0 \sqrt{\frac{2\,\lambda_c\,\Phi_{max}}{\pi\,\delta Y\,\cos^3\left(\frac{\Delta\theta}{2}\right)}} \tag{5.60}$$

where δY is the deviation from the straight line.

The contribution of rms deviation from a straight line to the extent of the focused scene is important. The data were fitted to the inverse square root of the deviation by two parameters. Values of these proportionality constants will serve as an analytic tool for specifying both the MOCOMP processing error and the size of the scene that will be appropriately compensated by the RMA algorithm.

If the deviation is known, then a modification of the RMA algorithm has been developed to compensate for the effects of systematic movement. However, the computations to estimate the error and recalculate the image are extensive. For non-real-time operations, these computations have been successfully applied. The alternative would be to divide the image into sub-apertures and update the RMA image after forming the sub-array components.

The points in Figure 5.30 fall along a curve that is approximately inversely proportional to the square root of the RMS motion deviation. The best curve fit to the data was estimated from UHF collection parameters as

$$L_{out} = \frac{\alpha}{\sqrt{\alpha_{SP}}} + \beta \tag{5.61}$$

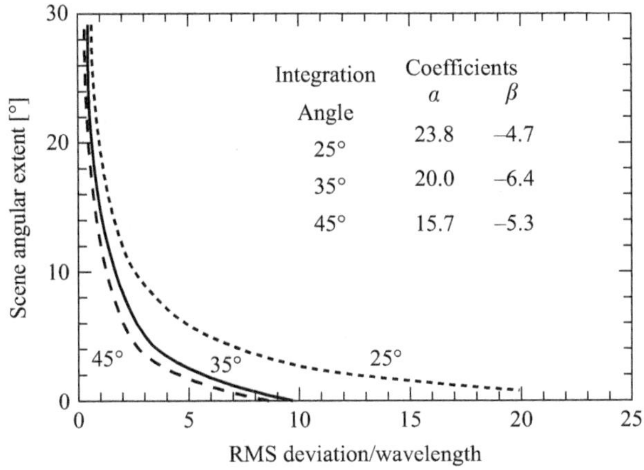

Figure 5.31 Angular scene extent versus platform motion [12]

This fit produces a good match to the data for σ_{SP} values of less than 5 m. For σ_{SP} values that are larger than 5 m, the best-fit curve significantly underestimates the achievable output scene length. When the output image length is below L_{out}, the focus will be within the design guidelines. This provides an estimate of how my subimages are required in the synthetic aperture collection to maintain the RMA algorithm focus over both range and cross-range.

In order to provide a general result that can be used in wide-angle UHF SAR design, the output scene length is expressed in terms of its angular extent:

$$\Delta\theta_{SL} = 2\ \tan^{-1}\left(\frac{L}{2R_0}\right) \tag{5.62}$$

The angular extent of the output scene was computed for integration angle of 25°, 35° and 45°. Based on 60 passes of P-3 UWB data, best-fit curves for each integration angle were generated. These data are summarized in the inset Figure 5.31. The figure also provides design data to illustrate how the angular extent of the output scene depends on the deviation of the platform motion from a straight line and on the SAR integration angle and wavelength.

5.5 Summary

Several algorithms have existed for several decades for UWB SAR processing depending on the spatial resolution and wavelength of the waveform. The early examples of UWB SAR emphasized the computational complexity for obtaining real-time images over wide collection angles, with severe challenges in obtaining wide swath image formation.

Because of the relative simplicity of RMA, it was used in early real-time image formation due to lower cost computation. The image MOCOMP assumed an approximation to straight-line flight path. And by the use of FFT processing, there was processing cost on an order of $N^2 \log_2 N$ signal processing cost. However, it is not always practical to maintain a straight-line collection path due to weather and flight dynamics. As a result, the RMA MOCOMP needs to be updated often during the SAR image collection. This in itself reduces some of the algorithm simplicity.

The competitive BPA was shown to provide better focus over the very wide-beam VHF and UHF SAR collections. The temporal-based algorithm provides an exact solution over wide swaths, if the real-time MOCOMP is accurate. However, the BPA algorithm has a complexity of the order of N^3, which drives higher computational cost. This is especially true for large image collections of 2 and 4 km on a side. However, with the rapid development of small, low-power hig-performance computer devices, the computational cost of BPA has been significantly reduced.

MOCOMP for wide area focus is important to reduce the need for image defocus, where the position and orientation of the platform or the knowledge of the image surface is well known. The techniques of adjusting image formation for depth of focus continue to be a research challenge.

References

[1] Soumekh M. *Synthetic Aperture Radar Signal Processing with MATLAB Algorithms*. New York, NY: Wiley and Sons; 1999.

[2] Carrara W.G., Goodman R.S., Majewski R.M. *Spotlight Synthetic Aperture Radar*. Norwood, MA: Artech House; 1995.

[3] Davis M.E. 'Foliage Penetration Radar – Detection and Characterization of Objects Under Trees'. *FOPEN SAR Image Formation, Chapter 4*. Raleigh, NC: Scitech Publishing; 2014.

[4] Hellsten H. 'CARABAS - an UWB Low Frequency SAR'. *IEEE MTT-S Int*. 1992;**3**(1-5):1495–8.

[5] Mahafsa B.M. *Radar Signal Analysis And Processing Using Matlab*. Boca Raton, FL: CRC Press; 2009. pp. 89–91.

[6] Yegulalp A.R. 'Fast Backprojection algorithm for synthetic aperture radar'. *Proceedings 1999 IEEE radar Conference*. Boston, MA; 1999. pp. 60–5.

[7] Wahl D.E., Eichel P.H., Ghiglia D.C., Jakowatz C.V. 'Phase gradient autofocus-a robust tool for high resolution SAR phase correction'. *IEEE Trans. Aerosp. Electron. Syst*. 1994;**30**(3):827–35.

[8] Madsen S.N. 'Motions Compensation for Ultra Wide Band SAR'. *Proceedings 2001 IEEE International Geoscience and Remote Sensing Symposium*. 3. Sydney, Australia; 2001. pp. 1436–8.

[9] Goodman R.S., Tummala S., Carrara W.F. 'Issues in Ultra-wideband, Widebeam, SAR Image Formation'. *Proceedings 1995 IEEE International Radar Conference*. Washington, DC; 1995.

[10] Clynch J.R. *"Navigation Coordinates", Naval Post Graduate School,* Monterey, CA. 2002. Available from https://www.oc.nps.edu/oc2902w/coord/coord.pdf.

[11] Kirk J.C. 'Motion compensation for synthetic aperture radar'. *IEEE Trans. Aerosp. Electron. Syst.*. 1975;**AES-11**(3):338–48.

[12] Kirk D.R., Maloney R.P., Davis M.E. 'Impact of Platform Motion on Wide Angle Synthetic Aperture Radar Image quality'. *Proceedings 1999 IEEE Radar Conference*. Boston, MA; 1999. pp. 41–6.

[13] Kirk D.R., Maloney R.P. 'Autofocus techniques for Wide-Band wide angle synthetic aperture radar'. *Proceedings SPIE*. Vol. 3370. Orlando, FL; 1998.

Chapter 6
Interferometric radar modes

Interferometric synthetic aperture radar (InSAR) is the process of combining the return from a transmitted SAR signal into two or more receive apertures. If the antennas are separated vertically, the processingwas often called cross-track, and the application was for terrain height measurements. If those apertures are along the velocity vector of the platform or platforms, the stationary background clutter can be cancelled and moving targets can be detected in the wideband SAR images. This mode is generally called along-track interferometry (ATI) and was first published in 1974 [1]. The early demonstration of InSAR and the science of this mode was demonstrated in 1990 by the National Aeronautics and Space Administration (NASA) Jet Propulsion Laboratory on their AirSAR platform, as shown in Figure 6.1 [3].

Previous displaced phase center antenna systems have cancelled the clutter return magnitude under the conditions that the moving target is correlated between the looks. This "simple" subtraction also attenuates the target return, lowering the probability of detection. ATI exploits the return signal phase information on each resolution cell and uses this phase to detect Doppler motion with respect to the surface background return. By coherently combining the signals from the two antennas, an interferometric phase difference can be formed for each imaging point.

The phase for an elevation difference between ground resolution cells or moving objects near the stationary ground in the scene is dependent on the difference in separation between the two apertures and the signal-to-noise ratio (SNR) of the signal. The phase difference between the returns determines the accuracy of either terrain height or target Doppler. For a small separation between the antennas, higher target speeds are needed to discriminate the phase from the background. At longer separations, an ambiguity in the phase of modulo 2π affects the ability to determine the target Doppler [1]. In a single-pass interferometric collection, the most critical condition for accurate terrain height measurement or slow-moving target detection is the accurate knowledge of the relative position and orientation between the two antennas. The position and relative insertion phase of the two antenna centers must be known within a fraction of a wavelength. This condition is generally met on a single platform where the multiple phase centers are stabilized or there is a baseline measurement system to determine the position and orientation of the antennas. A major technical accomplishment on the GeoSAR system was its laser baseline measurement system (LBMS), where the relative location of the antenna was within a

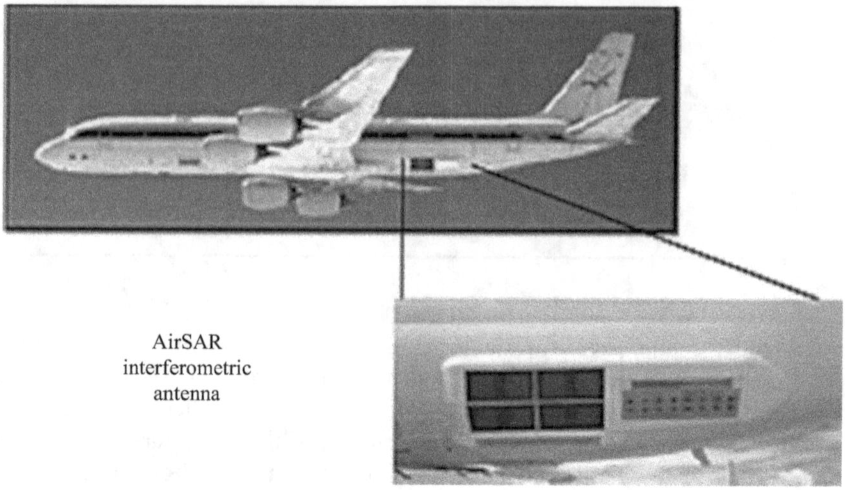

Figure 6.1 InSAR antennas on JPL AirSAR platform [2]

few millimeters, under dynamic flight conditions. These system requirements are detailed in the following sections below.

6.1 Cross-track InSAR

One of the initial cross-track InSAR testbeds was the interferometric SAR-elevation (IFSAR-E) radar developed in 1994 by Environmental Research Institute of Michigan (ERIM) and NASA Jet Propulsion Laboratory. ERIM designed the radar and integrated it into a LearJet platform, and JPL developed the ground processing system to enable high-resolution digital elevation model (DEM) for civilian and military applications [4]. The radar platform shown in Figure 6.2 housed two antennas in a center line radome, along with the inertial measurement unit (IMU). When

Figure 6.2 IFSAR-E radar platform [4]

coupled with a differential global positioning system (GPS) receiver, the navigation data for the InSAR image formation was adequate for the resolution objectives. The antennas were approximately 1.8 m long and were solidly mounted on a temperature-compensated structure, as shown in Figure 6.3. This rigid structure was developed to maintain the separation and orientation between the two radar channels under the temperature and flight conditions. The X-band radar and recording equipment were housed within the fuselage of the LearJet, along with high-speed digital recording for the I,Q data from the radar receivers [1].

IFSAR-E is a single-pass interferometric SAR system, which can collect data during day or night, and through cloud cover and in reasonable weather. Terrain height data were generated over featureless terrain. The system was designed to generate three-meter (one sigma) absolute accuracy for three-dimensional digital terrain elevation data (DTED) over large terrain areas for a –20 dB reflectivity, commensurate with arid terrain. The sensor data were collected for a 10-km stripmap swath from 12,300-m altitude at a platform speed of 12 km/min [4].

There are many civilian and military applications for IFSAR-E-derived elevation data. The greatest interest to date includes generation of DEMs, contour or ortho-quad data in areas that had never been mapped to ISFAR-E accuracy, providing rapid elevation data in areas where recent changes have occurred and providing terrain visualization data for operation planning and preparation. Figure 6.4 illustrates the processig flow for IFSAR-E, where the processing is separated between airborne collections of data and ground-based processing of the DEM products.

The raw SAR imagery resolution is 0.75 m along-track by 2.5 m cross-track. It is normally processed with four-sample along-track averaging, so the ground plane images are approximately a 3 × 3 m, 4-look imagery. The system looks orthogonal to the flight path and covers a 10-km ground swath, approximately centered about a 45° depression angle. The ground processing begins with motion data demultiplexing, differential GPS processing and Kalman filtering of the position, navigation and timing data. Both image formation and height extraction rely on the precise navigation solution and benefited from the advent of differential GPS.

The conversion of complex imagery to elevation data is accomplished by phase extraction via complex conjugate multiplication of the image pixel pairs, phase

Figure 6.3 IFSAR-E antenna [5]

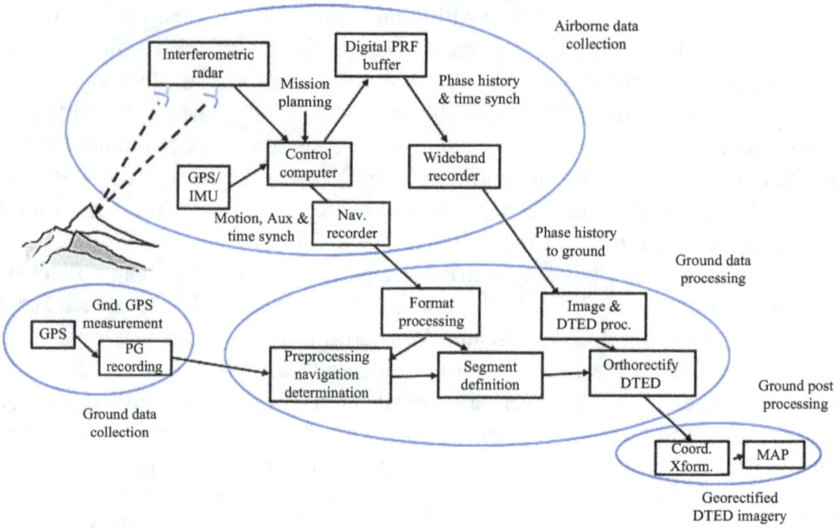

Figure 6.4 IFSAR-E high-level block diagram [4]

ambiguity unwrapping and scaling phase to height estimation as a function of geometry. The absolute phase determination algorithm was employed to determine the ambiguity of the unwrapped interferometric phase before scaling phase to height. An example of this development was a high-resolution digital elevation map of the Death Valley California area shown in Figure 6.5 with a significant height differential of over 1,700 m. After flight tests and performance verification, the IFSAR-E equipment and capability was transferred in 1997 to a commercial mapping company by the US Department of Defense under an Other Transactions procurement process. This matured rapid collection of DEM data and orthorectified mapping products was continued by Intermap Technologies Company, as the STAR-3i system [5].

The techniques for IFSAR system development and signal processing are widely reported. This section provides the basic cross-track image collection and DEM formation [6]. The SAR image measures the amplitude and phase as the composite of all scatterers in a resolution cell $A_b e^{j\varphi_b}$, as shown in Figure 6.6 for cross-track

Figure 6.5 Perspective view grayscale image of Death Valley [5]

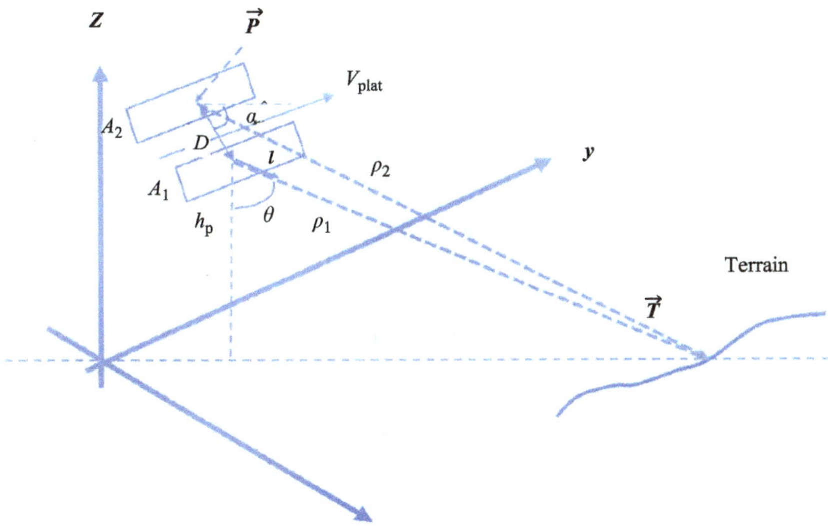

Figure 6.6 Cross-track interferometric SAR geometry [6]

interferometry. The backscatter from a resolution cell on the ground surface is measured at each aperture, and the interferometric amplitude and phase from the resolution cell are anticipated to be the same if there are no differences in the transmission path. The difference in the two returns is determined by the range from the SAR apertures to the resolution cell, based on the accurate GPS/IMU navigation measurement and radar timing. These signals at the two apertures are [6]

$$s_1 = A_b e^{j\varphi_b} e^{-j\frac{4\pi}{\lambda}\rho_1} \tag{6.1}$$

$$s_2 = A_b e^{j\varphi_b} e^{-j\frac{2\pi}{\lambda}(\rho_2 + \rho_1)} \tag{6.2}$$

If A_1 is the reference aperture, the interferometric phase is given

$$\phi = -\frac{2\pi}{\lambda}\left(\rho_1 + \rho_1 - (\rho_2 + \rho_1)\right) = \frac{2\pi}{\lambda}(\rho_2 - \rho_1) \tag{6.3}$$

It is important to remember that the phase return is modulo 2π. So, if the difference between the two paths is more than 2π, a phase unwrapping process is needed to remove the phase ambiguity and present a constant phase over a flat plain on the surface.

For terrain mapping, there is also a process of "ping-pong" between the two antennas, where the transmitter illuminates the surface alternately between the two SAR sensor antennas. This in-effect doubles the interferometric baseline for the ping-pong collection and the interferometric phase ϕ_{pp} is expressed by

$$\phi_{pp} = -\frac{2\pi}{\lambda}\left(\rho_1 + \rho_1 - (\rho_2 + \rho_2)\right) = \frac{4\pi}{\lambda}(\rho_2 - \rho_1) \tag{6.4}$$

The interferometric phase is measured by the vector separating the two antennas to a common point on the Earth's surface. This phase will vary with range and azimuth position of the sensor during the SAR integration flight. Therefore, the baseline D is defined by the location and orientation of the two antennas during the SAR collection. The scattering position on the Earth is obtained by the vector $\vec{\boldsymbol{T}}$, which is obtained by the vector addition between a reference position one of the SAR antennas $\vec{\boldsymbol{P}}$ and the unit look vector l along the distance ρ ($i = 1$ or 2) [6]:

$$\vec{\boldsymbol{T}} = \vec{\boldsymbol{P}} + \rho_i \hat{l} \tag{6.5}$$

As the aircraft moves along the SAR path, the terrain height is measured through the variation of the interferometric phase between the two antenna phase centers to each position on the Earth surface.

Assuming $\vec{P}$ and $\rho_i \hat{l}$ are known, the interferometric phase is calculated by

$$\phi \approx -\frac{2\pi}{\lambda} \left\langle \hat{l}, \vec{\boldsymbol{D}} \right\rangle \tag{6.6}$$

This can be written as

$$\phi = -\frac{2\pi\rho}{\lambda} D \sin\left(\theta - \alpha\right) \tag{6.7}$$

As the InSAR image is formed, there will be a cycle of phase change over a 2π radian or "fringe." This fringe frequency is the result of the natural variation of the terrain height and the surface height's effect on the interferometric phase. As the phase between the two paths is subtracted, there will be interference fringes due to the phase ambiguity, otherwise referred to as an interferogram. The phase variation in the interferogram can be flattened by subtracting the expected phase from the surface of constant elevation. The flattened interferometric phase can then be given by

$$\phi_{\text{flat}} = -\frac{2\pi\rho}{\lambda} \left(\left\langle \hat{l}, \vec{\boldsymbol{D}} \right\rangle \right) - \left\langle \hat{l}_0, \vec{\boldsymbol{D}} \right\rangle \tag{6.8}$$

where

$$\hat{l}_0 = \left(\sin\theta_0, \cos\theta_0\right) \tag{6.9}$$

and $\cos\theta_0$ is given by the law of cosines

$$\cos\theta_0 = \frac{\rho_0^2 + \left(r_e + h_P\right)^2 - \left(r_e + h_0\right)^2}{2\left(r_e + h_P\right)\rho_0} \tag{6.10}$$

The details of the cross-track processing algorithms for terrain height estimation will be given in the following sections, based on advances in the algorithms from the follow-on InSAR program Geoscience Synthetic Aperture Radar (GeoSAR).

6.1.1 GeoSAR processing

Under the GeoSAR program, JPL developed an end-to-end, dual-frequency (P-band and X-band) interferometric and polarimetric radar system. The GeoSAR platform

Table 6.1 System parameters for GeoSAR dual-frequency InSAR [7]

Parameter	P-band value	X-band value
Peak transmit power	4 kW	8 kW
Bandwidth	80/160 MHz	80/160 MHz
Pulse length	40 μs	40 μs
Antenna gain	11 dBi	26.5 dBi
Look angle range	22°–60°	22°–60°
Center frequency	350 MHz	9.755 MHz
Baseline length	20 and 40 m	2.6 and 5.2 m
Baseline tilt angle	0°	0°
Platform altitude	5,000–10,000 m	5,000–10,000 m

with the antenna location and the LBMS is shown in Figure 6.7. Both radar bands are designed to create interferometric image returns and to map simultaneously on the left and right sides of the aircraft. The key parameters for this system are summarized in Table 6.1. This radar system was the first ultra-wideband (UWB) interferometric radar system. As such, it required some major changes in the InSAR processing to handle the P-band wide-angle collection, along with the high percentage bandwidth at both X-band and P-band. It was also a first for having two different InSAR instruments on the same platform for DEM generation and forest characterization. The GeoSAR system has a center line pod with ranging laser measurements to surveyed reflectors on each antenna structures. These measurements are essential in correcting for any orientation of the antennas during in-flight operation [8].

The GeoSAR processing subsystem consists of the following major programs shown, as in Figure 6.8: motion measurement processing, X-band processor to generate DEMs and P-band processor to generate DEMs. The interferometric part of

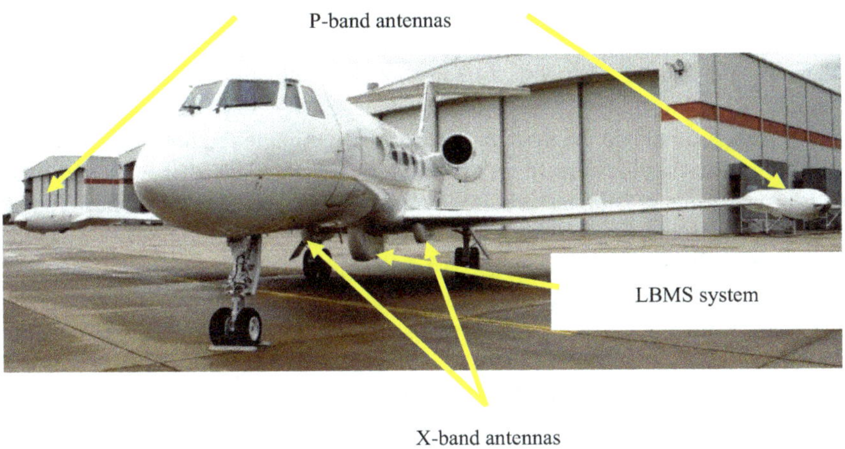

Figure 6.7 GeoSAR flight operation for terrain mapping [7]

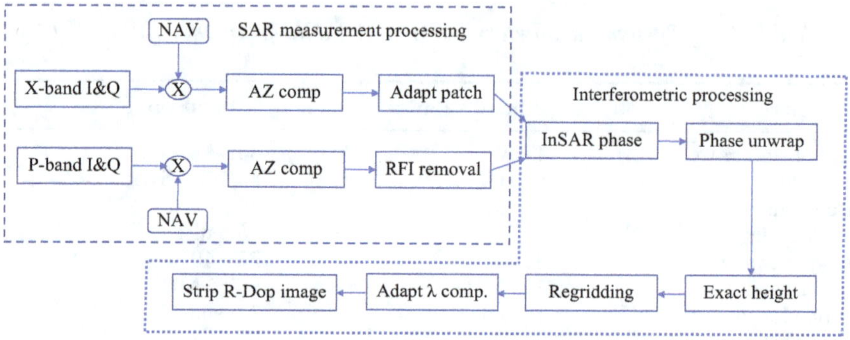

Figure 6.8 GeoSAR image DEM processing flow [9]

the processing is subdivided into two main programs: the SAR measurement processor (SMP) and the interferometric processor. The primary function of the SMP is to reformat and condition time-varying radar parameter data in a manner suitable for IFSAR processing. The on-board data collection records navigation data from the aircraft GPS inertial navigation unit for motion compensation (MOCOMP) of the complex data. And all of the data were processed in an *sch* coordinate system. This unique coordinate system was developed at NASA Jet Propulsion Laboratory to provide a geodetically accurate interferometric map [10]. In addition, the LBMS measures the range and angle variation for each of the four antennas, in order to correct assumptions of the parallel antennas within the X- and P-band radar [7].

The processors' main function is to generate strip maps from the raw radar signal history. Modifications for the P-band radar were necessary to handle the UWB SAR processing and reduce processor-induced mapping errors to the centimeter level. There are two algorithms for azimuth compression depending on X- and P-band wavelengths. In addition, there are improved algorithms for phase unwrapping and an absolute phase determination algorithm, exact height reconstruction equations and adaptive wavelength computation based on transmitted waveform and signal processing [7].

The P-band data are processed using a ω-k processing algorithm (covered in Section 5.2.4) followed by a special second-stage MOCOMP algorithm, which is designed to accommodate the large synthetic aperture angle required for P-band. The process of presuming and first-stage MOCOMP are identical to the X-band InSAR system. And azimuthal compression uses a modified ω-k processing, and both channels are processed to zero Doppler. Weighting of the azimuth reference function is a cosine on a pedestal with a user-specified pedestal height. Since the first-stage MOCOMP corrects the range delta along the fan beam direction (a narrow beam approximation), the imager is not well focused after the first stage. For each sub-patch, a best-fit plane to the reference DEM is used for the MOCOMP surface, and they are then focused in azimuth. The data are then re-skewed to the fan beam plane squint geometry.

The combination of X- and P-band InSAR was chosen specifically to characterize the terrain and any man-made structures under dense forests. X-band is historically the frequency of choice to image open-terrain and objects with high resolution.

P-band SAR has been used in many applications to detect objects under trees. The combination of the two frequencies was selected to measure terrain slopes and forest characteristics. X-band gives an excellent measure of the tops of the trees, and P-band penetrates the trees to approximate the DEM under the trees. Section 6.1.7 will summarize some of the results of GeoSAR for geophysical explorations.

6.1.2 Motion compensation

MOCOMP is the process where the radar signal data are resampled from the actual path of the antenna to an idealized path called the referenced path. As a result, the slant range error as a delta range $\Delta\rho$ due to the antenna tilt is compensated, as shown in Figure 6.9. GeoSAR employed an LBMS with the measurement system located on a center line structure, as shown in Figure 6.7. Each antenna on the aircraft had an optical reflector imaged by the LBMS. The range to each reflector was measured to a few millimeters, and the orientation was updated at 50 Hz to provide necessary relative location to the motion measurement processing. The reference path to the antenna location d, along the line of sight vector $\hat{l}_R$, is used to compute a range and phase correction to be applied to each sample using the expression [7]

$$\Delta\rho = \left\langle \bar{d}, \hat{l}_R\left(\rho, h_{ref}\right)\right\rangle, \ \Delta\phi = \frac{4\pi}{\lambda}\Delta\rho \tag{6.11}$$

where ρ is the range, λ the wavelength and h_{ref} the MOCOMP reference height. MOCOMP is carried out based on the position, navigation and time measurement of each reference antenna phase center relative to the local refence paths f. Next, MOCOMP is applied to a common reference path, which assures the data are coregistered in range after image formation. The processing applies the appropriate range spectral shift for a flat surface and flattens the fringes. For P-band data, a more sophisticated MOCOMP algorithm is employed. Range weighting is done after MOCOMP to avoid sidelobes and trim the nonoverlapping portions of the range spectra.

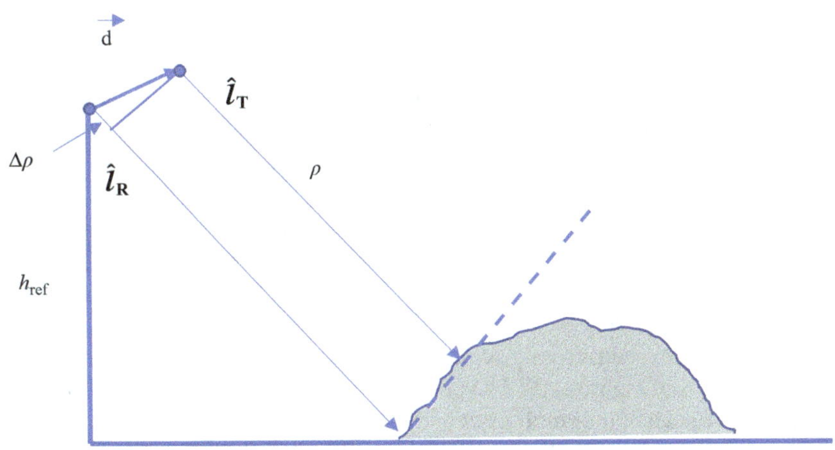

Figure 6.9 Motion compensation geometry [7]

Proper interferometric calibration of the GeoSAR system was essential in obtaining a DEM with the required submeter level of planimetric and vertical accuracies. Calibration begins with the determination of the yaw and pitch biases for the two embedded GPS inertial (EGI) units. To measure the angle biases, the Doppler centroid is estimated as a function of range using the raw range line data. By comparing the measured and expected Doppler centroids, the EGI biases can next be estimated. Finally, the delays were estimated based on the time delay difference between the channels due to different cable lengths. At this point, a cross-correlation of the time delays was carried out through the SAR images, and no identifiable targets were required [11].

The common range delays, the corrections for each mode for unmeasured electrical path lengths in the radar, are solved concurrently with the platform position error for each pass. The error of the range to the target is given by

$$\frac{\partial\rho}{\partial P_s}\Delta P_s + \frac{\partial\rho}{\partial P_c}\Delta P_c + \frac{\partial\rho}{\partial P_h}\Delta P_h + \frac{\partial\rho}{\partial\tau}\Delta\tau = \Delta\rho \tag{6.12}$$

where

$$\frac{\partial\rho}{\partial P_s} = \cos(\beta),\ \frac{\partial\rho}{\partial P_c} = \mu\sin(\theta),\ \frac{\partial\rho}{\partial P_h} = -\cos(\theta),\ \text{and}\ \frac{\partial\rho}{\partial\tau} = 1 \tag{6.13}$$

where β is the angle between the aircraft velocity and the look direction, θ the look angle, τ the common range delay, and P_s, P_c and P_h are the platform position in *sch* coordinate system. The parameter μ is given as

$$\mu = \sqrt{1 - \left(\frac{\cos(\beta)}{\sin(\theta)}\right)^2} \tag{6.14}$$

Finally, the location error in range of each target is measured, and (6.12) is solved using singular value decomposition to determine the platform position errors and range error. Data from multiple passes with different squint angles are synthesized to produce robust estimates of all the components of the baseline [11].

6.1.3 *Interferogram formation*

There are two SAR images formed to provide the process for forming an interferogram. One of the images is considered to be the reference image, and the second image is used for creating the interferometric correlation between the two SAR images. The interferometric phase is obtained by multiplying each corresponding pixel in the first image by the complex conjugate of the corresponding pixel in the second image. In order to reduce the amount of phase noise in the interferogram, a box car filter is used to average the phase. The size of this box car is determined empirically, typically using between a 3×3 or 7×7 averaging filter.

The variation in interferometric phase in a region of significant terrain height variation is shown in Figure 6.10. The benefit of power spectral filtering on the interferogram can be seen in the small inset box. A two-dimensional FFT of the complex image shows that there is well-defined peak at the spatial frequency. The areas in the flat lands have large separation of the phase fringes, while in the mountainous

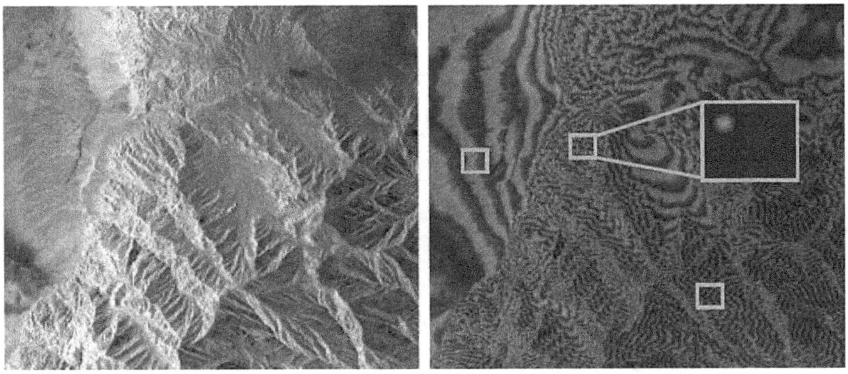

Figure 6.10 GeoSAR interferometric phase fringes in mountainous terrain [7]

terrain, the fringes have a higher spatial variation. This significant terrain variation was a challenge in the phase unwrapping. The use of a low-resolution terrain height map, such as commercial DTED, would often be required for improving the phase unwrapping process [7].

Coherent radar echoes will be correlated with each other, if each represents nearly the same interaction with a scatterer or set of scatterers. If the scatterers are randomly positioned within a resolution cell, the phase of each will be random and the sum will be characterized by a zero mean, complex Gaussian random number. The variance of this return will be proportional to the average radar cross-section of the surface. The variance due to sensor geometry will be spatial in nature and those due to target change due to motion within the two channels will be temporal. For an interferometric radar with two signals, the signals and noise (thermal effects) will be correlated as a function of noise given by ρ_{thermal} [12]:

$$\rho_{\text{thermal}} = \frac{\langle s_1 s_2^* \rangle}{\sqrt{\langle s_1 s_1^* \rangle \langle s_2 s_2^* \rangle}} \tag{6.15}$$

This relationship will be related to the SNR by

$$\rho_{\text{thermal}} = \frac{1}{1 + \text{SNR}^{-1}} \tag{6.16}$$

The radar returns can be divided into two parts: correlated signals, co, and decorrelated signals, de. The correlation of the two signals at infinite SNR can be expressed due to the spatial variation ρ_{spatial} given by

$$\rho_{\text{spatial}} = \frac{|\text{co}|^2}{|\text{co}|^2 + |\text{de}|^2} \tag{6.17}$$

The correlation of the combined spatial and thermal signal effects will be given by

$$\rho_{\text{thermal}} \cdot \rho_{\text{spatial}} = \frac{|\text{co}|^2}{|\text{co}|^2 + |\text{de}|^2} \frac{1}{1 + \text{SNR}^{-1}} \tag{6.18}$$

A similar argument holds for the temporal change in the returned signal ρ_{temporal}, so that

$$\rho_{\text{total}} = \rho_{\text{temporal}}\, \rho_{\text{spatial}}\, \rho_{\text{noise}} \tag{6.19}$$

The temporal decorrelation normally occurs in repeat pass InSAR collections and when the surface of the ground has changed from one pass to the next. This situation has been observed when there are earthquakes or evidence of disturbance of the Earth. For a single-pass InSAR, this temporal decorrelation is not expected, except in the case that there is evidence of different temporal scattering in the scene. For a typical radar where the impulse response is approximated by the two-dimensional function

$$W(x, y) = \text{sinc}\left(\frac{x}{\delta_x}\right)\, \text{sinc}\left(\frac{y}{\delta_y}\right) \tag{6.20}$$

where δ_x and δ_y are the azimuth and range resolutions, respectively. In this case, W represents a spatial decorrelation function.

The geometry for an interferometric look at a resolution cell is shown in Figure 6.11. The two antennas $A1$ and $A2$ illuminate a patch on the surface at incident angles of θ_1 and θ_2, respectively. The distance from the sensor to the center of the resolution cell is given by ρ_1 and ρ_2. When δ_x and δ_y are the cross-range and range resolutions respectively, the spatial baseline decorrelation due to a baseline separation of D can be expressed by

$$\rho_{\text{spatial}} = 1 - \frac{2\,|D|\, \delta_y \cos^2(\theta)}{\lambda\, \rho} \tag{6.21}$$

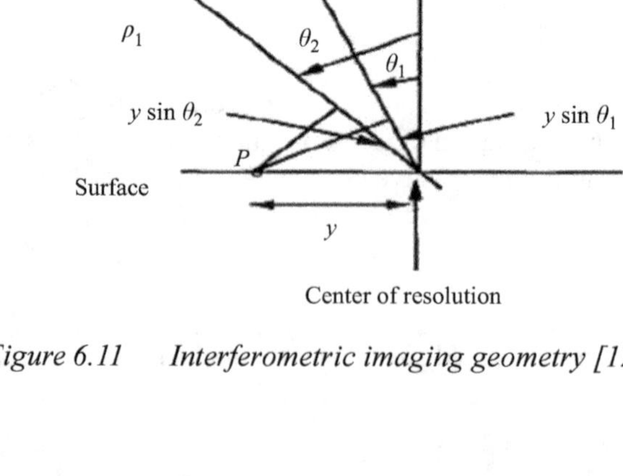

Figure 6.11 Interferometric imaging geometry [12]

There is a critical baseline where δ_{spatial} equals zero, and where the two returns are completely decorrelated. Under this case, the critical baseline D_c is given by

$$D_c = \frac{\lambda \rho}{2\,\delta_y \cos^2(\theta)} \tag{6.22}$$

For a single-pass interferometric radar, the range ρ_i and incident angle θ_i are the average between the two paths, as illustrated in Figure 6.11.

6.1.4 *Height reconstruction and regridding*

Interferometric height reconstruction is the determination of a target's position vector from known planform ephemeris information, baseline information and interferometric phase. Prior to height reconstruction, the interferometric phase altered during the MOCOMP process is restored to the true phase, as sensed by the antenna at the time the point was imaged. This process is called inverse MOCOMP and is applied by a spatial correction to the data. The phase screen applies a range dependent correction to the interferometric paths. After the phase and range have been corrected for propagation errors, the reconstructed target position vector $\overline{T}$ is obtained by solving the intersection locus of the range sphere. This sphere centered at the platform position vector P, the Doppler cone with generating axis given by the velocity vector $\overline{v}$ and cone angle as a function of the target Doppler centroid and the phase hyperboloid. The simultaneous set of equations is given by [7]

$$|\overline{P} - \overline{T}| = \rho \tag{6.23}$$

$$f = \frac{2}{\lambda}\left\langle \bar{v}, \hat{l} \right\rangle \tag{6.24}$$

$$\varphi = \frac{2\pi}{\lambda}\rho\left(\left(1 - \frac{2\left\langle \hat{l}, \bar{d}\right\rangle}{\rho} + \left(\frac{d}{\rho}\right)^2\right)^{1/2} - 1\right) \tag{6.25}$$

where φ is the interferometric phase, ρ the range from the platform to the target, b the baseline vector between the two antennas and λ the wavelength. After a choice of coordinate systems, this nonlinear set of equations can be readily solved in closed form. The orthorectified magnitude data are corrected for antenna pattern and range effects and scaled to a radiometrically correct values. The *sch* coordinates for a map pixel (sample, line) are obtained by the following expression:

$$\begin{bmatrix} s \\ c \\ h \end{bmatrix} = \begin{bmatrix} s_0 + (\text{line} - 1)\,\Delta s_{\text{post}} \\ c_0 + (\text{sample} - 1)\,\Delta c_{\text{post}} \\ h \end{bmatrix} \tag{6.26}$$

The GeoSAR processor had a set of regridding algorithms to correct the location of the data points on the *sc* plane. Since there may be a significant variation between the terrain height and the expected ground plane, the phase needs to be resampled to the collection grid. For regions that are uniform in planarity, a simple nearest-neighbor regridding could be used. And for the more significant variation in the terrain (especially in mountainous regions), either a convolutional or surface fit interpolation technique was needed. Each of these algorithms presents increased computational complexity to handle the DEM variation [7].

6.1.5 *Effective interferometric wavelength*

An interferometric radar that requires high-resolution image to either create a DEM or provide improved resolution on a structure or feature in the scene will be considered as UWB, as described in Section 1.2. Under these UWB cases, the processing cannot use the center frequency approximation, as was discussed in Section 5.2. In these high-accuracy applications, it is necessary to account for how these waveform differences affect the interferometric phase and correlation. The effective carrier frequency f_{eff} is a relation between the changing range difference and the corresponding interferometric phase shift. Since the interferometric range difference is proportional to the parallel component of the interferometric baseline $d_{\|}$, the effective interferometric wavelength λ_{eff} can be derived from the relation [9]

$$\lambda_{eff} = \frac{c}{f_{eff}} = \frac{c}{\frac{-1}{2\pi}\frac{c}{2}\frac{\partial\phi}{\partial d_{\|}}} = \frac{-4\phi}{\frac{\partial\phi}{\partial d_{\|}}} \tag{6.27}$$

Notice that the effective interferometric wavelength depends on the derivative of the phase with respect to the parallel component of the baseline and hence is range dependent. For high-precision topographic mapping systems, the range-dependent nature of the interferometric wavelength must be accounted for during the height reconstruction process.

In order to estimate the effective interferometric wavelength, it is necessary to determine the derivative of the interferometric phase with respect to the parallel baseline, $\frac{\partial\phi}{\partial d_{\|}}$. This is accomplished by assuming the downconversion frequencies are different for the two channels to account for wave number shift processing. The wavenumber k_0 of the carrier frequency f_c is

$$k_0 = \frac{2\pi c}{f_c} = \frac{2\pi}{\lambda_c} \tag{6.28}$$

which is assumed to be the center of the transmitted bandwidth. The downconversion is given by

$$k_i = k_0 \pm \frac{\delta k}{2} = k_0 + \mu_i \frac{\delta k}{2} \tag{6.29}$$

where $\mu_1 = 1$ and $\mu_2 = -1$. Let $g_i(\rho,s)$ for i = 1,2 be the impulse response functions at the carrier frequency for the two antennas forming the interferometric pair that is related to the baseband impulse response functions $w_i(\rho,s)$

$$g_i\left(\rho, s\right) = e^{i\,2k_0\rho}\, w_i \tag{6.30}$$

For high-accuracy applications, it is necessary to account for how these waveform differences affect the interferometric phase and correlation. The three-dimensional scattering location T is solved for using

$$\vec{\boldsymbol{T}} = \vec{\boldsymbol{P}} + \frac{\lambda\,\rho f}{2\,v}\vec{v} + \frac{\dfrac{d^2}{2\,\rho} - \dfrac{\lambda\,\phi}{2\pi\,p}\left(1 + \dfrac{\dot{\lambda}\,\phi}{4\pi\,p\rho}\right) - \langle\vec{d},\vec{v}\rangle\dfrac{\lambda\,\rho f}{2\,v}}{d\sqrt{1-\langle\vec{d},\vec{v}\rangle^2}}\;\frac{\left(\vec{v}\ x\ \vec{d}\right)\ x\ \vec{v}}{\left|\left(\vec{v}\ x\ \vec{d}\right)\ x\ \vec{v}\right|}$$

$$\pm\sqrt{1-\langle\hat{l},\hat{v}\rangle^2 - \langle\hat{l},\hat{n}\rangle^2}\;\frac{\vec{v}\,x\,\vec{d}}{|\vec{v}\,x\,\vec{d}|} \tag{6.31}$$

where $\hat{n} = \frac{(\vec{v}\ x\ \vec{d})\ x\ \vec{v}}{|(\vec{v}\ x\ \vec{d})\ x\ \vec{v}|}$. And the sign of the last term is determined by the requirement that the scatterer location have height less than the platform altitude and be located to the side of the platform to which the radar pointing.

The major impact of channel decorrelation occurs when there is a difference in the noise in the two radar echoes. This effect will add noise on the measurement of terrain height. The related phase standard deviations to height errors σ_h based on phase errors σ_ϕ are

$$\sigma_h = \frac{\lambda\ R\tan\left(\theta\right)}{4\pi\ D}\ \sigma_\phi \tag{6.32}$$

These corrections were made in the GeoSAR processing of P-band interferometric images to create the DEMs [9].

6.1.6 Phase unwrapping

Two-dimensional phase unwrapping is an important point in the InSAR phase use for terrain mapping. The phase of the resulting complex samples is proportional to elevation, and the magnitude is proportional to radar cross-section. For the placement of a pixel in three-dimensional space, however, the absolute phase measured by the interferometer must be known. Because the phase measurements are obtained modulo 2π, the collected values need to be unwrapped to absolute phase to obtain accurate terrain height estimates. There are two main approaches to phase unwrapping. The first class of algorithms is based on the integration with branch cuts approach. A second class of algorithms is based on a least-square fitting to the unwrapped solution to the gradients of the wrapped phase. This work will concentrate on the branch-and-cut algorithm to illustrate the processing [6, 13].

The classical branch-and-cut method starts with the localization of the phase residues. This is normally accomplished by an integration around a closed path. The result of such an integration is either zero or a multiple of 2π. To locate individual residues, the integration path must be made as small as possible which results, for example, in a path around four neighboring pixels for a sampled phase image [13].

$$\Delta^x_{i,j} = W\{\phi\,(i+1,j) - \phi\,(i,j)\} \tag{6.33}$$

$$\Delta^y_{i,j} = W\{\phi\,(i,\,j+1) - \phi\,(i,j)\} \tag{6.34}$$

where W is the wrapping operator that wraps all interferometric phase values into the range $[0, 2\pi]$. Using this notation, the condition for a residue is

$$\Delta^x_{i,j} - \Delta^x_{i,j+1} + \Delta^y_{i+1,j} - \Delta^y_{i,j} = \pm 2\pi \tag{6.35}$$

A simple approach to phase unwrapping would be to form the first differences of the phase at each image point in each image dimension as an approximation to the derivative and then integrate the result. Direct application of this approach allows local errors due to phase noise to propagate causing errors across the full SAR scene. Branch cut algorithms attempt to isolate sources of error prior to integration. The basic idea is to unwrap the phase by choosing only paths of integration that lead to self-consistent solution [6].

This phase unwrapping strategy is to connect nearby residues of opposite charge with cuts which interdict the integration paths, such that no net residues can be encircled and no global errors are generated. However, local errors in the vicinity of residues may occur. For any given residue, the one with opposite charge which is closest to it is chosen and cut-lines, are always taken as straight lines. Layover regions are masked prior to the residue search operation and remaining residues, for which no residue partner was found, are connected to the boundary or masked areas. After placement of the cut-lines the absolute phase can be determined by integration from an arbitrary point, which is not surrounded by cut-lines. While the absolute phase value obtained by the cut-and-branch method is consistent, global errors are sometimes propagated from isolated residues [13].

The interferogram coherence places an upper bound on the expected unwrapped-phase elevation change through spatial or baseline decorrelation [14]:

$$\rho_{\text{spatial}} = \max\left\{0,\ 1 - \frac{2\ |D_\perp|\ \delta_y}{\lambda\ \rho\ |\tan\theta_i|}\right\} \tag{6.36}$$

where $D_\perp$ is the interferometer perpendicular baseline, δ_y the slant range resolution and ρ the slant range to the surface. Several factors contribute to the decorrelation, but as θ_i becomes smaller, correlation will decrease due to multiple scattering mechanisms within the range and cross-range samples of the two radar paths. The observed local correlation is an important measure of the elevation change from one pixel to another.

Phase unwrapping is treated as a constrained optimization problem, as given by

$$\text{minimize}\left\{-\textstyle\sum_k \log\left(f(\Delta\phi_k|\Delta\psi_k,\ I,\ \rho)\right)\right\} \tag{6.37}$$

where I is the average SAR intensity, ρ the interferogram coherence magnitude and $f(.)$ the probability density function (PDF) of the variables in the argument. The parameters $\Delta\phi_k$ and $\Delta\psi_k$ are the unwrapped and wrapped differences within the scene, respectively. Since these two values are related to each other, except for the modulo 2π, the function to be minimized can be expressed as $f(\Delta\phi_k, I, \rho)$. The statistics of the scattering are based on many phenomena such as terrain slope, layover of rapid changes of topology and deformation of the surrounding terrain. For the majority of the DEM production, topology will dominate the phase unwrapping differences based on local slopes.

The phase difference $\Delta\varphi$ can be separated into true topographic scattering $\Delta\varphi_{topo}$ and an approximate Gaussian phase noise depending on the local correlation between the two scattering paths. The probability density function $f(\Delta\varphi_{topo}| I)$ is related to the image brightness through the backscatter coefficient σ^0, and the area of the scattering surface A:

$$E(I) = C\,\sigma^0 A \tag{6.38}$$

where C is a scaling factor and can be ignored. Both σ^0 and A depend on the local incidence angle θ_i, so the SAR image intensity contains a great deal of information about the topography.

The interferometric geometry is illustrated in Figure 6.12, where the terrain height z varies as a function of the distance in the local horizon to each pixel, y. The change in topology can be seen with many scattering surfaces. These surfaces, especially if there are many in a resolution cell, give way to image speckle [15].

This speckle directly affects the estimation of interferometric phase in an SAR image. As a result, it is common to use multilook processing, where the phase is estimated by multiple interferometric pairs of equal mean phase (or height). For homogeneous targets or images, the phase return is obtained by [15]

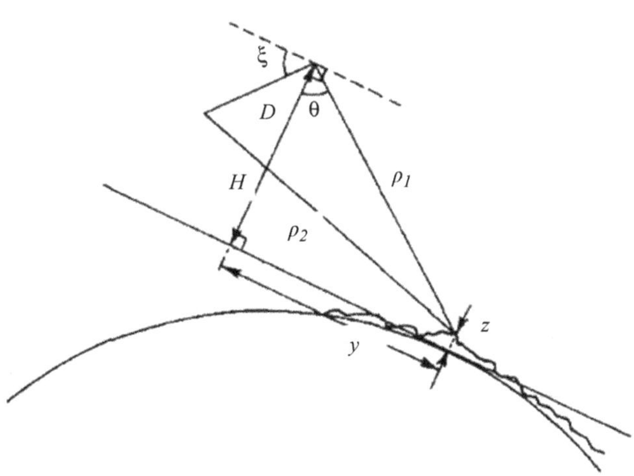

Figure 6.12 Geometry for interferometric phase estimation [15]

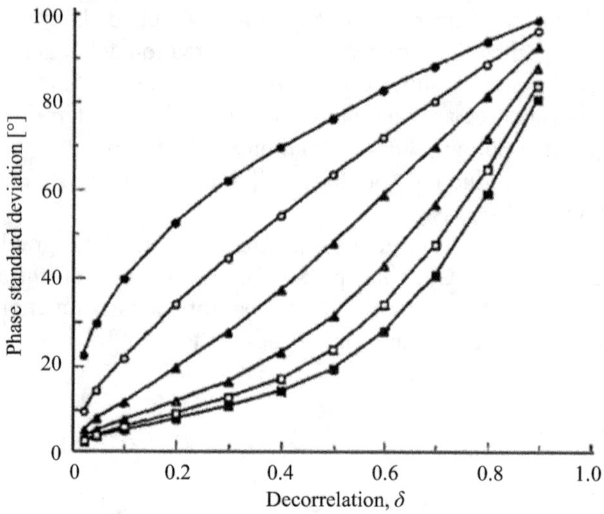

Figure 6.13 Standard deviation of phase based on signal decorrelation [15]

$$\hat{\Phi} = \arctan\left[\frac{\mathrm{Im}\left[\sum_{n=1}^{N_L} s_1(n)\, s_2^*(n)\right]}{\mathrm{Re}\left[\sum_{n=1}^{N_L} s_1(n)\, s_2^*(n)\right]}\right] \tag{6.39}$$

The maximum likelihood estimation of the phase is unbiased modulo 2π. As a result, the standard deviation of phase is estimated by

$$\langle\hat{\Phi}^2 - \langle\hat{\Phi}\rangle^2\rangle^{1/2} = \frac{1}{\sqrt{2N_L}}\frac{\sqrt{1-\rho^2}}{\rho} \tag{6.40}$$

where ρ is the correlation between the two signals ρ. Figure 6.13 shows the phase standard deviation as a function of decorrelation ($\delta=1-\gamma$), which is reduced by approximately $N_L^{-1/2}$. Standard deviation of the phase reduces significantly after four looks, especially if there is high correlation between the signals.

The height H of a target over the plane can be estimated by

$$z = H - R_s\cos(\theta) = H - \rho_s\left[\cos(\xi)\sqrt{1-\sin^2(\theta-\xi)} - \sin(\theta-\xi)\right] \tag{6.41}$$

$$y = \rho_s\sin(\theta) \tag{6.42}$$

where ρ_s is the slant range to the resolution cell, θ the incident angle and ξ the inclination angle of the baseline from the y-axis. The local incidence angle is found by

$$\sin(\theta-\xi) = \frac{(\rho_s+\Delta)^2 - \rho_s^2 - D^2}{2\,D\,R\rho_s} \tag{6.43}$$

The estimated interferometric phase difference is given by $\hat{\Phi}$, with the function $\Delta = -\hat{\Phi}\,\lambda/2\pi$ being the wavenumber of this phase variation to the slant range.

For the GeoSAR processing, unwrapping uses an updated version of the residue base unwrapping algorithm. The algorithm is patch based so that in principle an infinitely long strip of data could be unwrapped. For each patch, the correlation is used to mask out area of low correlation. Residues are computed to determine points where phase discontinuities exist in the unwrapped phase. Residues are connected in a tree groping process to ensure that the resulting phase can be recovered without phase discontinuities. Subregions of the interferogram that can be connected by a continuous path without crossing a branch cut are called connected components. Phase continuity between patches is preserved by retaining a line of phase values in the overlap region that is used to bootstrap the phase from one patch to the next. Trees (branch cuts) are used to connect residue and any line that cannot be crossed during the phase unwrapping process. There are two types of residues: positive and negative. They must be connected in such a way that the net sum is zero [9].

A point is designated as an intensity neutron if the following inequalities are satisfied:

$$I\left(i,j\right) = \mu_T \sigma_I + I \tag{6.44}$$

$$\gamma\left(i,j\right) \leq \gamma_T \tag{6.45}$$

where $I(i,j)$, and $\gamma\ (j,j)$ are the image intensity and correlation at point (i,j), I and σ_I are the mean and standard deviation of the image intensity for the patch, and μ_T and γ_T are the user-specified intensity and correlation neutron thresholds, respectively [7].

After determining the relative phase between the pixels, an overall constant multiple of 2π may need to be added to the phase to reconstruct the height correctly. Determining this multiple of 2π is called the absolute phase determination problem. Two algorithms for determining the absolute phase are included in the GeoSAR ground processing. One method uses a low-resolution DEM and the other attempts to estimate the absolute phase directly from the data. Sometimes, either of these algorithms may fail and therefore the user is provided with the capability to adjust the processor determined absolute phase either globally or on a patch-by-patch basis. Absolute phase determination is not done on every patch if it is possible to use the unwrapped phase from the previous patch to determine the absolute phase. This process is called phase bootstrapping. Bootstrapping compares the unwrapped phase along selected lines in the overlap region to determine the absolute phase constant for the next patch. How closely this bootstrapping must be applied depends on the closeness to a multiple of 2π wrapping point and is determined by the phase variance threshold specified by the user.

Bootstrapping is the preferred method of determining the absolute phase for all patches after the first patch. Phase is compared along a set of lines in the overlap region of two patches. If the number of points on the bootstrap lines exceeds a threshold, then the variance of the phase difference on the bootstrap lines is checked to see whether I exceeds the phase variance threshold. If the phase variance threshold is exceeded, then the correlation threshold is incremented by a user-specified amount, and the unwrapping process is repeated. This continues until either the phase variance threshold is not exceeded or the maximum correlation threshold is reached. If the maximal correlation threshold is reached and the phase variance still exceeds the phase variance threshold, then the region is marked as unwrappable.

6.1.7 GeoSAR application results

GeoSAR has the capability to collect two bands of InSAR data out of both sides of the Gulfstream aircraft, as shown in Figure 6.14 [16]. The X-band image provides excellent images on open structures, and the P-band penetrates the forest to reveal structures or geophysical features under the canopy. At X-band, the products include orthorectified intensity models, surface reflectance elevation and detailed visualization and analysis of the image. At P-band, the DEM represents near bare Earth and provides derived layers such as slope and contours. By using both the X- and P-band imagery and DEM, the image processing will provide a representation of bare Earth. The GeoSAR radar system simultaneously collects high-accuracy elevation and image data at more than 288 km^2/min with swaths of 14 km for each band and on both sides of the flight line.

GeoSAR has been able to provide mapping of geologic structures in areas where forest canopy obscures the terrain. Figure 6.15 shows two SAR images of a region in the Colombian forest. Figure 6.15a is the X-band image of the area with significant forest cover on the left side. In contrast, Figure 6.15b shows the P-band image, with the contours and geologic structure of the hills below the trees. These measurements have been used to characterize the hydrological characteristics of the surface contours as well as map river, stream and flooding below the forest canopy.

When the DEMs for both frequencies are created and aligned, the benefit of two frequency InSAR can be seen. Figure 6.16 shows the X- and P-band images of a

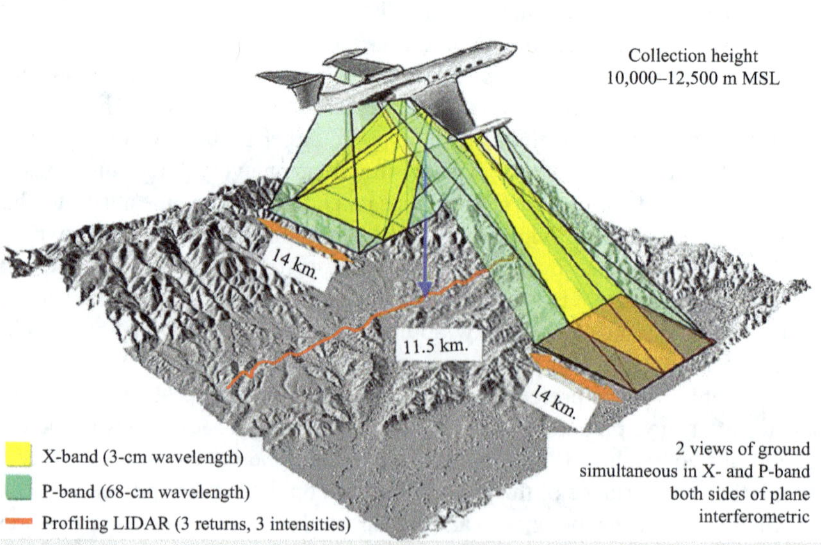

Figure 6.14 Flight characteristics of GeoSAR mapping system [16]

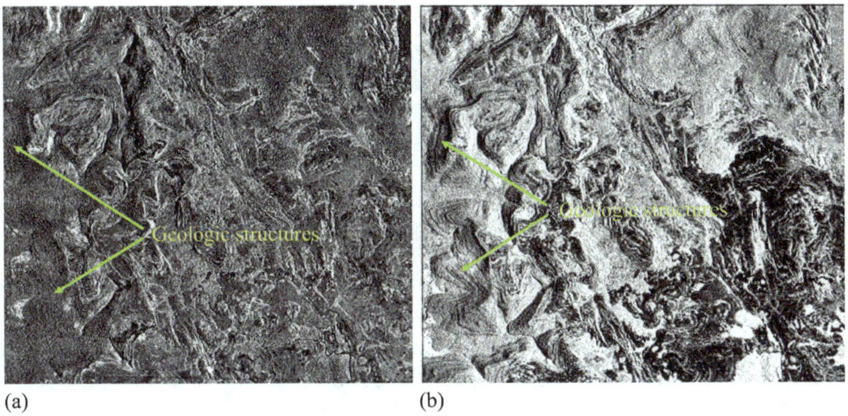

Figure 6.15 *Geologic mapping of Columbia forest [8]. (a) X-band 1-m image and (b) P-band 1-m image*

forested area of Columbia, near to an open field. The tops of the trees are very clear in the left image, with no indication of the terrain slope below the canopy. And for the open field in the center of the image, there is significant clutter return along with features of paths and isolated trees. In contrast, on the right image, the forested area has a speckled return with some bright returns from individual trees. Moreover, the return from the open field is very dense in intensity due to the greatly reduced clutter backscatter at P-band. This is very characteristic of foliage penetration radar.

Both images have a transect through the image going from the dense forest to the open field. The top trace is the DEM of the X-band image, while the lower trace

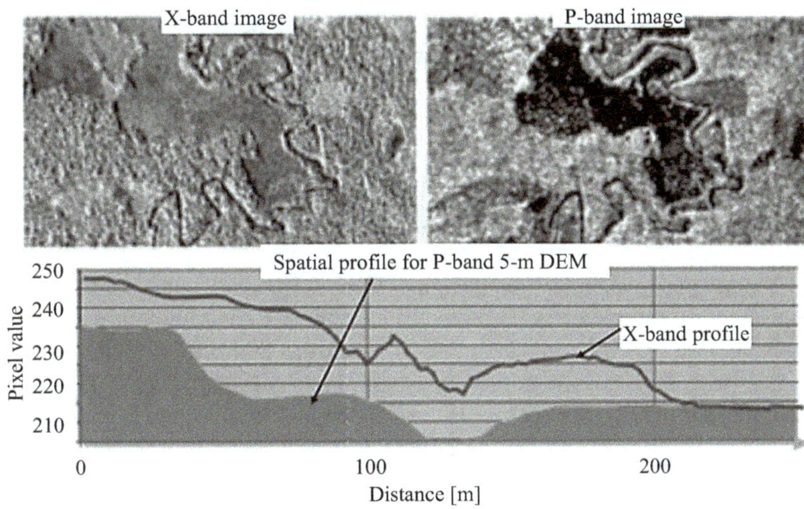

Figure 6.16 *GeoSAR DEM at X- and P-band in Columbia, South America [16]*

is from the P-band image. It can easily be seen that the difference between these two traces represents the estimation of the height of the trees. Moreover, the slope of the ground under the forest provides significant evidence for measuring the terrain slope.

6.2 Along-track InSAR

Ground moving target indication (GMTI) radars have a long history of innovation for detecting and tracking surface vehicles in open, rural and urban environments. As the clutter becomes more severe, competing with the vehicles due to radar platform motion and clutter spread, larger apertures and space time adaptive processing (STAP) have been developed. This competing clutter is normally in the vicinity of the vehicle and not obscuring the radar energy from detecting the object. In the case of vehicles moving in forests or along tree lines, the obscuration is severe, and both the attenuation and volumetric clutter in the detection cell must be considered.

UWB foliage penetration (FOPEN) radars have been developed to detect vehicles and other man-made objects under dense foliage. These FOPEN systems are generally designed in the very high frequency or ultra high frequency (UHF) bands to significantly reduce the effects of foliage attenuation [17]. With these longer wavelength designs, there is a problem for GMTI applications, where the detection is based on target Doppler being outside of mainbeam clutter (exo-clutter) or significant clutter cancellation from STAP (endo-clutter). In order to limit the clutter competing with the target detection, the resolution cell must be significantly reduced by high range resolution (high signal bandwidth) and fine cross-range resolution (long coherent integration times). These two criteria are synonymous with an SAR image collection.

ATI radars have been proposed for both space [6] and airborne applications [18] to detect very slow-moving vehicles and surface phenomena. Two SAR images are simultaneously formed with a baseline between the transmitter and the second aperture. By measuring the phase of each scattering cell, the fine Doppler content can be used to discriminate motion with respect to the background clutter. This approach has been shown to depend on the strong correlation of the return to the two receive apertures. When there is an attenuation of signal due to foliage volumetric scattering, or the target signature varies as a function of the angle to each receiver, this correlation criterion affects both the target detection and localization.

In this section, we consider the design of UHF ATI radar from several important aspects. The first parameter is the antenna baseline for discriminating very slow-moving vehicles. For ATI, this baseline is along the platform velocity vector, as opposed to vertical separation in the cross-track interferometry. This ATI geometry provides a discrimination in the Doppler return that is different from the background, stationary clutter. To measure this difference, it is important to assess the signal-to-clutter ratio (SCR) for the object. For detecting a moving vehicle, this difference provides a measure of the minimum discernable velocity (MDV). ATI requires a measurement of the phase between two antenna receivers. The final critical measure

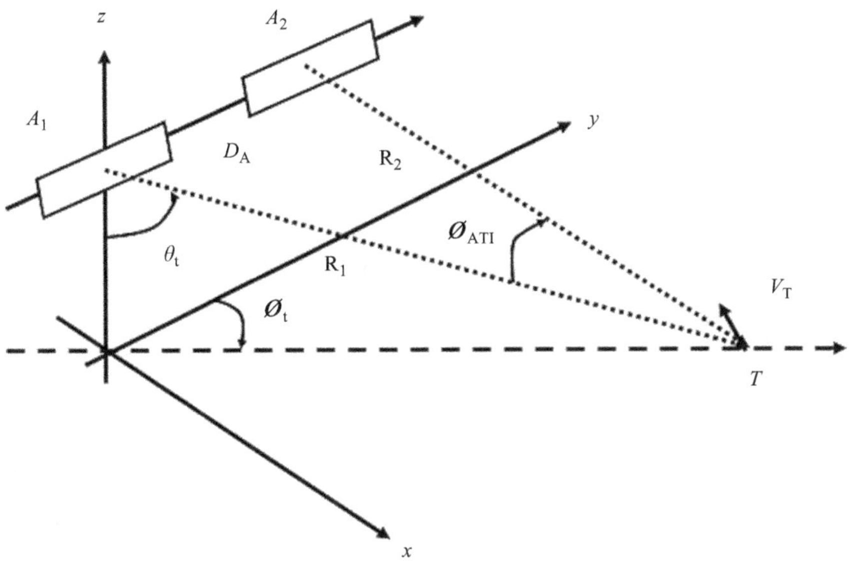

Figure 6.17 ATI SAR geometry [17]

is to determine the effect of signal decorrelation on the ATI ability to discriminate target from clutter and determine the target position and velocity.

ATI is an alternate GMTI architecture for detection of very slow-moving targets. The concept uses the along-track separation between phase centers, as shown in Figure 6.17, to measure the phase difference of a target between two different signal paths and to discriminate this phase from the background Doppler.

To develop the processing approach, one considers two independent receive signals S_1 and S_2 and the ATI measurement of these signals [18]:

$$\gamma_{tgt} \exp\left(j\, \phi_{tgt}\right) = \frac{\langle S_1\, S_2^* \rangle}{\sqrt{\langle S_1\, S_1^* \rangle\, \langle S_2\, S_s^* \rangle}} \tag{6.46}$$

where < > denotes expected value and * denotes complex conjugate. ATI processing relies on a high target correlation γ_{tgt} and a target phase ϕ_{tgt} distinguishable from clutter for detecting slow-moving targets with radial velocity:

$$v_T = \frac{\phi_{tgt}\, \lambda\, v_{plat}}{2\pi\, D_A} \tag{6.47}$$

where D_A is the separation between the two antenna phase centers and v_{plat} is the velocity of the platform. In this analysis, it is assumed that the two antennas are on the same platform.

For a target that has radial velocity different from the background clutter, the target will have a velocity measurement error given by [6]

$$\sigma_v = \frac{1}{\sqrt{2N_I}} \frac{\sqrt{1-\gamma^2}}{\gamma} \left(\frac{\lambda}{2\pi} \frac{v_{\mathrm{plat}}}{D_A \sin(\theta_t)} \right) \tag{6.48}$$

where θ_t is the nominal incidence angle and $\boldsymbol{v}_T$ the radial velocity of the target. The parameter γ determines the impact of the signal strength.

$$\gamma = \frac{\mathrm{SNR}}{\mathrm{SNR}+1} \tag{6.49}$$

The MDV then is related to the target radial velocity, the separation D between the phase centers and the additive noise of the receivers. This parameter is related to the SNR of the detection cell and the number N_I of independent cells averaged [18].

When there is a large separation between antenna phase centers, the potential exists for detecting small radial velocities, with a significant improvement in MDV over the conventional GMTI process. Figure 6.2 parametrically shows the MDV of two UHF arrays, as a function of the separation of the phase centers and additive noise σ [5]. These values represent the 1σ MDV and do not include the effects of internal clutter motion. They also assume that the SNR is noise limited and not clutter limited. Both these effects must be included in a more analytic approach. Due to the large potential separation between phase centers and panels of the UHF radar, the potential exists for detecting small radial velocities with a significant improvement in MDV over the conventional GMTI process. Figure 6.18a parametrically shows the MDV as a function of the separation of the phase centers along with the impact of additive noise σ. The values are the 1σ limits on MDV and will be increased by demands on either high probability of detection (Pd) or low false alarm rate [19]. Figure 6.18b shows the ATI ambiguous velocity versus the separation between the two antennas.

The predominant issue with GMTI is the target-to-clutter ratio. As the clutter cell is reduced by range and cross-range resolution, it is expected that the point target return will be larger than the distributed clutter. This measure of target resolution and improvement in SCR holds for both GMTI processing and ATI. However, there

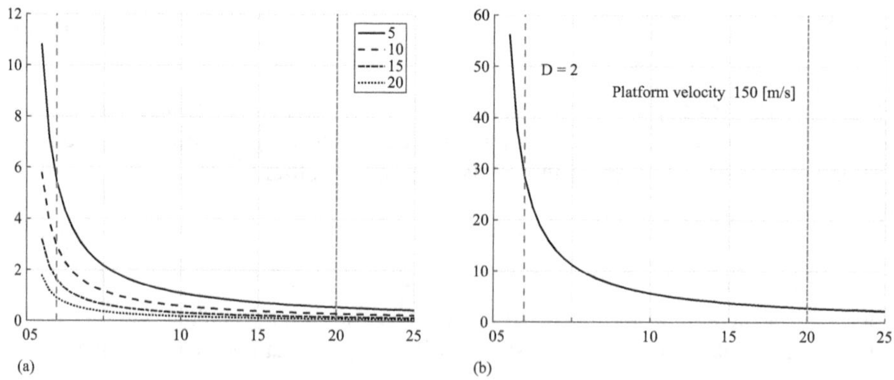

Figure 6.18 ATI performance versus baseline separation. (a) MDV and (b) ambiguous velocity

is a critical difference in the two modes in terms of the ability to maintain coherence during the wideband waveforms and long integration times. ATI utilizes SAR image processing and has a significant advantage in separating the targets from clutter. But the target motion within the SAR image collection will cause a displacement and defocusing of the targets if they have appreciable migration in either range or Doppler. These factors will be explored in this section in more detail.

Since a target within a clutter cell is in direct competition with the surrounding clutter, ATI is highly dependent upon a sufficient target-to-clutter ratio. An insufficient target-to-clutter ratio would produce a nonlinear target phase/velocity relationship after ATI processing. The resultant target phase after ATI processing as a function of the target-to-clutter ratio is given by [18]

$$\phi_{\mathrm{ATI}} = \frac{\mathrm{SCR}}{\mathrm{SCR}+1}\phi_{\mathrm{tgt}} \tag{6.50}$$

where SCR is the signal (target)-to-clutter ratio. This nonlinear relationship as a function of the target-to-clutter ratio is illustrated in Figure 6.19a. Target correlation also suffers from insufficient target-to-clutter ratios in ATI processing. Figure 6.19b shows that acceptable target correlation can be maintained for target-to-clutter ratios greater than 5 dB.

The relationship between aperture separation and either MDV or Doppler ambiguity can be examined. For a UHF system, Figure 6.18a provides the MDV versus separation at various SNR levels. Even at low SNR levels of approximately 5 dB, there is a significant improvement in MDV over a real aperture. Two baseline separations will be examined in greater detail. First, 2-m separation has a modest MDV of 0.75 m/s and large unambiguous velocity interval (30 m/s). This is a reasonable performance for targets that are moving with modest velocity over the ground. And it compares well with the objectives of GMTI systems. In contrast, a 20-m separation has a very low MDV under 0.2 m/s, but a very low-velocity ambiguity of 4 m/s. This approach would be applicable by mounting two antennas on a large aircraft and

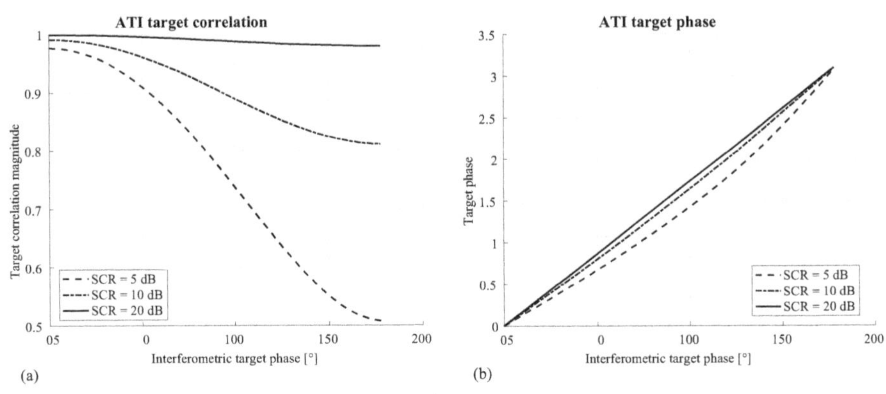

Figure 6.19 ATI target correlation with SCR effects. (a) Correlation and (b) phase

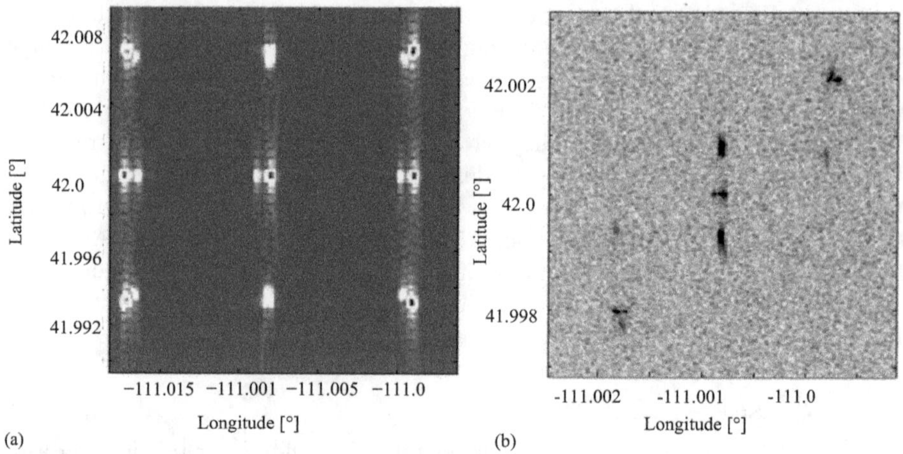

Figure 6.20 *Simulated scenario for evaluating ATI processing [19]. (a) Target scenario and (b) SAR image moving targets*

trying to measure very slow targets or differences in the movement of the surface. These have been used in earthquake assessment and detecting slow-moving boats on the water surface.

The ATI capability can be modeled with a simulation that combines multiple targets, with discrete scattering centers. The radar platform flies an SAR collection path with realistic motion over DTED terrain features. And the background clutter is simulated with land-use characteristics with clutter having the Doppler content as well as several clutter types in the scene. Details of this simulation approach to investigate new applications will be given in Chapters 8 and9. An example of how ATI can be evaluated will be given here.

The scenario shown in has nine targets arranged in an array (see Figure 6.20a). Each target is made up of ten 1-dBsm scattering centers emblematic of a truck, with eight scatterers at the front of the vehicle and two at the rear. The bright returns are shown with the eight moving targets moving away from the center at 5 m/s ground velocity. One target is stationary at the center position. In order to evaluate the velocity ambiguity effect, two ATI baselines were used: a 2-m baseline and a 20-m baseline. Figure 6.20b shows the SAR image with targets displaced in the cross-range (latitude) dimension. The radar characteristics and SCR simulated were varied from 5 to 20 dB. These examples are for the 20-dB SCR conditions [19].

The ATI effects are obtained by calculating the ATI magnitude and phase from (6.43) and is shown in Figure 6.21. The signal processing uses a 3×3 boxcar filter and constant false alarm ratio (CFAR) process to eliminate the stationary background clutter. However, it is clear that there are several detections of the targets during the SAR ATI processing. The cluster of detections at the center of the polar plot is from scintillating clutter. Details of the simulation, ATI processing and receiver-operating characteristics of this example will be presented in

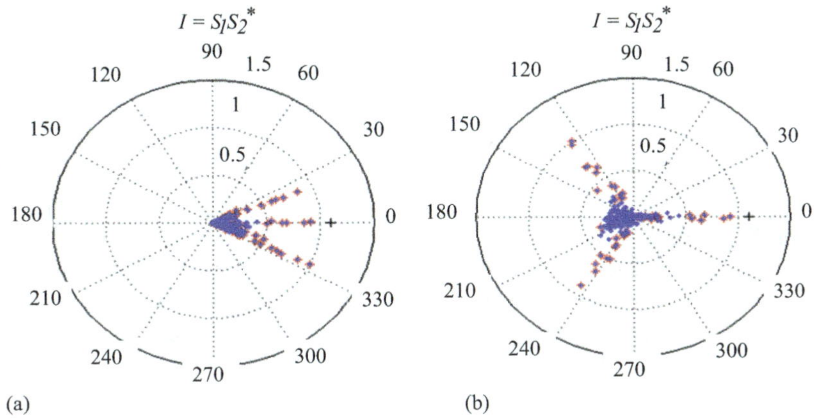

Figure 6.21 *Simulated ATI phase and target detections [19]. (a) 2-m baseline and (b) 20-m baseline*

Section 7.3. The main conclusion of this example is illustrating the ATI magnitude and phase measurement and obtaining a sense of the baseline effects on ATI phase ambiguities. For this UHF radar, the 2-m baseline target velocity ambiguity is at 23 m/s and the 20-m baseline ambiguity is 2.3 m/s.

The complexity of processing to handle Doppler ambiguities must be explored. In addition, the limitations of SCR will affect the ability to detect slow-moving targets and determine their velocity vector. Modeling and simulation is reasonably straight forward to assess the practical aspects of these applications. Chapter 7 will provide examples for both X-band and UHF cases of detecting slow-moving targets in two clutter regimes: open areas with low reflectivity surface and areas within strong clutter from forested regions.

Moving targets are defocused in the SAR images in both ATI channels. This defocus is caused by the motion of the target through resolution cells. If the moving target has a velocity in the along-track direction v_x, the phase history of the moving target will not match the reference phase history ϕ_{ref} necessary for SAR azimuth compression [18]:

$$\phi_{ref} = -\frac{4\pi}{\lambda}\frac{v_{plat}^2\,t^2}{2\,R_s} \tag{6.51}$$

Neglecting the cross-track velocity component, the target phase history is given by

$$\phi_{tgt} = -\frac{4\pi}{\lambda}\frac{\left(v_{plat}-v_t\right)^2 t^2}{2\,R_s} \tag{6.52}$$

The phase error ϕ_{err} at the ends of the synthetic aperture is

$$\phi_{err} = -\frac{2\pi}{\lambda\,R_s}\left[v_{plat}^2-\left(v_{plat}-v_t\right)^2\right]\left(\frac{T_{int}}{2}\right)^2 \tag{6.53}$$

A target will be smeared in the along-track dimension due to velocity v_t in the slant range direction. This target velocity gives an additional component of the target phase history due to change in the Doppler centroid of the target. This fractional Doppler shift Q_{Dop} is given by

$$Q_{Dop} = \left| \frac{v_t \, L}{v_{plat} \, \lambda} \right| \tag{6.54}$$

It should be noted that range migration for targets moving in radial velocity will not be affected since the image match filter will focus that contribution of velocity. However, if the velocity exceeds modulo 2π, then the uncompensated range migration $\Delta\rho_n$ between pulses will appear in a different point in the scene.

$$\Delta\rho_n = \eta_{Dop} \frac{\lambda}{2} \tag{6.55}$$

The total apparent range extent ρ_n traversed by the target over the synthetic aperture time is

$$\rho_n = \eta_{Dop} \frac{\lambda^2 \, R_s \, \mathrm{PRF}}{2 \, L \, v_{plat}} \tag{6.56}$$

where PRF is the pulse repetition frequency.

It is important to note that the geolocation of the targets depends on the detection of the target and an accurate measurement of the ATI phase. A moving target in an SAR image will appear in a different along-track location and is associated with stationary clutter at that image point. This along-track shift is due to the target Doppler frequency being the same as the Doppler in the SAR image. The ground point in the beam θ_{az} can be determined by

$$\frac{-2 \, v_t}{\lambda} = \frac{2 \, v_{plat} \, sin\theta_{az}}{\lambda} \tag{6.57}$$

And the apparent offset of the moving target x_{off} is given by

$$x_{off} = R_s \sin\theta_{az} = -R_s \frac{v_t}{v_{plat}} \tag{6.58}$$

The preceding analysis assumed that the clutter was homogeneous over large areas of the ground. This cannot be expected in the real world where there are boundaries between roads, fields and forested regions. It will be important to collect data from an ATI radar over a reasonable variation of scenes. One early ATI experiment by JPL was carried out during an experiment with the AirSAR platform and various slow-moving targets on the road. The moving target was a person pushing an electronics cart with a GPS receiver, as shown in Figure 6.22a. The horizontal velocity of the person was only 1 m/s, clearly a slow-moving target. Figure 6.22b shows two InSAR images collected at two different baseline separation. The fully processed SAR images have 3.75 m range resolution and 85 cm of cross-range resolution. The along-track baseline is 19.8 m and 39.6 m (ping-pong), producing the interferometric images in Figure 6.22b. The target is easily detected due to the longer SAR

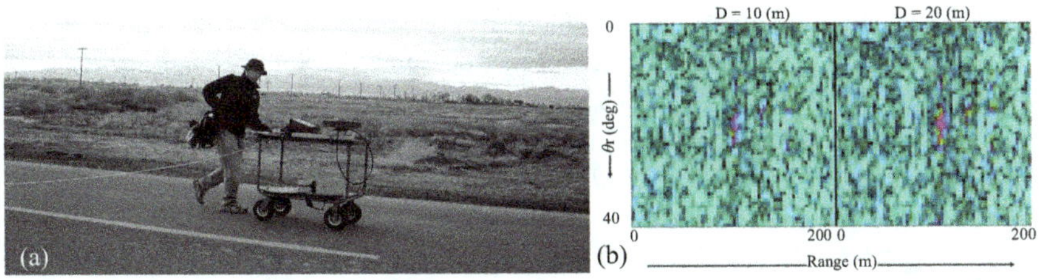

Figure 6.22 *L-band interferogram imaging person with cart [20]. (a) Control target and (b) Interferogram*

integration times. The ping-pong resulted in higher signal to clutter due to the finer cross-range resolution and two looks (interleaved pulses) [20].

References

[1] Graham L.C. 'Synthetic interferometer radar for topographic mapping'. *Proceedings of the IEEE*. 1974;**62**(6):763–8.

[3] NASA Jet Propulsion Laboratory. *AIRSAR: Airborne Synthetic Aperture Radar*. Available from https://airsar.jpl.nasa.gov.

[3] Van Zyl J.J., Chu A., Hensley S., Lou Y.K. 'The AIRSAR/TOPSAR integrated multi-frequency polarimetric and interferometric SAR processor'. *Proceedings 1997 IGARS Symposium*. Vol. 3. Singapore; 1997. p. 3.

[4] Adams G.F., Ausherman D.A., Crippen S.L., Sos G.T., Williams B.P., Heidelbach R. 'The ERIM interferometric SAR: IFSARE'. *IEEE Aerospace and Electronic Systems Magazine*. 1996;**11**(12):31–5.

[5] Bullock M.E., Lawrence F., Dams R.V., Tennant K. 'Map generation utilizing IFSARE imagery and digital elevation models from the Intermap STAR-31 system'. *Proceedings 1997 IEEE International Geoscience and Remote Sensing Symposium*. Vol. 3. Singapore; May 1997. pp. 1350–7.

[6] Rosen P.A., Hensley S., Joughin I.R., *et al*. 'Synthetic aperture radar interferometry'. *Proceedings of the IEEE*. 2000;**88**(3):333–82.

[7] Hensley S., Chapin E., Freedman A., Michel T.R. *Improved Processing of AIRSAR Data Based on the GeoSAR Processor*. Available from https://www.semanticscholar.org/paper/Improved-Processing-of-AIRSAR-Data-Based-on-the-Hensley Chapin/c412ce410857bdb358556049fdd9ef1101c55da9.

[8] Jones J.J. GeoSAR Radar Mapping, Fugro Earth Data, Inc Brochure. Available from http://www.latornell.ca/wp-content/uploads/files/presentations/2015/Latornell_2015_T3C_Joe_Jones.pdf.

[9] Hensley S., Madsen S.N. 'Interferometric radar waveform design and the effective interferometric wavelength'. *Proceedings 2007 Waveform Diversity & Design Conference*. Pisa, Italy; June 2007.

[10] Zebker H.A., Hensley S., Shanker P., Wortham C. 'Geodetically accurate InSAR data processor'. *IEEE Transactions on Geoscience and Remote Sensing*. 2010;**48**(12):4309–21.

[11] Chapin E., Hensley S., Michel T.R. 'Calibration of an across track interferometric P-band SAR'. *2001 IEEE International Geoscience and Remote Sensing Symposium*. Sydney, Australia; 2001. pp. 502–4.

[12] Zebker H.A., Villasenor J. 'Decorrelation in interferometric radar echoes'. *IEEE Transactions on Geoscience and Remote Sensing*. 1992;**30**(5):950–9.

[13] Just D., Adam N., Schwabisch M., Bamler R. 'Comparison of Phase Unwrapping Algorithms for SAR Interferograms'. *Proceedings International Geoscience and Remote Sensing Symposium*. 1. Florence, Italy; 1995. pp. 767–9.

[14] Chen C.W., Zebker H.A. 'Two dimensional phase unwrapping with statistical models for nonlinear optimization'. *Proceedings International Geoscience and Remote Sensing Symposium*. 7 vols. Honolulu, HI; 2000. pp. 3213–5.

[15] Rodriguez E., Martin J.M. 'Theory and design of interferometric synthetic aperture radars'. *IEE Proceedings F Radar and Signal Processing*. 1992;**139**(2):147–59.

[16] Reis J.J., Hensley S. 'GeoSAR P-band and X-band performance in southern California and Colombia, South America'. *ISPRS International Workshop*. Banff, Canada; June 2005.

[17] Davis M.E. 'Chapter 8 FOPEN Ground Moving Target Indication'. *Foliage Penetration Radar - Detection and Characterization of Objects Under Trees*. Raleigh-Durham, NC: Scitech Publishing; 2011.

[18] Chen C.W. 'Performance assessment of Along-Track interferometry for detecting ground moving targets'. *Proceedings 2004 IEEE Radar Conference*. Philadelphia, PA; April 2004. pp. 99–104.

[19] Kapfer R.M., Davis M.E. 'Along track interferometry for foliage penetration moving target indication'. *Proceedings 2008 IEEE Radar Conference*. Rome, Italy; 2008.

[20] Chapin E., Chen C.W. 'Along-track interferometry for ground moving target indication'. *IEEE Aerospace and Electronic Systems Magazine*. 2008;**23**(6):19–24.

Chapter 7
Ultra-wideband ground moving target detection

This chapter investigates ground moving target radar modes for surveillance from air-to-ground radar systems. This mode has been important, starting with both Multiple Aperture Surveillance Radar (MASR) and Standoff Target Acquisition System (SOTAS) to detect slow-moving targets in a battlefield environment. Its technical challenges are the detection of slow-moving targets that are masked by the clutter Doppler within the main beam of the radar antenna. A wide range of algorithms have been developed over the past 50 years, as the radio frequency (RF) components and high-speed signal processing techniques have been improved [1]. But as the technology advances, several new algorithms and system architectures were invented, based on the advances in component technologies. These techniques include obtaining high resolution of the targets for discrimination, and moving to longer wavelength radar systems to counter the masking and techniques that the adversary uses to avoid detection [2]. In addition to military ground moving target indication (GMTI) techniques, there are several civilian and geoscience applications for characterizing terrain, forests and ecological measurements using radar images.

The first section will present a simplistic assessment of modes for detecting ground moving targets. The motion of the airborne platform carrying the radar limits the ability to detect the target Doppler return due to the Doppler spread of the clutter. The primary system requirement for moving target detection is characterized by minimum detectable velocity (MDV), or the ability to discriminate slow-moving targets from the background clutter Doppler. The primary measure of effectiveness is the subclutter visibility of a radar system and its associated signal processing.

Improvements in MDV have been well reported in the literature based on space-time adaptive processing (STAP) [3, 4]. The improvement in GMTI MDV is obtained by adding additional spatial channels to the radar, and extensive Doppler signal processing to cancel the background clutter. Early STAP was developed for surveillance radar systems with low- and moderate-range resolution. However, as image resolution requirements are improved, the bandwidth of the radar is increased. A wideband system further complicates the desire for subclutter visibility, when the target motion through range and Doppler cells affects the assumptions on target correlation for discriminating between stationary clutter and moving targets. Recent radar techniques for handling this wideband requirement include digitally sub-banding the radar returns and processing multiple narrow-band adaptive processing in parallel. After the signal adaptation and clutter cancellation, the individual sub-bands are recombined to recover the target resolution. These approaches are covered in Section 7.2.4 based on recently published radar research.

Finally, this chapter will introduce interferometric synthetic aperture radar (SAR) to measure displacement of the targets or terrain feature returns between two channels, exploiting the phase difference between two channels that are separated in either cross-track or along-track [5]. Interferometric radars have been used to improve spatial resolution of terrain height measurement by comparing two channels with angle diversity. The interferometry can also be used in along-track interferometry (ATI) to improve the ability to detect and geolocate moving targets in SAR images [6]. Depending on the requirements for improved spatial resolution, the moving targets experience both a Doppler displacement and migration in range and cross-range during the image formation. These effects are covered in Section 7.2.

A new radar signal processing area of research is "focus before detection" where target or scene characteristics are complicated by the motion of target and features during the coherent integration timeframe. When targets or platforms have very long integration dwells, and the motions cannot be adequately measured or predicted, targets will be defocused. This will cause a reduction in the signal-to-noise ratio (SNR) and lower the detection probability. Inverse SAR techniques as well as radon Fourier transform techniques have been reported in the radar literature [7]. This is particularly important to ultra-wideband (UWB) surveillance radar system, where the objective is to simultaneously detect and characterize both fixed and moving targets. The focus-before-detection techniques can overcome the SNR effects for motion across range cells, Doppler cells and even between beams in a multiple beam radar system. A second effect of long coherent dwells is the consideration of both range and Doppler ambiguities in extracting the unambiguous parameters needed for improved focusing of the targets. Several techniques of autofocus, minimum entropy (ME) focusing and multiple look estimation of target motion parameters are required [8]. This chapter introduces the impact of UWB processing on moving target detection. Chapter 9 will address the issues of simultaneous detection of both fixed and moving targets with a common, long-dwell waveform.

7.1 GMTI mode

The GMTI mode has become a major capability for surveillance radars. And the problem of detecting a target via Doppler in a real beam radar is well known. When the platform has a velocity v_p and there is a finite beamwidth, the clutter will spread in azimuth within the beam, as illustrated in Figure 7.1 [6]. This spread in clutter is minimum when the azimuthal angle φ_p is exactly 90°. For any angle off of zero creates a range rate of the clutter compared to the platform velocity, which will compete with slow-moving ground clutter. Targets will need to have a radial velocity greater than the clutter spread within the beam to be detected. Many techniques have been developed to reduce the effective main beam spread through MTI pulse Doppler and STAP. MTI will normally cancel the clutter to the half beamwidth, and STAP down to a tenth of a beamwidth, if there is no internal clutter motion (ICM).

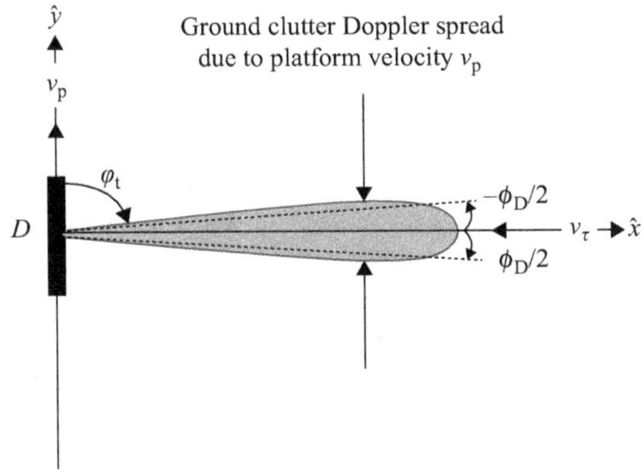

Figure 7.1 Antenna footprint on ground [6]

However, there is going to be a residual Doppler width to compete with low MDV targets. More importantly, MTI and STAP adaptive processing will attenuate slow-moving targets, making it more difficult to detect the signal. As a result, the development of adaptive processing to improve GMTI requires an estimation of the target discrimination from the clutter backscatter, under the assumption that the clutter is stationary with a zero mean backscatter coefficient. If there is motion within the clutter returns, often referred to as ICM, then this clutter motion will contribute to increasing the target MDV.

The basic ground moving target detection starts with an understanding of the background Doppler characteristics. The ground clutter Doppler around platform is given by :

$$f_{DP} = -\frac{2v_P}{\lambda}\cos(\varphi_P)\cos(\gamma_P) \tag{7.1}$$

where φ_P is the angle from the velocity vector v_P and γ_P is the grazing angle [6].

This illustrates that a negative rate of change of a target phase results in a positive Doppler shift. That is when a target is moving toward the radar, the Doppler frequency increases. And when the target moves away from the platform ($v_P > 0$), the target Doppler frequency is reduced. The separation of the target Doppler from the clutter is obtained by looking at the spread of the clutter within the beamwidth of the radar. This beamwidth θ_D is approximated by the ratio of the wavelength λ to the width of the antenna D:

$$\theta_D = \frac{k\lambda}{D} \tag{7.2}$$

The problem of detecting a target via Doppler in a real beam radar is well known. When the platform has a velocity v_P and there is a finite beamwidth θ_D, the clutter will spread in azimuth within the beam. To detect a target with velocity v_T, it has to have a Doppler value greater than that of the ground clutter spread within the beamwidth.

$$v_T \geq v_P \sin \frac{k\lambda}{2D} \cos \left(\gamma_P\right) \tag{7.3}$$

By using an MTI pulse Doppler canceller, the main beam clutter can be partially cancelled by using a multiple pulse cancellation, as shown earlier in Section 3.2.1. Many techniques have been developed to reduce the effective main beam spread through MTI pulse Doppler and STAP. MTI will normally cancel the clutter to the half beamwidth, and ideal STAP can cancel the clutter down to a smaller fraction of a beamwidth, if there is no ICM. However, there is going to be a residual Doppler width to compete with low MDV targets. More importantly, MTI and STAP processing will attenuate slow-moving targets making it more difficult to detect the signal.

Figure 7.2 illustrates the MDV with antennas at X and UHF bands, on a platform moving at between 50 and 200 m/s. The most effective technique to reduce main-beam clutter spread in MTI radar systems is to increase the aperture, shown in the figures between 2 and 10 m in length. At microwave frequencies (e.g., X-band), it is possible to get main-beam Doppler spread below 0.5 m/s. This has been demonstrated on several systems as effective for wide-area surveillance, even when the vehicles are not moving toward the radar.

However, at UHF frequencies, the antenna must be excessively long to eliminate the main-beam Doppler clutter spread. Even with antennas of over 10 m in length, the MDV is limited between 5 and 10 m/s. Little can be done with MTI or STAP processing with a real-beam antenna traveling at appreciable speeds to detect very slow-moving targets. However, there are two approaches that have been considered in the past. One is to tether the antenna on an aerostat or helicopter to minimize the platform motion. The second approach is to form a synthetic aperture to provide narrow azimuth resolution cells and then cancel the ground clutter with a displaced phase center antenna [9].

The SOTAS radar, as shown in Figure 7.3a, was designed at microwave frequencies and was demonstrated on a helicopter at longer detection ranges [1]. The second approach has been demonstrated in the early US Camp Sentinel FOPEN system from an MTI radar on a tall tower (Figure 7.3b), albeit with limited detection range and no mobility [2]. But it is well known that microwave frequencies do not provide acceptable foliage penetration and it gives poor performance against tactical targets hiding under trees and camouflage nets. The following sections will summarize improved GMTI adaptive processing, afforded by advances in multiple channel radar architectures and high-speed radar signal processing.

7.2 Space-time adaptive processing

STAP for detection of slow targets by cancelling clutter has been developed for large, airborne GMTI radar systems over the past three decades [3]. The early adaptive processing to remove clutter was referred to as spatio-temporal adaptive processing [10]. By adding the spatial diversity in addition to the Doppler processing in the radar architecture, there are independent looks at both the ground clutter and

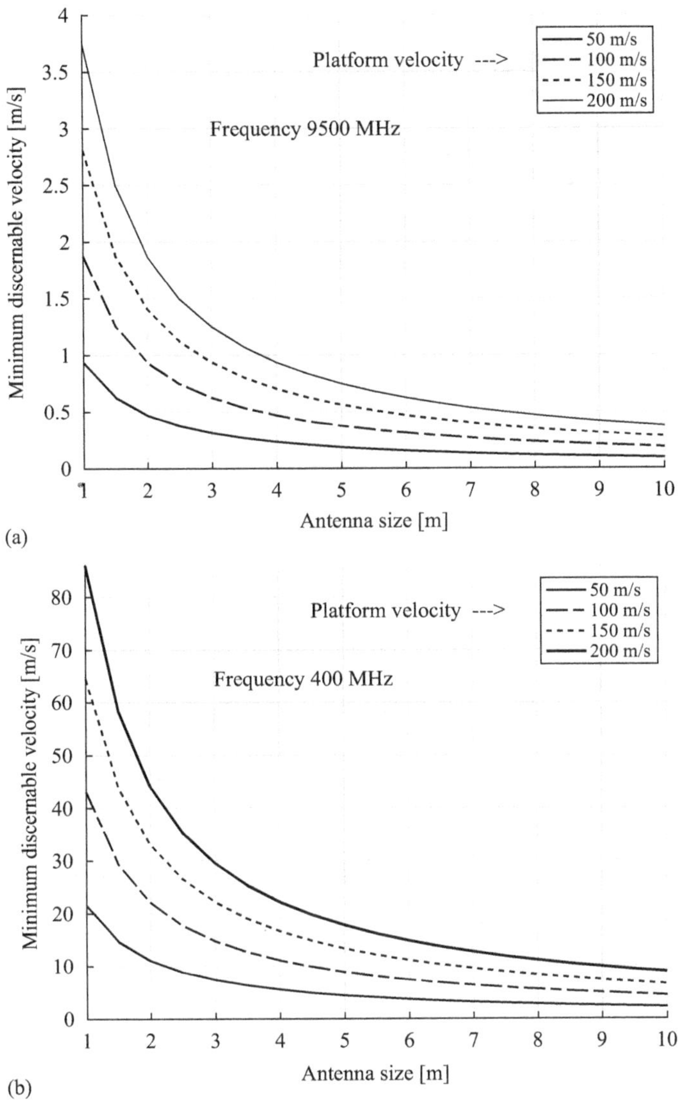

Figure 7.2 MDV versus antenna size. (a) X-band radar and (b) UHF radar

targets. The clutter spatial and spectral return is limited primarily by the size of the aperture, where the spatial directivity will limit ground clutter extent by the velocity of the platform, as illustrated in Figure 7.4. Similarly, the range and Doppler characteristics of the target will be determined by the waveform being used in GMTI modes. It should also be noted that STAP can also assist in the removal of discrete RF interference along an angular sector of the surveillance space.

Figure 7.3 *Early GMTI systems employing Doppler processing. (a) X-band SOTAS radar [1] and (b) UHF camp sentinel radar [2]*

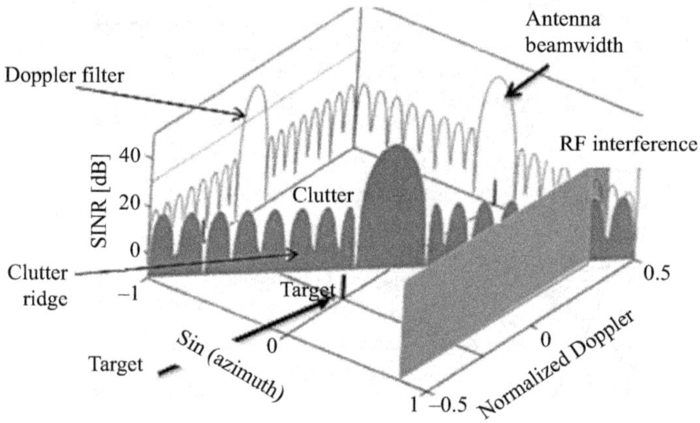

Figure 7.4 *Target in clutter and RF interference on moving platform [3]*

STAP combines signal in both the spatial (multiple channels on the antenna) and temporal (Doppler processing within the pulse repetition frequency). The design of the sidelobes of the antenna and Doppler filtering will provide the ability to discriminate the moving target from distributed clutter and interference. Providing high-efficiency detection of moving ground targets requires the ability to cancel the stationary clutter, while minimizing the attenuation of desired targets. This ability to improve the MDV of targets is affected by the signal decorrelation within the main beam and sidelobes of the antennas. As a result, channel match and balance are important in implementing STAP processing and obtaining the desired low MDV.

The geometry for analyzing STAP is given in Figure 7.5. The unit vector from the center of the antenna is given by $\hat{k}(\phi_t, \theta_t)$, measured in the azimuth φ_t and elevation θ_t dimensions from the face of the antenna. The returns from the antenna illumination of an iso-range ring is along R_s.

The concept of spatial and temporal degrees of freedom (DOFs) is illustrated in Figure 7.6. The antenna is depicted by multiple spatial channels, in the form of a uniform linear array of N elements or subarrays. The pulse-Doppler waveform is a

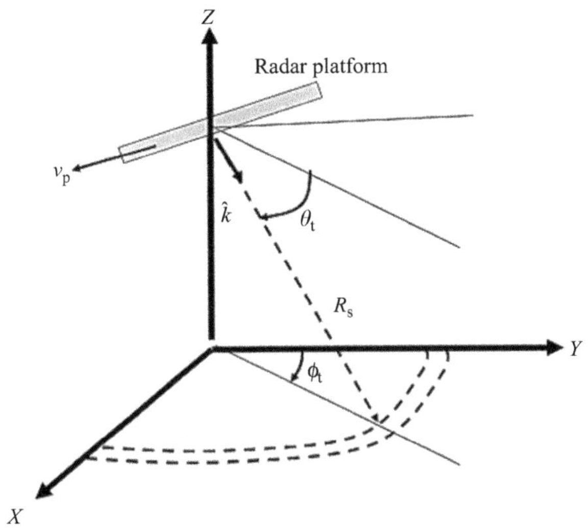

Figure 7.5 STAP platform coordinate system [6]

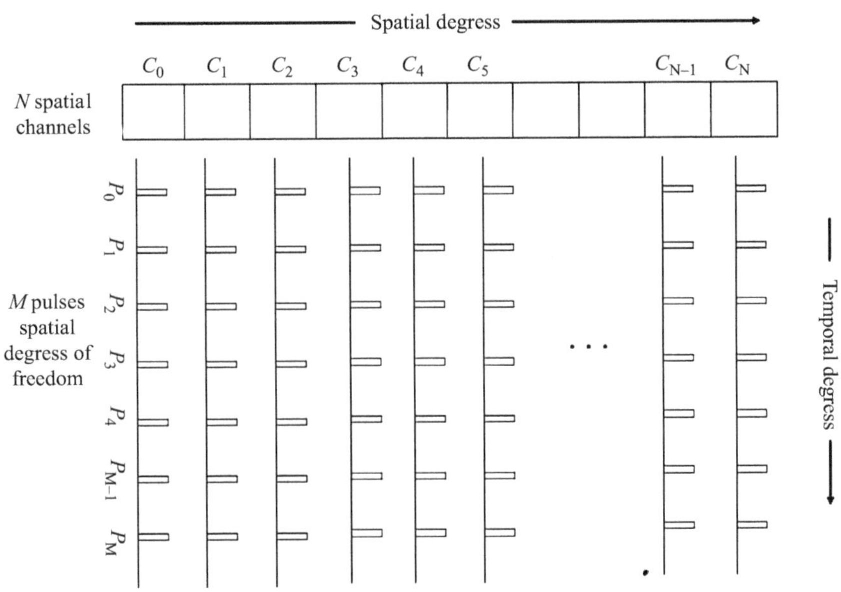

Figure 7.6 Pulse collection for STAP processing [6]

sequence of M pulses, which are transmitted at a constant pulse repetition interval (PRI) of T_r. The spatial effects of the individual element or subarrays are characterized by a normalized spatial frequency, indicative of the time delay of the received signal between each phase center of the antenna. The temporal characteristics are measured by the normalized Doppler, i.e., the target Doppler return divided by the pulse repetition frequency (PRF) of the waveform.

The spatial DOFs can be generalized for the array having a uniform separation between channels of d. When it is oriented such that the nth element has a position at a reference time along the x-axis of $\vec{\boldsymbol{r}}_{\mathbf{n}}$, the location of each spatial DOF is given by [3]

$$\vec{\boldsymbol{r}}_{\mathbf{n}} = n\, d\, \hat{x} \tag{7.4}$$

where $\hat{x}$ is the unit vector in the plane of the antenna. The target echo received on each of the N receive channels is given as

$$s_{\mathrm{n}}(t) = a_{\mathrm{r}}\, u\left(t - \tau_{\mathrm{n}}\right) \exp\left[\,\mathrm{j}2\pi\ \left(f_0 + f_{\mathrm{t}}\right)\left(t - \ \tau_{\mathrm{n}}\right) + \mathrm{j}\psi\right] \tag{7.5}$$

where a_{r} is the echo amplitude. The incremental times τ_{n} represent the temporal DOF, and f_{t} is the Doppler frequency imparted by the target motion.

Measured at the instantaneous frequency λ_0, the signal function is defined as

$$u(t) = \sum_{m=0}^{M-1} u_{\mathrm{p}}\left(t - m\ T_{\mathrm{r}}\right) \tag{7.6}$$

It is useful to define a normalized target Doppler frequency by the PRF, f_{r} (i.e., $1/T_{\mathrm{r}}$):

$$\varpi = f_{\mathrm{t}}\, T_{\mathrm{r}} = \frac{f_{\mathrm{t}}}{f_{\mathrm{r}}} \tag{7.7}$$

The target delay to the nth element consists of two components:

$$\tau_{\mathrm{N}} = \tau_1 + \tau_{\mathrm{n}} = \frac{2R_{\mathrm{T}}}{C} - n\frac{d}{C}\cos\theta_{\mathrm{t}}\, \sin\varphi_{\mathrm{t}} \tag{7.8}$$

The first term in (7.8) is the round-trip delay from the radar to the target, and the second is the relative delay measured from the phase reference to the nth element. It is also useful to define a target spatial frequency by:

$$\vartheta_{\mathrm{t}} = \frac{\hat{k}\left(\phi_{\mathrm{t}}, \theta_{\mathrm{t}}\right) \bullet \vec{\boldsymbol{d}}}{\lambda_0} = \frac{d}{\lambda_0} \cos\theta_{\mathrm{t}} \sin\phi_{\mathrm{t}} \tag{7.9}$$

The matched filter output of the nth receiver is the signal:

$$x_{\mathrm{n}}(t) = a_{\mathrm{t}} \exp\left[\mathrm{j}n2\pi\, \mathrm{v}_{\mathrm{t}}\right] \sum_{m=0}^{M-1} \exp\left[\mathrm{j}m2\pi\, \varpi_{\mathrm{t}}\right] \chi\left(t - \tau_{\mathrm{t}} - mT_{\mathrm{r}}, \varpi t\right) \tag{7.10}$$

where $\chi(r,f)$ is the waveform ambiguity function. This equation can be normalized such that $\chi(0,0) = 1$. Under these assumptions, the target samples are expressed by the matrix elements:

$$x_{nm}(t) = \alpha_{\mathrm{t}} \exp(\mathrm{j}n2\pi\vartheta_{\mathrm{t}}) \exp(\mathrm{j}m2\pi\varpi_{\mathrm{t}}) \ {}^{n=0,\ldots,N-1}_{m=0,\ldots,M-1} \tag{7.11}$$

The variable x_{nm} is normally referred to as the STAP data cube, which is processed to estimate and cancel the distributed clutter as well as any RF interference. The target data can now be assembled into a space-time matrix snapshot, written as [3]

$$\chi_{\mathrm{t}} - a_{\mathrm{t}}\, \mathrm{v}_{\mathrm{t}}\left(\vartheta_{\mathrm{t}}, \varpi_{\mathrm{t}}\right) \triangleq a_{\mathrm{t}}\, b\left(\varpi_{\mathrm{t}}\right) \otimes a\left(\vartheta_{\mathrm{t}}\right) \tag{7.12}$$

where

$$\vec{a}\ (\vartheta_t) = [1; \exp(j2\pi\ \vartheta_t); \ldots ;\ \exp(j2\pi\ (N-1)\ \vartheta_t)] \tag{7.13}$$

$$\vec{b}\ (\varpi_t) = [1; \exp(j2\pi\ \varpi_t); \ldots;\ \exp(j2\pi\ (M-1)\ \varpi_t)] \tag{7.14}$$

are the space-time steering vectors to the target. The term $\boldsymbol{v}_t\ (\theta,\varpi)$ in (7.12) is the target steering vector.

The clutter return is characterized by its separation from the target's Doppler frequency in angle from the velocity vector. When the clutter return is received at a slant range R_S and azimuth angle θ_c, the Doppler frequency from the clutter patch is given by

$$f_c(\phi_c, \theta_c) = -\frac{2\hat{k}(\phi_c, \theta_c) \cdot \vec{v_P}}{\lambda_0} \tag{7.15}$$

In this case, the angles θ_c and φ_c are continuous variables due to the distributed nature of the clutter. When the aircraft velocity vector is aligned with the antenna's longitudinal axis, there is a direct relationship between the normalized Doppler frequency and the spatial frequency. Equations (7.7) and (7.9) represent the orthogonal space-time dimensions to segregate the target from the clutter Doppler. The target can be seen in Figure 7.4, where the target is outside of both the angle and Doppler dimensions of the normalized space and time dimensions. For those targets that are outside of these two main beam dimensions, the target is considered to be exo-clutter.

The clutter return is characterized by dependency of the radar platform's velocity vector to the clutter patch on the surface. STAP processing considers both the target matrix elements and the clutter characteristics. The target will be considered as a point with discrete spatial and temporal characteristics. The clutter will have both normalized Doppler and spatial frequencies, due to its distributed nature. In fact, it is common to refer to the "clutter ridge" as the space-time effects of stationary ground clutter. In this case, the angles θ_c and φ_c are continuous variables due to the distributed nature of the clutter. The clutter Doppler frequencies along the clutter ridge are given by [3]

$$f_c(\theta_c, \phi_c) = -\frac{2\ \vec{\boldsymbol{k}}\ (\theta_c, \phi_c) \cdot \vec{v_P}}{\lambda_0} \tag{7.16}$$

When the aircraft velocity vector is aligned with the antenna's longitudinal axis, there is a direct relationship between the normalized Doppler frequency and the spatial frequency. The Doppler frequency from the clutter patch ϖ_c is expressed as

$$\varpi_c = f_c\ T_r = \left(\frac{2\ v_P\ T_r}{d}\right) \phi_c \tag{7.17}$$

With these normalized coordinates, the slope of the clutter ridge is given by

$$\beta_c = \frac{2\ v_P\ T_r}{d} \tag{7.18}$$

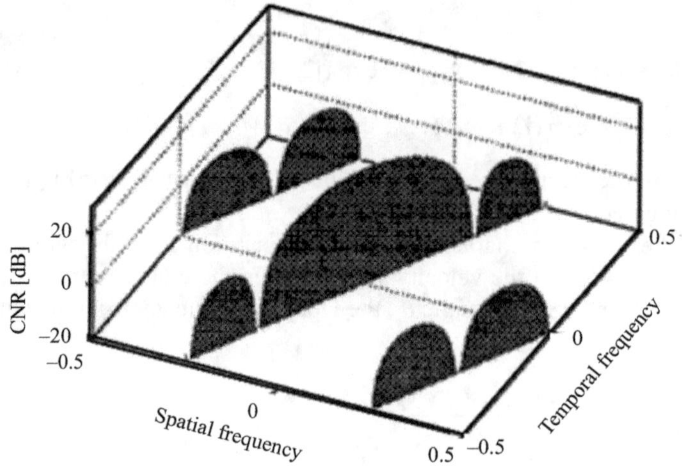

Figure 7.7 Normalized angle Doppler space for $\beta_c > 1$ [4]

This term β_c represents the number of inter-element spacings traversed by the platform during one PRI. For half-wavelength spacing, β_c is equivalent to the number of times the clutter Doppler spectrum aliases into the unambiguous Doppler space. When $\beta_c = 1$, the clutter ridge exactly fills the spatial dimension, as seen in Figure 7.4. Under the condition $\beta_c > 1$, the clutter ridge wraps around and multiple Dopplers will occupy the signal space as shown in Figure 7.7 [4].

It is important to recognize that STAP requires a high degree of correlation in both time and frequency. If there is temporal variation in the clutter, as in ICM, there will be a decorrelation of the clutter covariance matrix. Furthermore, the assumption is that the targets are highly correlated in order to obtain the target covariance matrix. Note that the normalized spatial and Doppler characteristics are expressed at the center frequency. If the antenna or target characteristics change within the waveform frequency, there will be a limitation in the ability to adaptively cancel the clutter. These factors are commonly referred to as a stationarity assumption. Since wavelength will change within the waveform, STAP will be limited in correlation properties and be a limiting factor in the STAP processing, as covered in Section 7.2.4.

The basis of a surveillance radar is to process signals coming from wide angles and range and adaptively cancel the clutter to detect the targets. The data available for a radar signal processor comes from M pulses and each of the N elements (or channels). As shown in Figure 7.6, the data will be collected by a space-time processor and utilize a linear filter to detect a target at a particular range gate. The space-time processor can be represented by an MN-dimensional weight vector w. The output of the data collection z can be represented by the inner product of the weight vector and the signal snapshot of interest x_{mn} as given in (7.11):

$$z = w^H \chi \tag{7.19}$$

Ideally, the space-time processor provides coherent gain on the target and spatially nulls the returns from distributed clutter and spatial RF interference. Since the geometry and characteristics of the interference is not known, it must be estimated by the space and time dimensions of the data cube. A particular weight vector is optimized for a specific angle and Doppler. Assume that the target signal is present in a background of clutter at a range of interest. The data snapshot can be written as [3]

$$\chi = \alpha_t v_t + \chi_u \tag{7.20}$$

where $v_t = b(\varpi_t) \otimes a(\vartheta_t)$ is the target steering vector and χ_u denotes the interference plus noise components of the data. It is well known in the STAP community that the optimum space-time filter is given by

$$\boldsymbol{w} = \boldsymbol{R}_u^{-1} v_t \tag{7.21}$$

where the covariance matrix $\boldsymbol{R}_u$ of interference plus noise is estimated by the snapshot in (7.19):

$$R_u = E\{\chi_u \chi_u^H\} \tag{7.22}$$

In order to estimate the interference effects, the covariance matrix must be estimated from the data. Depending on the DOFs in the problem, there needs to be 2 × DOF samples to estimate R_u [11].

A critical measure of the adaptive processing is the signal-to-interference plus noise ratio (SINR). By carrying out the space-time adaption shown above, the SINR is defined by

$$\text{SINR} = \frac{\sigma^2 \xi_t \left|w^H v_t\right|^2}{w^H R_u w} \tag{7.23}$$

where σ^2 is the noise power and ξ_t is the SINR at the range bin for a single pulse and a single element in the array. Substituting the optimum weight vector into (7.22), the optimum SINR is given by

$$\text{SINR}_0 = \sigma^2 \xi_t v_t^H R_u^{-1} v_t \tag{7.24}$$

It is instrumental to measure the SINR loss of the adaptation by comparing the noise only case with the adaptation of signal, interference and noise. The optimum SNR (without interference) is expressed as the gain of M pulses and N elements as

$$\text{SINR}_0 = MN\,\xi_t \tag{7.25}$$

Furthermore, the SINR loss of a space-time processing is defined to be its performance relative to the matched filter SNR in an interference free environment. Thus,

$$L_{\text{SINR}} = \frac{\text{SINR}(\varpi)}{\text{SINR}_0} \tag{7.26}$$

The estimation of SINR is useful to the surveillance radar, in that it will provide a loss curve versus Doppler of the target. This loss will be used to measure the detection range performance compared to the clutter-free region. Typically, the SINR loss curve is built into an analysis program to evaluate the ability to detect targets of varying Doppler. Examples of this will be given in Section 7.2.2.

7.2.1 Internal clutter motion

Clutter is typically modeled as a stationary (i.e., zero Doppler shift) scatter from the surface. However, motion internal to the background clutter resolution area of vegetation and trees will result in the clutter having nonzero bandwidth. This effect is generally referred to as ICM, and it will tend to increase the rank of the clutter in the STAP covariance matrix [11]. This results in a two-dimensional "ridge" in angle-Doppler space along which the clutter resides. This ICM decorrelating effect results in an increase in the rank of the clutter, which must be considered in order to accurately assess STAP performance, especially when eigen decomposition based.

In an extensive series of radar measurements across many geographic regions, MIT Lincoln Laboratory measured the spectra of ground scatter under a variety of wind conditions and frequencies. These measurements ranged from mid-VHF to X-band [12] over many terrain types and grazing angles. Analysis of these measurements resulted in a clutter spectrum model of simple mathematical form that is based on operating (carrier) frequency and wind speed. The ground scatterer was shown to be consistent with a model that has two components, a stationary (DC) term plus a noise (AC) term that has an exponential frequency spectrum. This model results in a clutter spectrum of the following form [13]:

$$P_c(f) = \frac{r}{r+1}\,\delta(r) + \frac{1}{r+1}\,\frac{\beta\lambda}{4}\,e^{-\frac{\xi_c\lambda}{2}|f|} \tag{7.27}$$

where f is the clutter Doppler shift in Hz, λ the wavelength and δ the Dirac delta function. The shape parameter ξ_c of the exponential spectrum was shown via the measurements to be generally invariant with carrier frequency, but a strong function of wind conditions. Note that the spectrum is normalized, so that $\int_{-\infty}^{\infty} P_c(f) = 1$. The ratio r between the power of the DC and AC components of the ground scatter was shown to be a function of both the carrier frequency and wind speed:

$$10\log(r) = -15.85\log(w) - 12.2\log f_c + 63.2 \tag{7.28}$$

where w is the wind speed (m/s), f_c the carrier frequency in MHz and log(–) the base-10 logarithm. The correlation function is obtained by the Fourier transform of (7.28):

$$\begin{aligned} r_c(\tau) &= \int_{-\infty}^{\infty} P_c(f)\, e^{j2\pi f\tau}\, df \\ &= \frac{r}{1+r} + \frac{1}{r+1}\,\frac{(\xi_c\lambda)^2}{(\xi_c\lambda)^2 + (4\pi\tau)^2} \end{aligned} \tag{7.29}$$

The variation of ac to dc components versus wind speed and frequency is given in Figure 7.8. The spectral width of the ac component of vegetation ICM varies as ξ_c^{-1}, and it varies approximately as the logarithm of the wind speed. Note that $v = \xi_c^{-1}$ is

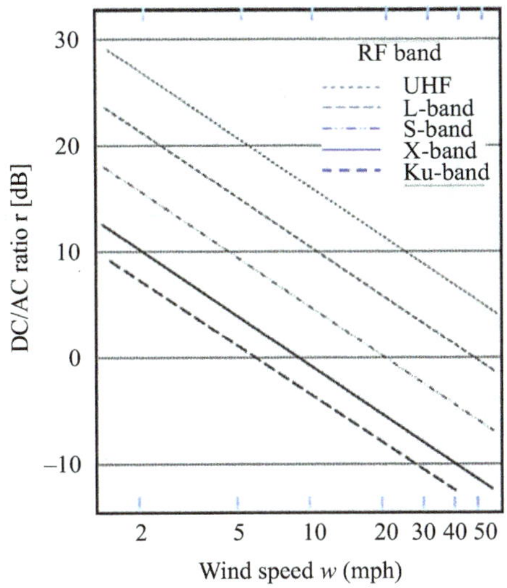

Figure 7.8 *Exponential model for ac clutter shape from windblown vegetation [12]*

the point where the spectrum is –4.34 dB down from its zero Doppler peak. Pac(v) is largely dependent on radar frequency, but the value of dc/ac ratio *r* is strongly dependent on both the wind speed and the radar frequency.

The extensive measurements by Billingsley provide the empirical representation of (7.29) as shown in Figure 7.9. The quantity *r* is also the ratio of steady to random average power describing the temporal amplitude of the statistics of the clutter. The spectral power resolved in the zero-Doppler bin of the radar return is

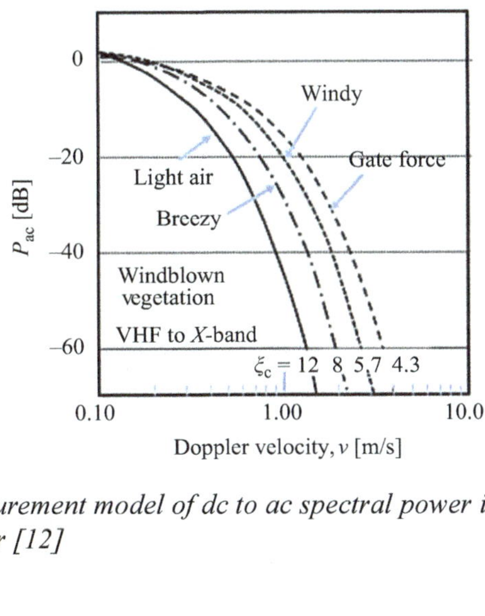

Figure 7.9 *Measurement model of dc to ac spectral power in windblown forest clutter [12]*

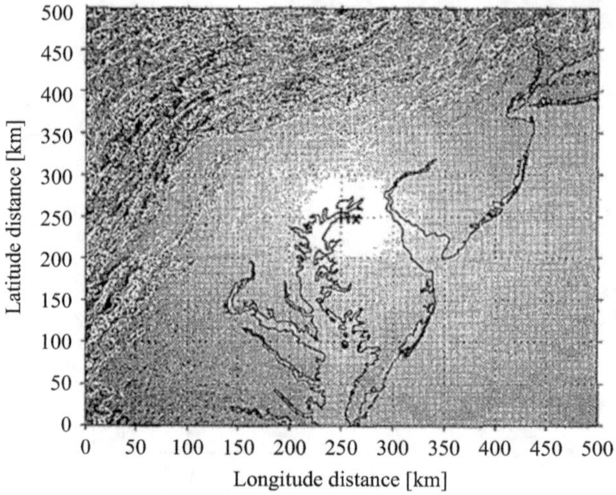

Figure 7.10 Region of DELMARVA peninsula [11]

provided by the relatively motionless parts of the forest and the ground surface. The higher spectral content is from limbs, leaves and, to some extent, the grass lands. At the lower frequencies, there is greater penetration of the tree-canopy and the trunks have a lower Doppler shift. Whereas, for microwave frequencies, the scattering is predominantly due to the top surface of the foliage, and the motion in high winds provides a more significant scattering of the energy.

The impact of ICM is considered next on the SINR loss performance of full DOF STAP. An area covering the Delaware--Maryland--Virginia (DELMARVA) peninsula is shown in Figure 7.10 [11]. The figure is a radar backscatter map simulated at L-band from a platform traveling at approximately 100 m/s. The remaining parameters of the simulation are found in Table 7.1. Figure 7.11a plots the eigenvalues for the ideal covariance for the range bin numbers 1 and 400 ("rbl" and "rb400," respectively). In addition, the estimated covariance matrices based on the first 400 range bins both with and without ICM is presented. The clutter-only eigenvalues

Table 7.1 L-Band STAP radar simulation parameters [11]

Parameter	Value
Carrier frequency	1,240 MHz
Bandwidth	500 KHz
Number of elements	16 at $\lambda/2$
Pulse repetition frequency	450 Hz
Clutter power (CNR)	50 dB/element and pulse @ Rmin
Minimum range	50 km
Maximum range	250 km
Platform heading	172.6° truc azimuth

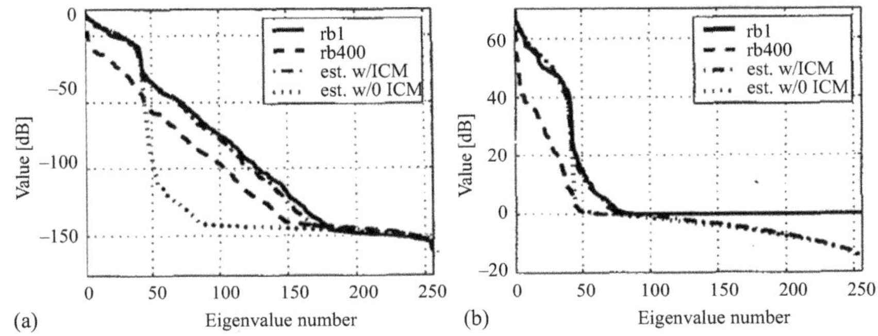

Figure 7.11 *Clutter eigenvalues impact on STAP [11]. (a) Clutter-only eigenvalues and (b) clutter and noise eigenvalues*

indicate that the ICM has greatly increased the rank of the clutter. Figure 7.11b shows the same results, but with the noise power added to each channel. It can be concluded that this effect is solely from the ICM (and not from finite sample effects) since the eigenvalues of the estimated covariance matrices do not show significant spread beyond that expected from the clutter scenario. The number of eigenvalues that are above the noise indicate that the effective rank is increased by the ICM [13].

The effect of the ICM on the clutter notch is a smoother increase in the width of the notch with decreasing range. This is expected when one considers the clutter power plot of Figure 7.12, where the DELMARVA scenario average CNR is plotted as a function of (one-way) range. The site-specific clutter is compared to a constant reflectivity coefficient of $\sigma^0 = -8$ dB. A CNR of 50 dB is assumed for the minimum range bin (50 km). The average of the diagonal elements of covariance matrices is shown. Since the clutter power variation is much smooth, there is a corresponding impact on the clutter notch width of Figure 7.12.

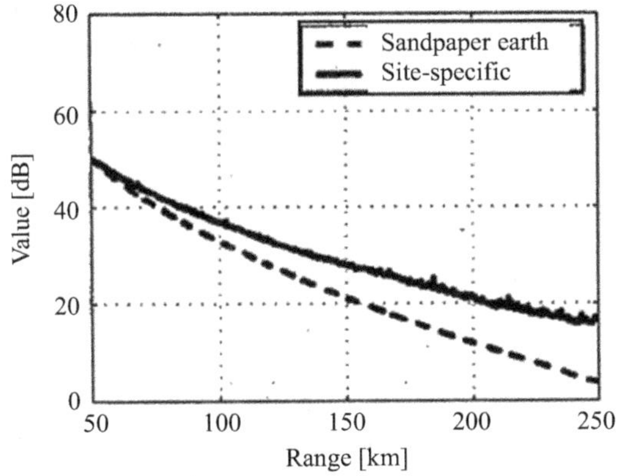

Figure 7.12 *Average clutter power from DELMARVA simulation [11]*

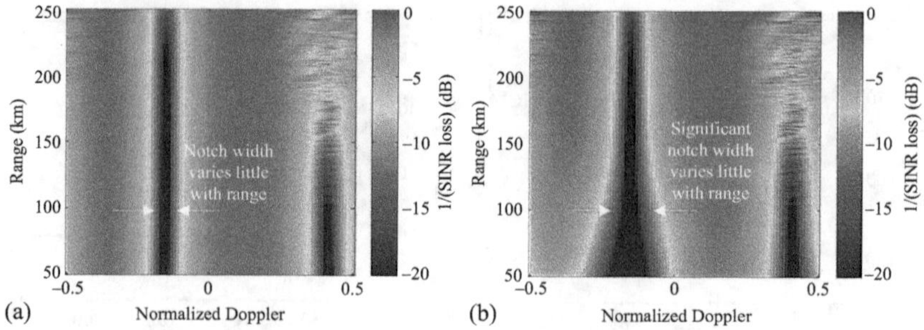

Figure 7.13 *Site-specific clutter variation in Doppler and range. (a) Clutter-only and (b) clutter and ICM [14]*

Results for this scenario are shown in Figure 7.13, showing the SINR performance from the Delmarva scenario. It should be noted that the 7.3° of crab is produced by the mismatch of the array axis and platform heading and the off-boresight look direction result in an additional notch (when compared to the previous scenario) from the clutter back lobe. SINR loss due to clutter is plotted for varying range and normalized Doppler for a look azimuth of 110°. It should be noted that if one were to use a global training strategy, particularly one based on training on the ranges with the strongest clutter power, that significant over-nulling of the clutter would result at many ranges leading to an overly broad clutter notch and consequent loss of radar sensitivity.

This effect is similar to that which is seen when considering wideband signals incident on an adaptive array. It has previously been noted that the effective rank of a nonzero bandwidth signal on an array is a function of SNR and bandwidth in addition to angle from the array boresight. This results from the decorrelation of the signal as it propagates across the array. The effect here is similar, except that in this case the nonzero bandwidth of the clutter results in decorrelation over time (i.e., from pulse-to-pulse). It is interesting to note that while the effects of ICM in this case would not be discernible using Fourier analysis (e.g., analyzing the Doppler spectrum by taking the discrete Fourier transform over pulses). However, they do have a very significant effect on the structure of the covariance matrix and the resulting STAP performance [15].

7.2.2 Narrow-band STAP

This section illustrates the impact of platform motion on a narrow-band radar for GMTI. This case is for a UHF system, operating at two velocities. The size of the antenna is assumed to be 7 m, very long for a small unmanned air system (UAS). However, it is representative of the FORESTER radar on an A-160 unmanned rotorcraft, as shown in Figure 7.14 [6]. A simulation of the platform motion and UHF radar operation is summarized in Table 7.2. The case of the platform motion at 50 m/s would be characteristic of a very slow-moving platform, and for 1 m/s would be a stationary platform on either a tethered aerostat or hovering helicopter, as shown in

Figure 7.14 *FORESTER on A-160 remotely piloted helicopter [6]*

Table 7.2 *Simulation parameters for UHF STAP example [6]*

Frequency	360	MHz	Antenna	7.1	meters
Bandwidth	15	MHz	N_{az}	16	
PRF	1,000	Hz	N_{el}	4	
Power	2,500	watts	d_{az}	0.53	Wavelength
Pulse width	25	μs	d_{el}	0.74	Wavelength
Mp	64		Altitude	7.5	km
Duty factor	2.56	percent	Velocity	[1, 50]	m/s
Target	−3	dBsm	R_s	25	km
Clutter	−8	dBsm/m^2			

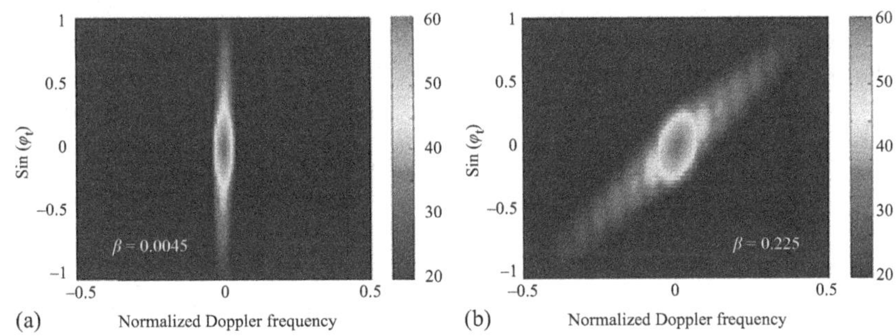

Figure 7.15 *STAP results for UHF radar (a) 1 m/s velocity, and (b) 50 m/s velocity [6]*

Figure 7.15. These two cases will be examined for first-order impacts on the MDV after STAP processing. It should be noted that these analyses are for 15-MHz bandwidth, which is representative of a moderate-bandwidth GMTI system.

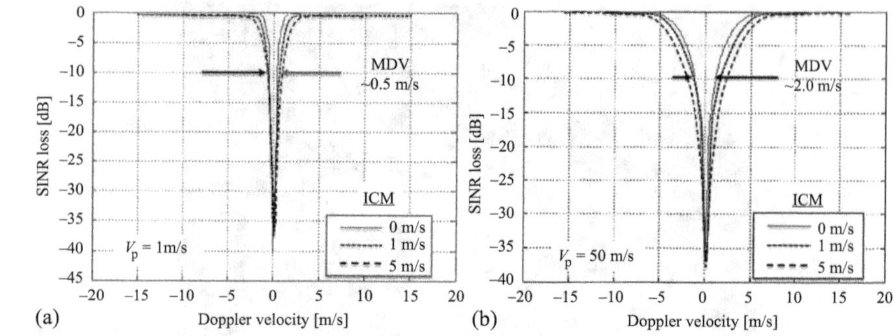

Figure 7.16 MDV for UHF STAP at 1000 Hz PRF. (a) v_P = 1 m/s and (b) v_P = 50 m/s [6]

The clutter ridge can be seen clearly for the two cases shown in Figure 7.15: for the 1 m/s "hovering" helicopter, the ridge has a slope of $\beta = 0.0045$ s^{-1}, and the 50 m/s moving platform with $\beta = 0.225$ s^{-1}. Figure 7.16 shows the corresponding SINR loss curves using ideal STAP, and where the MDV is measured at the –10 dB loss point. There are three curves, depending on the ICM of 0, 1 and 5 m/s. In this instance, the MDV is less than 0.5 m/s for the hovering platform, which is comparable with a person walking. However, when the platform is moving at 50 m/s, the MDV is approximately 2 m/s. This is sufficient to detect slow-moving vehicles, but not persons.

These analytic results give the estimate of an ideal GMTI platform with stationary clutter. When there is significant movement in the clutter, or the clutter spatial variation is great between the spatial channels, the MDV will be affected. A second point is that the bandwidth is low on these GMTI measurements, so that the moving target cannot transition either range or Doppler cells during the CPI. One needs to understand the effects of wideband operation on the STAP performance.

As a comparison with the UHF STAP, an X-band GMTI performance on a moving platform is simulated given the parameters in Table 7.3. There are two cases of platform velocity at 50 and 125 m/s, as shown in Figure 7.17, corresponding to a small UAV and a business jet, respectively. The major differences is that at X-band,

Table 7.3 Simulation parameters for X-band STAP example

Frequency	9,500	MHz	Antenna	[1.0, 0.19]	Meters [az, el]
Bandwidth	15	MHz	N_{az}	64	
PRF	500	Hz	N_{el}	8	
Power	2,400	W	d_{az}	0.53	Wavelength
Pulse width	25	μs	d_{el}	0.74	Wavelength
Mp	64		Altitude	3	km
Duty factor	5.12	%	Velocity	50, 125	m/s
Target	–3	dBsm	R_s	25	km
Clutter	–8	dBsm/m^2	ICM	[1, 5]	m/s

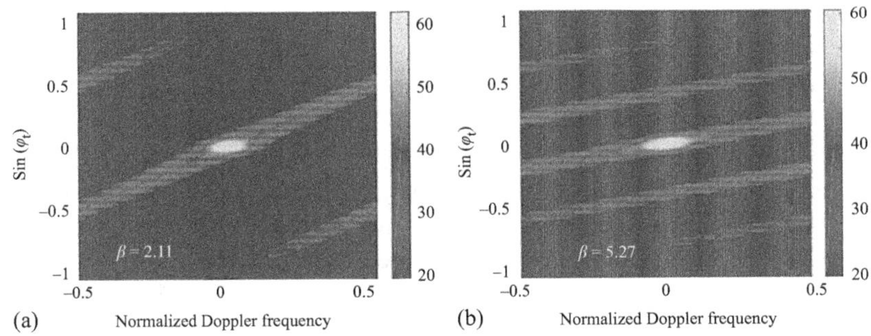

Figure 7.17 X-band clutter ridge. (a) 50 m/s and (b) 125 m/s

the antenna is 1-m long and the PRF is raised to 3,000 Hz. At the shorter wavelength and higher PRF, the clutter ridge β slope is significantly reduced from the UHF example. The clutter ridge at 50 m/s gives $\beta = 2.11$, and for the 125 m/s velocity, $\beta = 5.27$. These are large slopes on the clutter ridge, which are known to increase the eigenvalues due to Doppler wrap in STAP processing.

The SINR loss is similarly calculated from the simulated covariance matrix and for three ICM parameters, 0, 1 and 5 m/s, as shown in Figure 7.18. The MDV is evaluated at −10 dB SINR loss as well. With these inputs, the MDV at 0 m/s ICM is 0.47 m/s at 50 m/s platform velocity and 0.95 m/s at 125 m/s platform velocity. X-band has a significant advantage for GMTI target detection with moderate platform speeds. When compared with the 7.1-m antenna at UHF, the 1-m X-band antenna has comparable MDV with a moderate-speed platform velocity. However, the STAP processing will be corrupted if there is significant ICM. For the longer wavelength, the ICM is reduced based on the Billingsley clutter model versus frequency.

It is clearly seen that any appreciable wind velocity on the X-band clutter will decorrelate the signal during a CPI and negate the benefit of STAP. For efficient

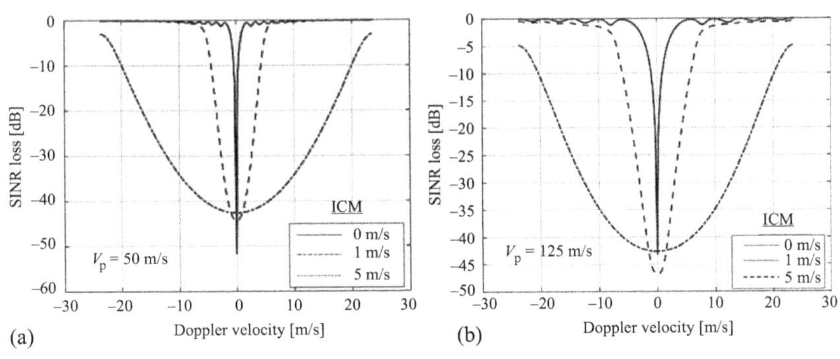

Figure 7.18 X-band SINR loss performance at 3000-Hz PRF. (a) 50-m/s velocity and (b) 125-m/s velocity

STAP processing, the PRF needs to be adjusted based on the platform velocity to manage the clutter ridge effects on the DOFs for cancelling the clutter.

In summary, STAP algorithms have been widely described in the published literature, but those works are primarily used in narrow-to-moderate bandwidth GMTI systems. However, with the need for high-resolution imaging for target characterization, UWB waveforms have increasingly been developed and published. In narrow-band STAP, when there are sufficient independent samples to train for adaptive weights, the detection of low-MDV targets can be carried out with minimal attenuation of the target SINR. For wideband STAP, a new approach is required to provide multiple narrow-band solutions, where the correlation of signals is maintained. The following section will introduce the channelized filtering needed for many UWB applications.

7.2.3 *Channelized filters*

UWB systems need to be processed in sub-bands to maintain coherence on object features, and adaptive processing for reduction of clutter and RF interference. Several developments have been reported giving evidence of the need for wideband radar operations, often as a result of innovations in the digital communications world. However, radar processing is significantly different than digital communications, because of the need to equalize channels over wide bandwidth, to provide match filtering to increase channel cancellation ratio and to achieve required interference mitigation metrics.

Digital filter banks are used in a large number of communications applications [16]. Figure 7.19 illustrates a typical M-channel decimated filter bank, where $H_k(z)$ and $F_k(z)$ are analysis and synthesis filters, respectively. The analysis filters $H_k(z)$ channelize the input signals into M sub-band signals, which are in turn decimated by M. At the receiving end, the M sub-band signals are decoded, interpolated and recombined using a set of synthesis filters $F_k(z)$.

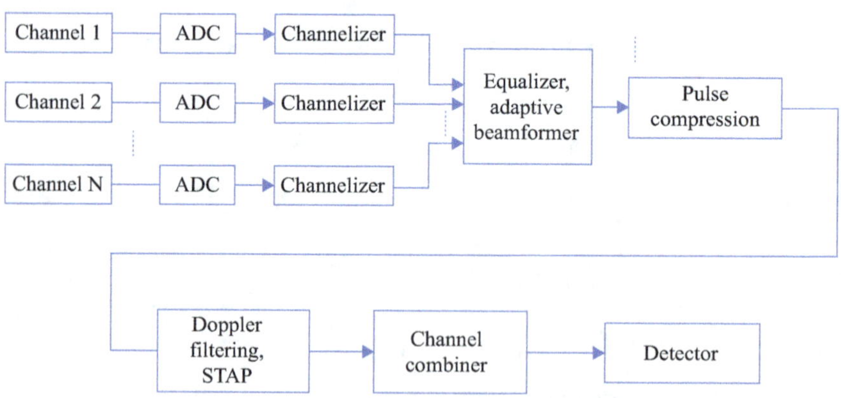

Figure 7.19 Multiple channel radar with sub-banding [17]

The impulse responses of the analysis and synthesis filters $h_k(n)$ and $f_k(n)$ are cosine-modulated version of the prototype filter $h(n)$ given by [16]

$$h_k(n) = 2h(n)\cos\left\{(2k+1)\frac{\pi}{2M}\left(n - \frac{N-1}{2}\right) + (-1)^k \frac{\pi}{4}\right\} \tag{7.30}$$

$$f_k(n) = 2h(n)\cos\left\{(2k+1)\frac{\pi}{2M}\left(n - \frac{N-1}{2}\right) - (-1)^k \frac{\pi}{4}\right\} \tag{7.31}$$

where $0 \le n \le N-1$ and $0 \le k \le M-1$ and N is the length of $h(n)$

A modern radar has multiple, digitized channels and needs sub-banding to handle the processing complexity to maintain coherence or channel balance. Figure 7.19 shows the generic flow of such a system. After digital conversion on each channel, it is beneficial to sub-band the signals for preparation of the next signal processing steps.

Each radar channel is then processed as shown in Figure 7.20. The channelizer decomposes an input signal $x(n)$ into N components $x_k(m)$. And within the channel combiner, the N components are recombined to form a recovered signal $\hat{x}(n)$. In the channelizer, a single filter is used within a bank of modulators to subdivide the signal spectrum into similar overlapping segments. The response of the FIR filter $h(n)$ is low pass and corresponding to the spectrum of the zeroth sub-band. The filter is tuned into a passband filterbank by the modulation of the sub-bands by a bank of FFT exponential sinusoids W_N^{-kn}, which are the trigonometric constant coefficients given by

$$W_{N_{SUB}} = e^{j\frac{2\pi}{N}} \quad \text{and } k = 0, \ldots, N_{SUB} - 1Nd \tag{7.32}$$

7.2.4 UWB STAP processing

With UWB waveforms, the target dispersion in both range and Doppler appears as loss across the Doppler clutter-free region [18]. This has a major impact on the detection probability as well as target location. It is necessary that alternative algorithms be employed to reduce the target signal loss. One approach is to use a filter bank such as polyphase filters shown in the previous section, to provide narrow sub-bands of the measurements in the spatial channels. Within the narrow sub-bands, it is sufficent to run STAP to cancel the clutter since the target correlation is high in the narrow-band situation. Once the clutter is removed from the channels, the UWB

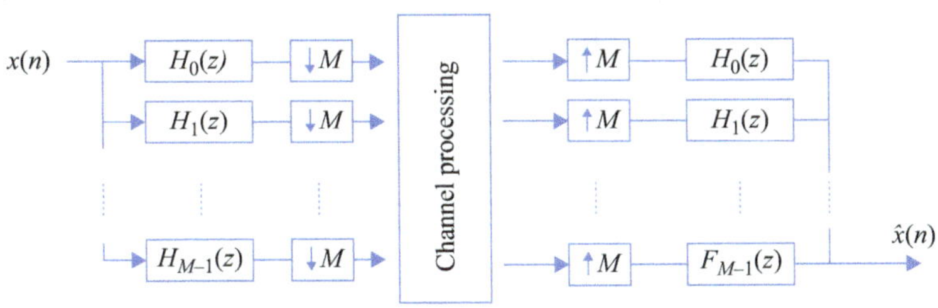

Figure 7.20 Individual channel sub-banding approach [17]

signal is recovered by providing an inverse transformation back into the UWB frequency span. The SINR loss is mitigated, and the target can be processed for detection and characterization [19].

Narrow-band STAP has efficient detection of slow-moving targets, given sufficient DOFs to separate the moving target from the background clutter. However, UWB STAP suffers significant losses due to decorrelation of the independent channels due to dispersion in time and Doppler. UWB STAP losses can be circumvented by decomposing the wideband signal into a bank of narrow sub-bands. Narrow-band STAP is applied to each narrow sub-bands, followed by recombining into wideband signal for targets after clutter suppression. Losses are greatly reduced, and MDV is maintained.

There are two concerns in applying conventional narrow-band STAP to a wideband system. First, the mismatch in the target steering vector is important because the target Doppler frequency is a function of wavelength; therefore, a constant Doppler frequency across the entire bandwidth is not valid in wideband STAP processing. Second, there is a dramatic increase of clutter rank in wideband systems because of the mismatch of antenna patterns and frequency responses between channels across the system bandwidth.

Using the sub-banding technique, the target Doppler frequency at the kth sub-band can be expressed as [19]

$$f_k^t = \frac{2\,v_t}{\lambda_k} \tag{7.33}$$

where λ_k is the wavelength of the central frequency of the kth sub-band. The spatial frequency for the kth sub-band (assuming an equivalent linear array) becomes

$$f_k^s = \frac{d}{\lambda_k}\cos(\theta_t)\sin(\phi_t) \tag{7.34}$$

Using Equations (7.33) and (7.34), the steering vector for the kth sub-band can be written as

$$s_k = s_t\left(f_k^t\right) \otimes s_s\left(f_k^s\right), \quad k = 0, \ldots, K-1 \tag{7.35}$$

where K quantifies the number of sub-bands and index 0 refers to the reference sub-band.

The assumption is made that the sub-banded signal is sufficiently narrow that a simple steering vector can be used for the target signal in each sub-band. Thus, the signal at the kth sub-band in the rth range cell can be expressed as

$$x_k(r) = \alpha\, s_k\left(f_k^t, f_k^s\right) + d_k \tag{7.36}$$

where d_k is the disturbance component. A transformation matrix $T\left(f_k^t, f_k^s; f_0^t, f_0^s\right)$ can be simply constructed in terms of steering vectors, given by

$$T\left(f_k^t, f_k^s; f_0^t, f_0^s\right) = \text{diag}\left\{\frac{s_0(1)}{s_k(1)}; \frac{s_0(2)}{s_k(2)}; \ldots; \frac{s_0(JN)}{s_k(JN)}\right\} \tag{7.37}$$

where J and N are the number of antenna elements and pulses, respectively, and $s_k(m)$ is the mth entry of the vector s_k.

Using (7.36), the combined signal at the reference sub-band can be expressed as

$$y(r) = \sum_{k=0}^{K-1} T\left(f_k^t, f_k^s; f_0^t, f_0^s\right), x_k(r) = y_s(r) + y_d(r) \tag{7.38}$$

where $y_s(r)$ and $y_d(r)$ are the target and disturbance components, respectively. The disturbance covariance matrix can be estimated by using Q secondary data from adjacent range gates as

$$\hat{R}_y = \frac{1}{Q} \sum_{q=1}^{Q} y_d(q)\, y_d^H(q) \tag{7.39}$$

The adaptive STAP weight vector is then estimated by

$$\hat{w} = \hat{R}_y^{-1} s_0\left(f_0^t, f_0^s\right) \tag{7.40}$$

It should be emphasized that the training of $\hat{R}_y$ should not use any range cells that contain the desired target. This should be the same training cells for each sub-band, as long as the targets are not spread over the range extent for the total bandwidth of the adaptive processing. Figure 7.21a illustrates the processing for wideband STAP

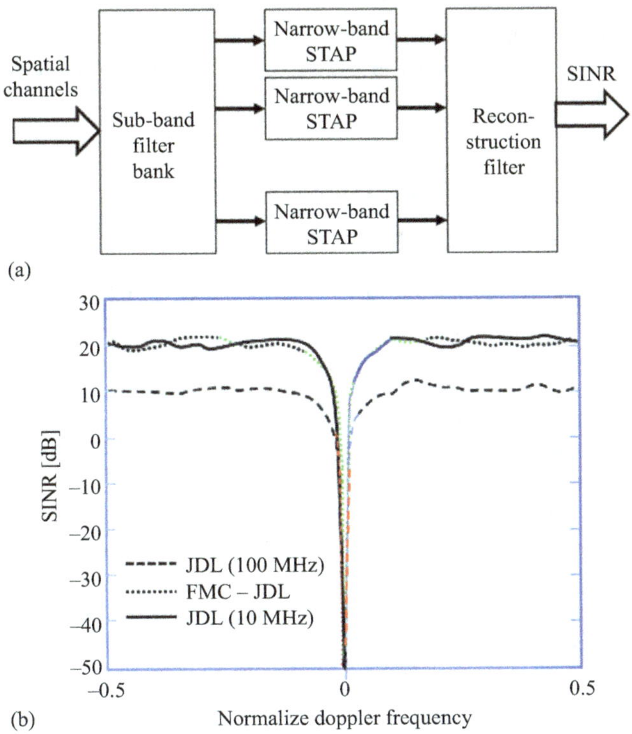

Figure 7.21 *UWB STAP processing [19]. (a) STAP sub-band processing and (b) SINR results*

to improve the low-loss adaptive processing in the narrow bands. After the narrow bands were run to detect the targets, the reconstruction filter was applied to the resulting narrow-band data cubes. Figure 7.21b compares the SINR loss for a narrow-band (10 MHz) STAP processing compared to the same data cube with 100 MHz bandwidth. The degradation of the SINR was approximatey 10 dB. When the UWB STAP was run on the 100-MHz bandwidth data cube, the loss was generally recoved across the Doppler band.

In summary, for narrow-band GMTI, it was observed that the radar platform velocity should be small compared to the radial velocity of any moving target. In addition, the range resolution needs to be sufficiently coarse so that the target cannot move appreciably in the CPI. These limitations are not a problem when detecting moving vehicles in the scene. However, for detecting persons, or characterizing vehicles based on size or shape, the resolution in both range and Doppler needs to be finer. This makes a significant impact on the design of the radar. A second conclusion is that STAP is directly affected by the motion of either the targets or internal motion of the clutter. Channel match is a major driver in determining the clutter cancellation of the system. With modest bandwidth and presence of ICM, a conservative limit on the MDV will be correlated with half of the mainbeam clutter cancellation.

It should be noted that there is an effective signal processing technique for detecting a moving target, while simultaneously obtaining the spatial resolution on the target for discrimination. That technique is ATI approach, which is covered in the following section.

7.3 Along-track interferometry

Generally, when the radar waveform has a coarse resolution, there is significant clutter in the same resolution cell as the tartet. This lowers the target to clutter plus noise ratio, reducing the detection probability. With higher resolution processing, the targets are resolved from distributed clutter. During the HOWLS collectons, the most effective approach make the spatial resolution smaller than the target to be detected. We have seen that SAR is the most effective approach for focusing surface scatterers. If two SAR images can be formed with a separation along the velocity vector of the platform, the stationary background clutter can be cancelled by coherent cross-correlation of the two images. However, moving targets will be discriminated due to their interferometric phase, but they are displaced in the ground plane as a result of the Doppler with respect to the SAR platforms. If these targets can be focused, the displacement can be measured and the subclutter visibility can be improved over the SAR image [9]. This is the essence of the ATI mode as previously introduced in Section 6.2.

This section now presents an ATI approach for detection and geolocation of UWB moving targets, along with the waveform limitations on moving target detection. UWB waveforms are required for fine range and cross-range resolution of targets. However, when there are moving targets present in an SAR image, they are displaced and defocused in the synthetic aperture image formation processing. These displacements need to be detected to restore location accuracy, and the targets need to be focused to restore the signal-to-clutter ratio (SCR).

ATI has been developed for detection of slow-moving targets by imaging from fast radar platforms. There are several issues that exist for detecting fast-moving targets in SAR images. First, the ATI waveform ambiguity needs to be evaluated for velocity estimation. Next, the design of multiple channel operation is required for forming multiple SAR images to obtain high correlation between channels in order to detect the moving targets. After the images have been formed and detected, defocused targets need to be focused for restoring the SCR.

This section illustrates the UWB ATI design based on detailed simulation of an UHF multichannel SAR system. The process is to form two SAR images from different channels of the radar. Figure 7.22 illustrates two subapertures moving together along the flight path. One aperture is used to measure the primary SAR image, and the second aperture forms a scene that is either delayed or precedes the main aperture. The process is to measure phase changes between the two apertures, which will indicate a movement during the SAR integration time. This ATI phase is directly related to the velocity of the image motion, as related to the separation between the pixels in the two images.

The detection problem is further complicated in forested environments by severe signal attenuation, volumetric scattering and ICM. As a result, an UWB waveform is required for adequate range resolution, and large real or synthetic apertures are required for adequate cross-range resolution. The speed of the platform and the size of the aperture dictate the achievable MDV performance [20]. Because of the non-linear nature of the dual SAR image collection and the interaction with forest volumetric clutter, a detailed modeling and simulation approach is required to quantify the system performance.

The two signals from the moving target are given by

$$S_1 = A_b e^{j\phi_{tgt}} e^{-j\frac{2\pi}{\lambda}\rho_1} \tag{7.41}$$

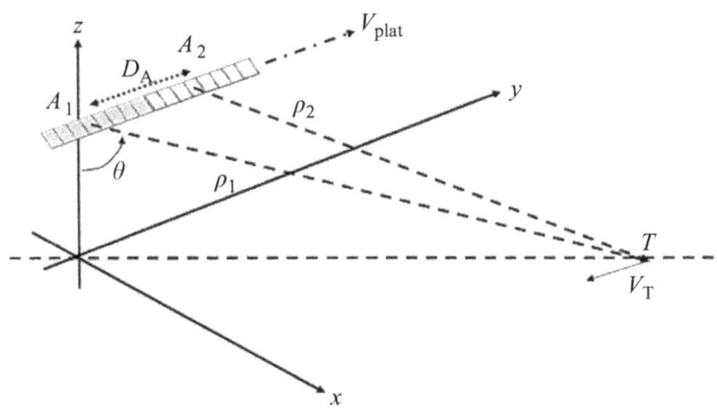

Figure 7.22 ATI signal collection [6]

$$S_2 = A_b e^{j\phi_{tgt}} e^{-j\frac{2\pi}{\lambda}\rho_1 + \rho_2} \tag{7.42}$$

The return S_1 is the direct return from the target to the transmitting phase center of aperture A_1. And the indirect return S_2 is return from the target to interferometric aperture A_2. The two phase centers are separated by a distance of D_A. In this example, the two antenna phase centers are aligned along the velocity vector V_{plat} and are separated by 2 m.

In this section, the design of an UHF ATI radar is described from several important aspects including the effect of SNR on target MDV and radial velocity estimation. Next, the analysis will explore the benefit of SAR image focusing algorithms to ATI detectability and velocity estimation. Finally, the effect of foliage attenuation and ICM on ATI ability to discriminate target from clutter and determine the target position and velocity.

7.3.1 ATI moving target analysis

Several efforts have been carried out to illustrate the ability of detecting slow-moving targets in ATI SAR images. Section 6.2 developed the theory of ATI. This section will extend this theory to a particular example to verify the applicability to UWB moving target indication. Figure 7.22 provides the geometry of an UHF ATI applications where the two antennas are aligned with the velocity vector. There are two phase centers of a multiple channel-phased array, where the transmit signal is from aperture A_1, and the received signals are received by both A_1 and A_2. Depending on the frequency and phase center characteristics, the antenna channel can be a single element or a subarray of several elements. As was shown in (6.43), the target amplitude and phase are based on the cross-correlation between each pixel in the two images, given by [20]

$$\gamma_{tgt} \exp\left(j\phi_{tgt}\right) = \frac{\langle S_1 S_2^* \rangle}{\sqrt{\langle S_1 S_1^* \rangle \langle S_2 S_2^* \rangle}} \tag{7.43}$$

The along-track separation between the two antenna phase centers provides interferometric phase due to the difference between the slant ranges along the two paths ρ_1 and ρ_2 due to the target motion v_T and the time difference Δ_t in the SAR signal propagation:

$$\begin{aligned}\Phi_{12} &= \phi_2 - \phi_1 \\ &= \frac{2\pi}{\lambda}\left[\rho_1(t) - \rho_2(t+\Delta t) =\right]\frac{2\pi}{\lambda}\upsilon_T \Delta t\end{aligned} \tag{7.44}$$

The use of ATI phase for target detection and discrimination from background clutter relies on a high target correlation γ_{tgt} and a target phase ϕ_{tgt} for detecting slow-moving targets with radial velocity:

$$v_T = \frac{\phi_{tgt}\,\lambda\, v_{plat}}{4\pi\, D_A} \tag{7.45}$$

A Doppler ambiguity in target phase state can be similarly defined, when the target phase is modulo 2π:

$$v_{\mathrm{amb}} = \frac{\lambda \, V_{\mathrm{plat}}}{2D_{\mathrm{A}}} \tag{7.46}$$

As a result, the wavelength and separation between the two antennas determines the range of radial velocities of ground targets that can be detected with a single pair of phase centers.

Figure 7.23 parametrically shows the MDV of two UHF arrays, as a function of the separation of the phase centers and additive noise σ [21]. Additive noise is a critical factor when there are separate channels that have different spatial or frequency characteristics, and these differences add noise to the signal processing chain. These values represent the 1σ MDV and do not include the effects of ICM or target decorrelation. They also assume that the SNR is noise-limited and not clutter-limited. Both these effects must be included in a more analytic approach. With a large separation between UHF radar element phase centers, the potential exists for detecting small radial velocities with a significant improvement in MDV over the conventional GMTI process.

The two most effective enhancements in ATI processing for GMTI are to mitigate SAR pixel migration (e.g., range-Doppler migration) and to accurately estimate radial velocity as they degrade detectability, velocity estimation and geolocation. These enhancements will be demonstrated on simulated I--Q data and their benefits will be quantified individually and together as a viable GMTI signal processing approach.

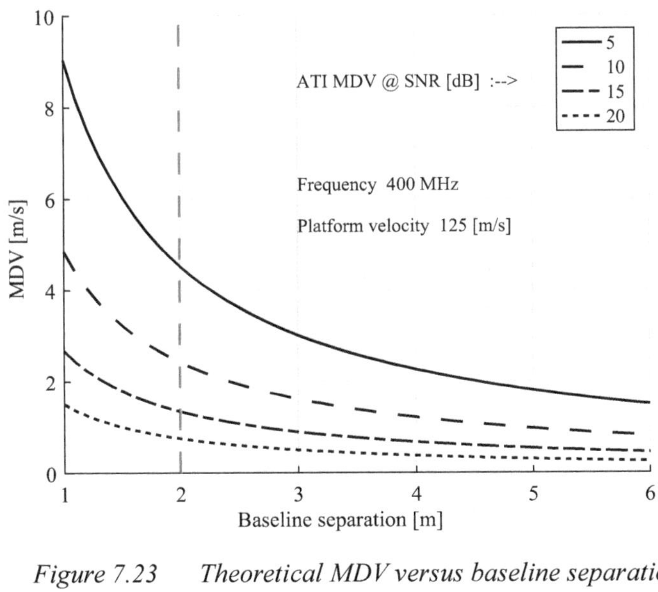

Figure 7.23 Theoretical MDV versus baseline separation [21]

Target coherency improvement

The most effective processing step is to integrate the multiple target detections that occur during the long SAR integration times. These multiple detections are localized to the centroid of the range-Doppler ATI "smear." In a normal SAR integration for fixed targets, the objective is to obtain the finest range and cross-range resolution for target detection or characterization. For an ATI mode to detect moving targets, the SAR integration time needs to be moderated such that the targets do not move significantly outside of a range or Doppler cell. Figure 7.24 illustrates the integration time versus cross-range resolution at various slant ranges to the image point. The UHF SAR is assessed at 400 MHz, and the X-band SAR is at a 9.5-GHz frequency. The X-band integration time provides a 2-m cross-range resolution with 2-s integration time at 30-km range. As a result, the moving targets will only move a few meters during a collection for ATI. In contrast with a UHF system, the SAR integration time is over 10 s for a cross-range resolution under 5 m at only 10-km range. This defocus of the moving targets due to the long SAR integration times is a significant challenge for ATI operation at long wavelengths. And the coherence of target features needs to be considered even at microwave frequencies when the motion of the vehicle can defocus the image. A focus-before-detect approach to the UWB surveillance image collection needs to be considered [8].

The first step in improved target detected due to target motion in the range and Doppler bins is using a box-car filter in the region around each segment of the ATI image. This has been shown to increase the SCR by approximately 5 dB, depending on the size of the box-car filter [9]. This gain has been estimated assuming nondestructive phase, for a UHF system as a function of cross-range resolution as illustrated in Figure 7.25. Once the detection region is determined and a target radial velocity is estimated, the pixels are coherently added using an autofocusing algorithm yielding even greater signal gain [22]. These processes are given below.

Radial velocity estimation

Target smearing and low SNR plague ATI radial velocity estimation. The inherent sensitivity of ATI to phase makes it a powerful slow-moving target detector; however, this same sensitivity creates problems for estimating the radial velocity of the ground mover and image focusing techniques. ATI's phase sensitivity is illustrated as a function of SCR in Figure 7.26, at different SCR values. When the SCR is greater than 20 dB, the interferometric phase error is under 1%. And with an SCR approaching 5 dB, the phase error is increased, making it more difficult to reposition the target in the SAR image.

7.3.2 ATI simulation

In order to evaluate the techniques for an ATI mode, a simulation scenario was used to determine the separability of point scatters in the scene, as well as the detection of moving targets. The scenario was modeled using Rome Laboratory Space-Time Adaptive Processing (RLSTAP), a detailed pulse-by-pulse simulation of the

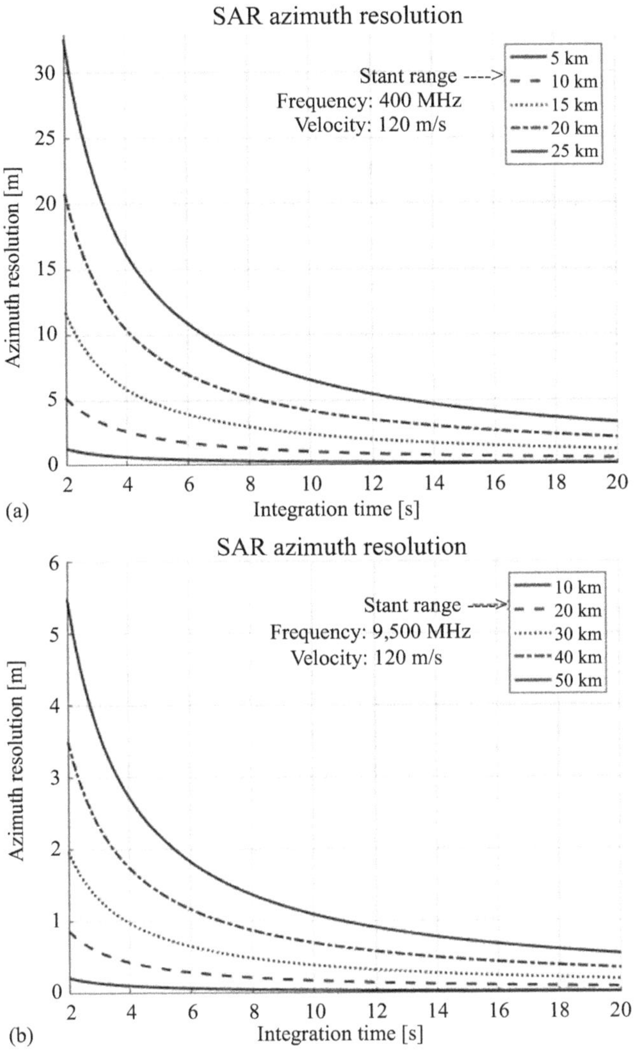

Figure 7.24 *Cross-range resolution as a function of integration time. (a) UHF SAR image and (b) X-band SAR image*

transmitter and receiver characteristics, clutter and propagation characteristics and target model motion. The following characteristics of the modeling are needed for UWB ATI evaluation [19]. First, the modeling of terrain type, terrain height and foliage types provides the variation of clutter backscatter over the change in clutter type and terrain slope. In a multiple channel system, the calculation of the position, orientation and velocity of distributing transmit and receive apertures is necessary, along with independent location and motion during the coherent integration times. As for the ground moving targets, providing multiple ground target motion, with

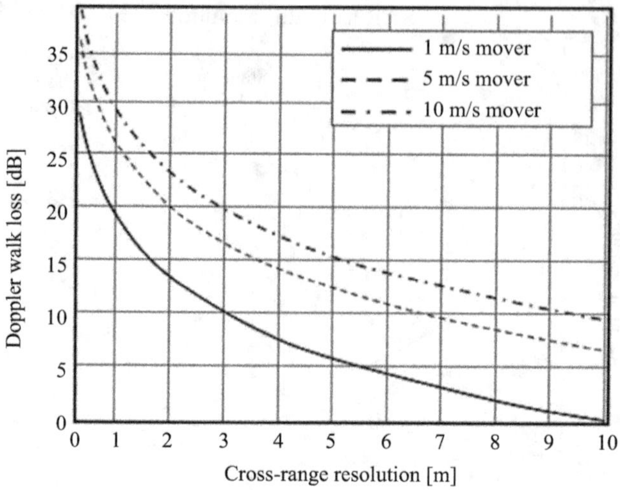

Figure 7.25 ATI "smear" loss as a function of cross-range resolution [21]

target model details provide real-world conditions for decorrelation studies. It is important to have the ground targets backscatter models that depend on not only the ground plane variation, but also the effect of multiple scattering centers that scintillate in the slant plane look at the target. Finally, the model should add noise and time errors to the signal, along with statistical ICM.

Because the modeling tool represents the position and velocity at each pulse, effects of navigation errors, range and Doppler walk, and waveform and beamshape contributions can be evaluated [23]. Figure 7.27 shows the RLSTAP data flow

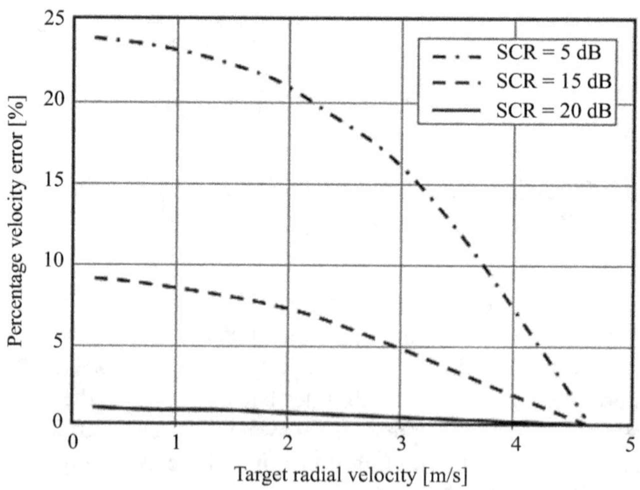

Figure 7.26 Percentage error in ATI radial velocity estimation versus SCR [21]

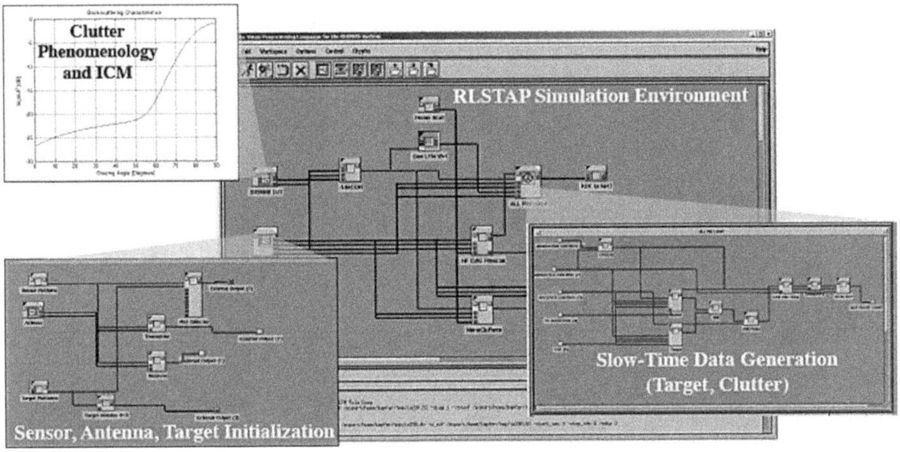

Figure 7.27 RLSTAP line-up for simulating ATI of multiple moving targets [23]

method for independent transmit and receive operation for parametric evaluation of FOPEN ATI operation.

The scenario being considered for UWB ATI evaluation is an SAR and GMTI collection from a platform flying at 120 m/s, from a 7-km altitude, and a slant range of 8.8 km to the scene. The simulation area is given in Figure 7.28 which shows the

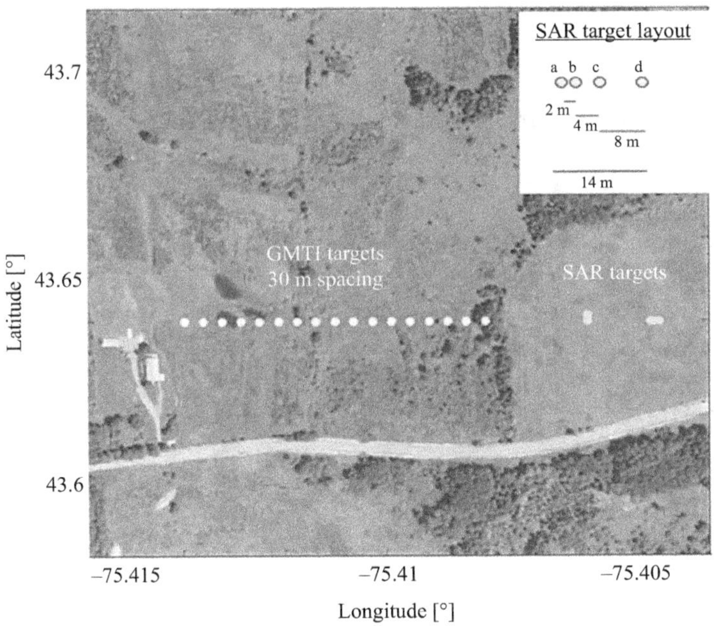

Figure 7.28 Scene for analyzing fixed and moving target detection [21]

Table 7.4 Simulation parameters for UHF ATI analysis [21]

Parameter	Value
Frequency	400 MHz
Bandwidth	160 MHz
Transmitter height	7 km
Stant range	8.8 km
Velocity	120 m/s
SAR length	1,200 m
ATI separation	2 m

scenario used in this analysis, which consists of 21 moving targets and 8 stationary targets on a 1 by 1.7 km region composed of forest and farmland. Table 7.4 summarizes the important scenario parameters used in the evaluation process. The scenario presents a nominal incidence angle of 51°. The separation between the two ATI phase centers is 2 m. And the SAR imaging path was offset from the symmetric line, with a squint angle of 3°.

The problem being considered is a 6-m UHF aperture with two phase centers separated by 2 m, shown in Figure 7.29. The antenna architecture consists of 6-m aperture with 16 elements spaced $\lambda/2$. There are two subarrays consisting of six horizontal and three elevation elements yielding a 2 m by 1 m uniformly weighted area. The fore and aft subarrays used for ATI present a 2-m baseline. The antenna pattern for various combinations of subarray phase centers is given in Figure 7.30. The subarray pattern for a single SAR image is shown as the blue line. If the two apertures were coherently combined, the array pattern is shown as the green line, where the uniform distribution gives a $(\sin x)/x$ pattern with nulls at half of the azimuth interval of the subarray. However, if the two apertures are coherently added, the sidelobe structure of the array will be modulated by the subarray pattern. These factors will provide both an amplitude and phase effect on the SAR images and any moving targets.

There are three regimes in processing moving targets in a surveillance radar. When the targets have a Doppler that is outside of the mainbeam clutter null, the processing is called exo-clutter. And when the targets fall (typically) within the 10 dB gain of the main beam, it is considered to be endo-clutter. In between these two regimes, the target Doppler is within the skirt of the mainbeam clutter, as illustrated in Figure 7.31. In order to examine these regimes, there are twenty-one 10 dBsm

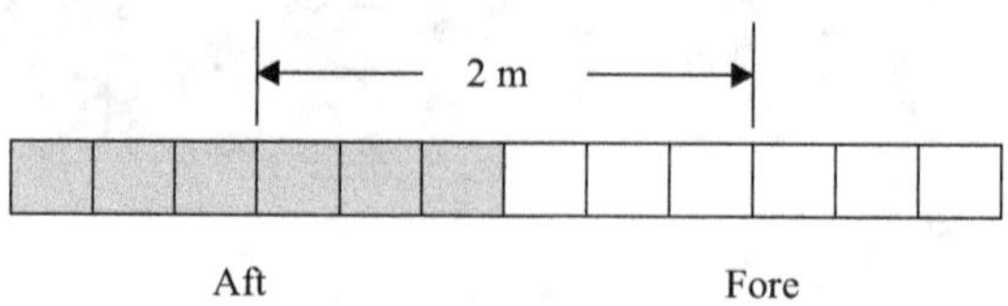

Figure 7.29 Antenna architecture

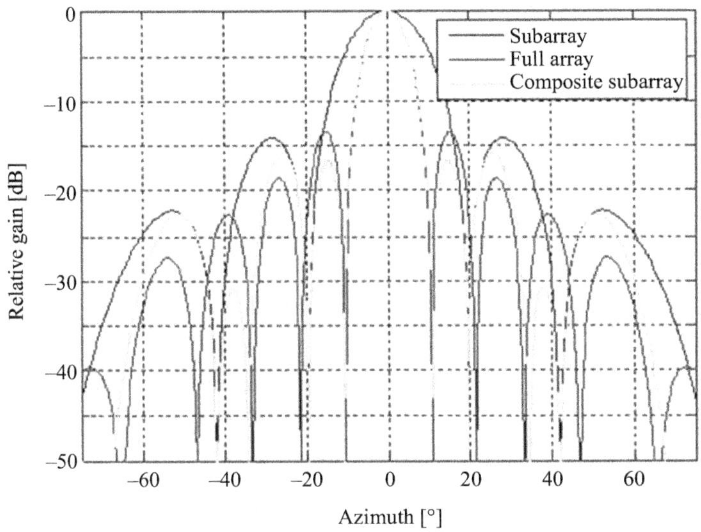

Figure 7.30 SAR/ATI antenna patterns

targets included in the scene to test the ATI displacement of the moving targets and verify the range and cross-range resolution of the stationary targets within the clutter Doppler spectrum. The 20 moving targets move radially toward and away from the RADAR at ground speeds of [±0.25 ±0.5 ±0.75 ±1 ±2 ±4 ±6 ±8 ±12 ±16] m/s, with one stationary target in the middle of the pattern. The target velocities were selected to provide both endo- and exo-clutter moving targets. Each stationary target is made up of four scatterers with separations of 2, 4 and 8 m between them. With these fiducials in the scenario, the simulation was designed to test SAR resolution, moving target P_d and MDV, and geolocation accuracy of the moving targets.

The flight path was not centered on the broadside, as shown in Figure 7.32, but with a squint angle of 32.5°. This causes a radial velocity of the image focal point with respect to the flight profile that changes during the collection. Figure 7.33a plots the variation of the slant range to the transmit phase center, and Figure 7.33b

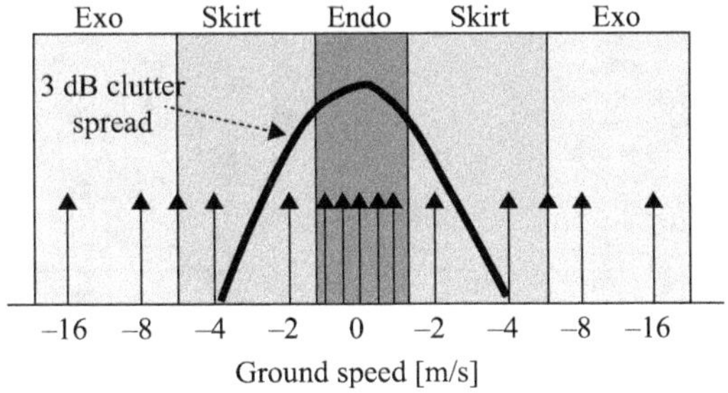

Figure 7.31 Moving target ground speed variation

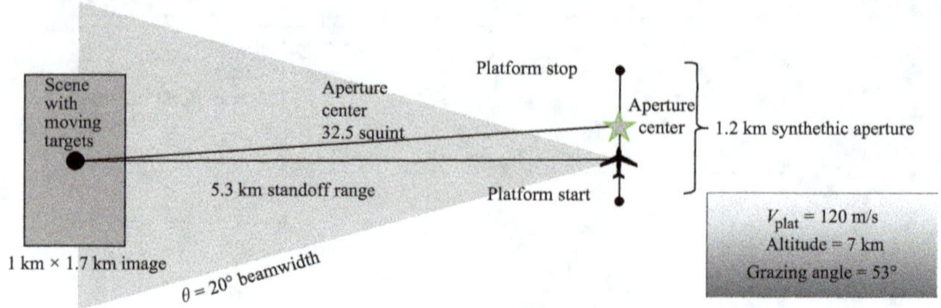

Figure 7.32 SAR collection geometry [21]

shows the comparable radial velocity between the antenna and the scene center. As a result of these changes in geometry during SAR image formation, the squint angle must be compensated during the ISAR image collection.

This 160-MHz UWB waveform was integrated for 10 seconds to achieve a cross-range resolution of 5.4 m. A 2.0-m SAR image was developed as shown in Figure 7.34, utilizing back-projection processing of the scenario in Figure 7.32. Note that the moving targets are displaced in cross-range proportional to the radial velocity.

The choice of speeds enabled assessment of target detection at exo-clutter with higher radial velocity and endo-clutter with slower radial velocity. Since they are closely spaced with the same scattering cross-section, the effect of ATI versus STAP detection can be evaluated. In addition to the moving targets, there are two groups of four stationary SAR targets in range and cross-range orientations to assess the SAR resolution. These stationary targets are separated by [1, 2, 4] m. The expansion of the SAR image around the stationary targets is shown in Figure 7.35, where the four "reflectors" indicate the image resolution was approximately 2 m in both range and cross-range.

The target location in the image magnitude is well localized. However, the phase is smeared due to the physical motion of the target during the integration time.

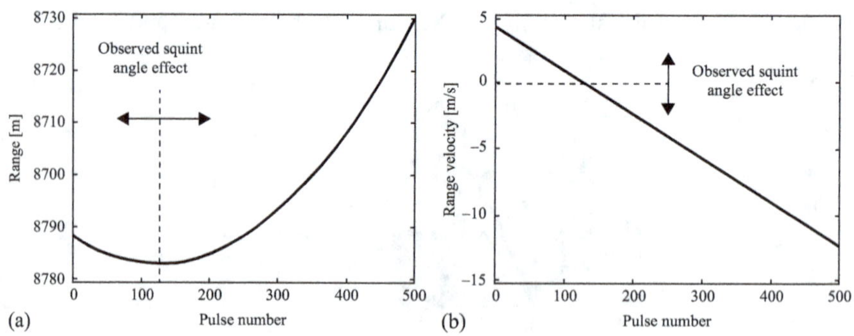

Figure 7.33 Range variation during SAR integration [21]. (a) Slant range during SAR and (b) radial velocity to scene center

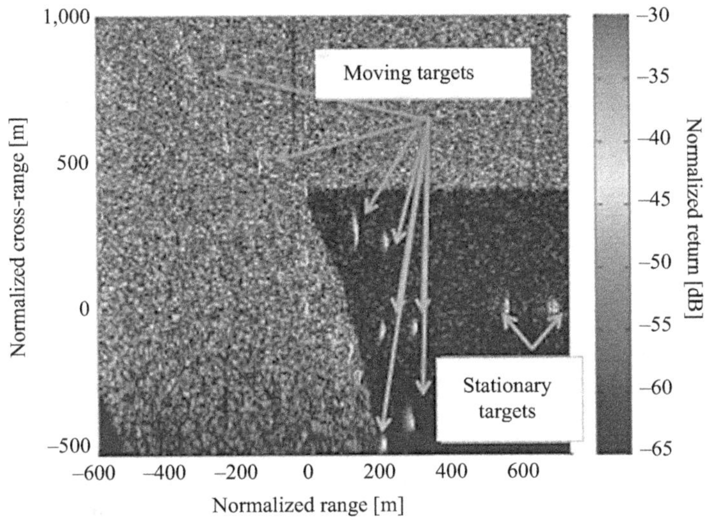

Figure 7.34 SAR images of fixed and moving targets [21]

Accurate location of the targets depends on the ability to measure the target velocity and compensate for the displacement. A multilook ATI image was formed by applying a 3 by 3 moving average window to the SAR image. The multilook process is an effective way to reduce unwanted speckle in SAR images.

The ATI processing consists of forming two SAR images as shown in Figure 7.36, but with the image focus at the receiver phase centers separated by 2 m. The ATI magnitude and phase are calculated from (7.32) and are shown in Figure 7.37a and b, respectively. A constant false alarm ratio (CFAR) detector has been used over the SAR images, and the results are indicated by "X" in Figure 7.37a. Since this analysis was only carried out in noise, there were few false alarms. With a 2-m baseline separation, Doppler ambiguities are expected to occur at ±22.4 m/s, which will impact high-speed target estimation. But at the maximum radial velocity of 16 m/s, no Doppler wrapping in the image is experienced.

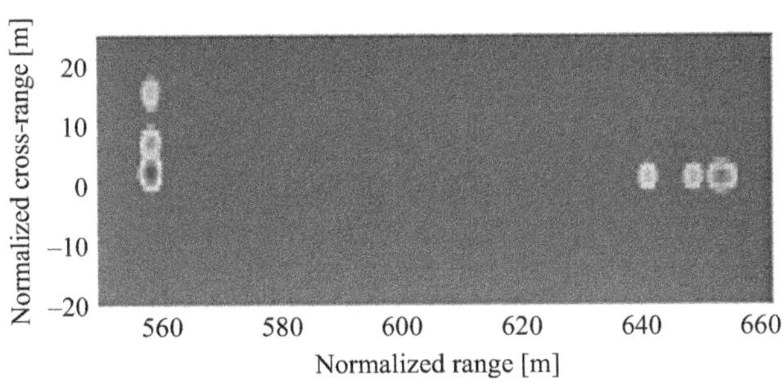

Figure 7.35 SAR image of fixed target for range and cross-range resolution [21]

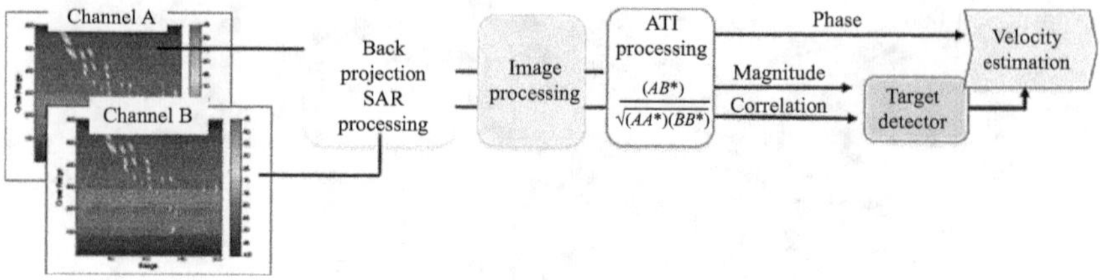

Figure 7.36 ATI processing flow for moving target detection [21]

Within the two SAR images, a cell-averaged CFAR was used to estimate the location of targets, albeit they were greatly defocused. However, the signal-to-clutter level of each target is degraded due to the motion during the SAR images. As a result, the ATI phase will be corrupted, and the geolocation accuracy will be seriously degraded for the faster targets. These will be shown below.

Table 7.5 summarizes the data for each of the 21 targets in the ATI evaluation array. It provides the slant range from the center of the SAR array, and the instrumented ground speed of each target. The impact of a squint in the SAR array is evident, since the center (stationary) target has a +4.13 m/s relative ground speed due to the SAR platform motion. For each of the detected targets, the ATI phase was estimated from the data.

It is clear that an improvement in the SCR is needed to detect and accurately locate the targets. The SAR images for each channel are generated using back-projection SAR algorithm, with the motion compensation from the phase center of each antenna. Image processing with several pixels integrated around a resolution cell are used to reduce speckle. The ATI processing is carried out on spatially the two separated radar images, with the transmit slant range from one aperture to each pixel in the image, and the receive slant range to the two ATI antennas. Next, a two-dimensional cell-averaging CFAR is applied to ATI magnitude for potential fixed or moving target detection. From the CFAR output, the detection cells are defined for localizing the ATI phase used to estimate target velocity and reposition the moving targets into the estimated geographic location.

The SAR along-track offset is due to the relative motion of the platform v_{Plat} and the targets v_{T}, and is given by:

$$\text{Along-track offset} = R_s \frac{v_{\text{T}}}{v_{\text{Plat}}} \tag{7.47}$$

It can be seen in Figure 7.37 that there is an asymmetry between the targets closer to the radar and those farther away. This effect is explained by (7.49) due to two effects. The radial velocity of the targets with respect to the platform is additive for targets moving away from the platform, and subtractive for targets moving toward. And the magnitude of the offset depends on the slant range to the target, which varies by 600 m in the ground plane. Furthermore, the incidence angle between the radar and the ground plane is 52°, which further dilutes the ground range geometry. The SAR offset column in Table 7.5 agrees with this theory. The ATI phases from Figure 7.37b are plotted versus the radial velocity with respect to the platform in

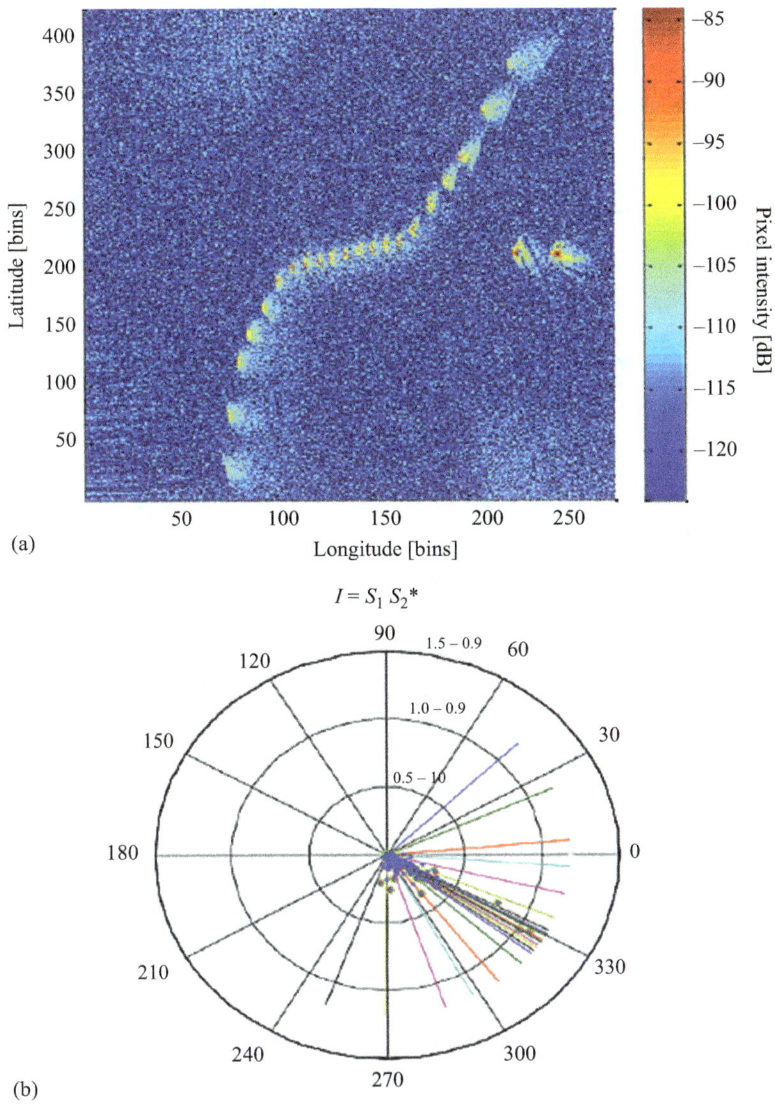

Figure 7.37 Results of ATI processing on 21 moving targets [21]. (a) ATI magnitude with detections and (b) ATI phase

Figure 7.38 and show the same offset in ATI phase angle as a function of the target ground velocities.

The loss of SNR due to target "smearing" can be evaluated qualitatively in Figure 7.39, where the signal level corresponds to the radial velocity shown in Figure 7.38. All of the slow targets clustered at 4.1 m/s have little or no SNR loss due to limited range or Doppler walk of the returns. And the movers over 10 m/s

Table 7.5 Summary of ATI target measurements [21]

Range (km)	Gound speed (m/s)	Radial velocity (m/s)	ATI phase (deg)	SNR (dB)
8.59	−16	−5.06	−44.08	26.36
8.61	−12	−2.77	−24.79	28.46
8.63	−8	−0.47	−5.5	27.45
8.64	−6	0.67	4.14	24.79
8.66	−4	1.83	13.79	31.62
8.68	−2	2.99	23.43	30.22
8.70	−1	3.57	28.25	29.45
8.71	−0.75	3.71	29.46	25.96
8.73	−0.5	3.85	30.66	29.20
8.75	−0.25	3.99	31.87	31.00
8.77	0	4.13	33.07	31.51
8.79	0.25	4.28	34.28	32.29
8.80	0.5	4.42	35.49	32.72
8.82	0.75	4.56	36.69	32.32
8.84	1	4.71	37.9	30.37
8.86	2	5.32	42.71	31.93
8.88	4	6.54	52.35	23.86
8.90	6	7.78	61.98	23.65
8.91	8	9.02	71.61	24.98
8.93	12	11.51	90.87	19.13
8.95	16	14.02	110.13	16.89

relative to the platform suffer over 15-dB SNR loss. This partially explains the loss of detections in Figure 7.37. If one expands the fast-moving targets, shown in Figure 7.39, it can be seen that the number of distinct detections in the ATI plot are diminished. This directly affects the ability of a simple CFAR to detect the targets.

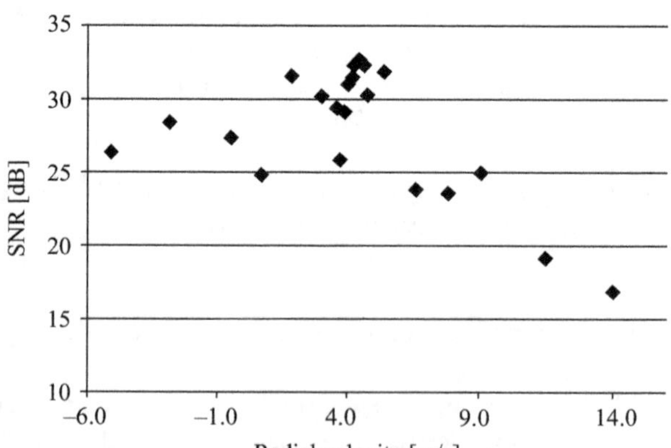

Figure 7.38 ATI phase versus radial velocity [21]

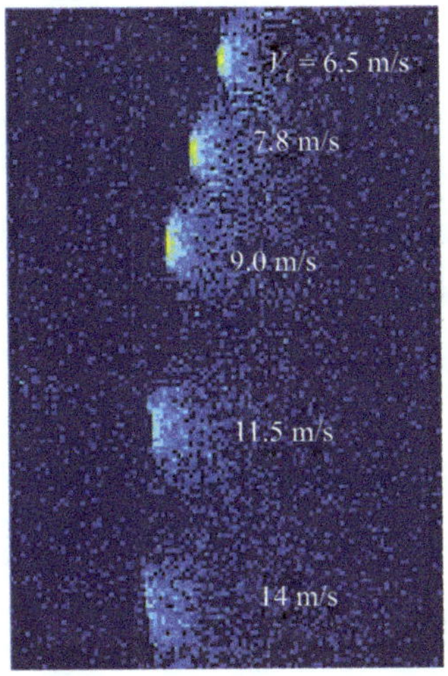

Figure 7.39 Poor image performance on fast-moving targets [21]

Research is continuing on coherent integration of the area returns (in the presence of clutter) to reduce this loss of signal [22].

The faster the moving target, the greater the defocus of the point spread performance in SAR images. At UHF, the integration times for cross-range resolution requires several seconds. During that time, the targets will walk in both range and Doppler, and the return will be smeared in the clutter return. This represents a loss of SCR, directly affecting the detection probability.

It is necessary that these defocused targets be refocused for detection and measurement of the velocity. Once the radial velocity has been accurately estimated, the targets can then be replaced on the SAR image. It is important that the ATI phase be accurately estimated for this geolocation process. Chen [9] has derived an error model for SCR in the ATI processing of targets. This can be used for estimating the velocity error as a function of the target radial velocity.

7.3.3 Focus before detection

Two autofocus techniques were applied to each of the movers in the image. A phase gradient autofocus provided modest improvement in the target SNR [23]. However, the moving baseline SAR approached the theoretical limit of the 10 dBsm target across the velocity band [24]. An improvement in the SNR and geolocation accuracy will be presented below.

The first step is to define a method for focusing the targets prior to the ATI processing and velocity estimation. A process of imaging moving targets in SAR causes a significant defocus due to range and Doppler walk. This target motion during long inegration times blurs the image and causes severe reduction of target detections with the long image collection times. Before any image autofocus, there is a reduction in the SCR for the individual targets, along with affecting the ATI phase. The

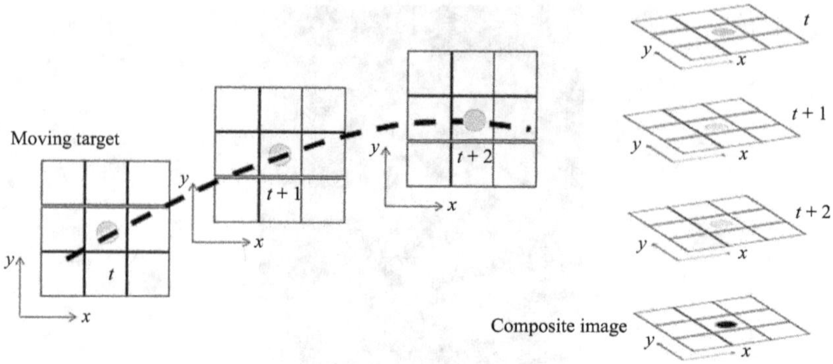

Figure 7.40 Moving back-projection SAR for target focus [21]

fundamental issue is long integration times required for cross-range resolution causing severe Doppler walk, which is increased with higher target velocity.

The moving targets will have a defocus effect depending on the radial velocity to the SAR platform. If the targets are not focused before detection, the SCR will directly affect the detection probability. A moving back-projection SAR autofocus algorithm was chosen to estimate the center of the displaced chip, as illustrated in Figure 7.40. During the collection time, the SAR chip is sequentially motion compensated to account for the range and Doppler motion, resulting in an improved focus of target scatterers.

One approach is to use spatial–temporal SAR focus point to enhance pulse-to-pulse alignment on moving target. This algorithm incorporates a prominent point tracker into detection process to estimate center SAR image for the "track before detect." The algorithm uses track prediction state to update SAR focus point temporally. This causes a challenging tracking problem due to along-track offset of moving targets. For this experiment, predicted states were assumed to be known perfectly providing an upper bound of performance.

The focus is accomplished in the frequency domain to emphasize correlation autofocus. Multiple short-time images are generated around the region of the target locations. A quad tree decomposition is used to determine the likely migration of the targets in range and Doppler. This is efficiently carried out in the frequency domain, as summarized in Figure 7.41. By using the back-projection superposition property, the spatial correction is done in the frequency domain, which maximizes the cross-correlation of the two channels. The early processing generates multiple images along short synthetic apertures (lower resolution) "Quick SAR." These subapertures

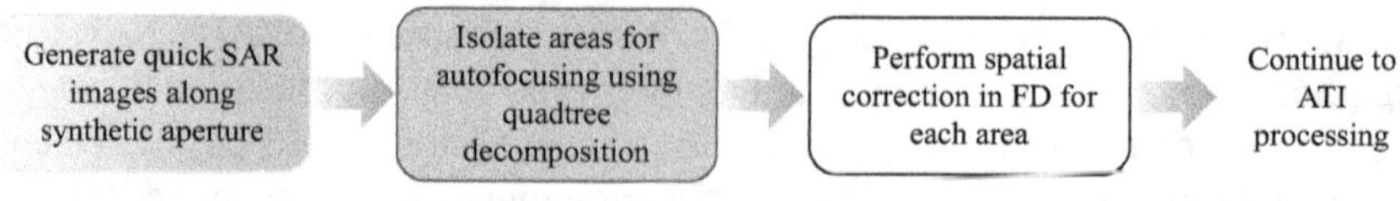

Figure 7.41 Moving back-projection focus of moving targets [21]

Figure 7.42 Quadtree on full image [21]

exploit the back-projection's superposition property, but introduces grating lobes. The algorithm utilizes a Quadtree decomposition to isolate areas for autofocusing. As illustrated in Figure 7.42, each region of the SAR image can be evaluated for regions that contain artifacts of moving target impact on the background clutter. In order to isolate these areas, a course threshold detector is applied to the individual SAR images, which greatly reduces the number of autofocusing operations. When there is a smaller region detected with a potential moving target, it can be assumed that the target is localized by less defocus of the target.

Next, the processor performs spatial correction in frequency domain using a computationally efficient approach, given by the transformation [20]:

$$f(x - x_0,\, y - y_0) \leftrightarrow F(u, v)\ \mathrm{e}^{-\mathrm{j}2\pi(ux_o/M+vy_0/N)} \tag{7.48}$$

To maximizing cross-correlation, the following spatial filtering is employed:

$$C = \frac{\sum_{x,y}\left(I_1(x,y) - \bar{I}_1(x,y)\right)\left(I_2(T(x,y)) - \bar{I}_2(T(x,y))\right)}{\sqrt{\sum_{x,y}\left(I_1(x,y) - \bar{I}_1(x,y)\right)^2 \sum_{x,y}\left(I_2(T(x,y)) - \bar{I}_2(T(x,y))\right)^2}} \tag{7.49}$$

With this approach, sub-pixel accuracy is obtained by pixel alignment with only a translation movement.

The subaperture processing utilizes a 0.5-s range-Doppler chip on the region of the ground mover, as shown in Figure 7.43. There is an improvement in the target SNR by this short integration. This was sufficient to detect the smeared target and use the ATI phase to estimate the cross-range migration. Sufficient resolution improvement and pixel structure were obtained for correlation. Using the quick SAR, the pixel gain was improved by approximately 2 dB. Furthermore, the improved resolution of the target location was improved from 100-m spread to under 10 m, a 10:1 improvement. This provided a geolocation accuracy by 3:1, as indicated in Figure 7.44a. However, the short quick SAR images will create grating lobes, as shown in Figure 7.44b, when the full integration is carried out.

Figure 7.45 shows the results of the two autofocus techniques on the ATI images. The unfocused detection SNR for each of the moving targets is given by the diamond data points. It clearly illustrates the fall off of target SNR as the moving

(a) (b)

Figure 7.43 Quick SAR image with autofocus improvement. (a) Unfocused target and (b) autofocused target [21]

targets experience the range walk during the full SAR integration interval. The phase gradient autofocus provides a moderate improvement in SNR on the fast movers given as the triangular points [25]. There is a wide variation in the SNR depending on the efficiency of the autofocus technique applied. It is surmised that the quadtree segmentation could be improved. Finally, the square data points give the results of the moving map autofocus. A significant improvement in the moving target SNR and hence the geolocation accuracy was achieved [21].

7.3.4 Receiver-operating characteristics

A final and more substantive measurement of the need for focus before detection is illustrated in a receiver-operating characteristics, as shown in Figure 7.46. The detection results for this ATI analysis are shown for the unfocused targets in SAR images, use of phase gradient autofocus and the moving baseline SAR focus of the targets. The solid lines are based on the simulation without ICM, and the dotted lines use the Billingsley model with a wind speed of 10 mph and a shape factor of

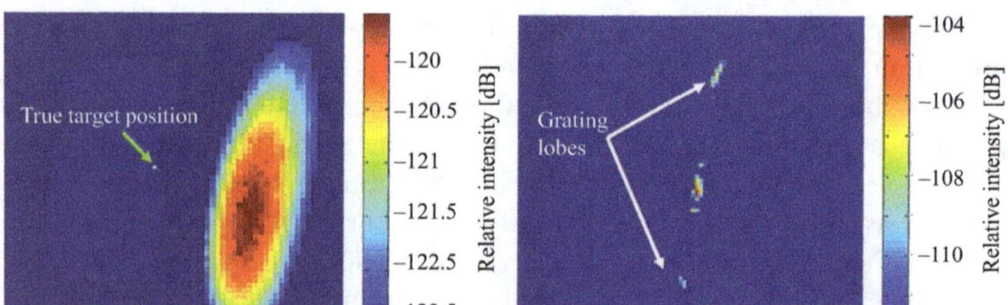

(a) (b)

Figure 7.44 Quick SAR image and focus of fast-moving target [21]. (a) Geolocation of target 34 m and (b) sub-image processing with grating lobes

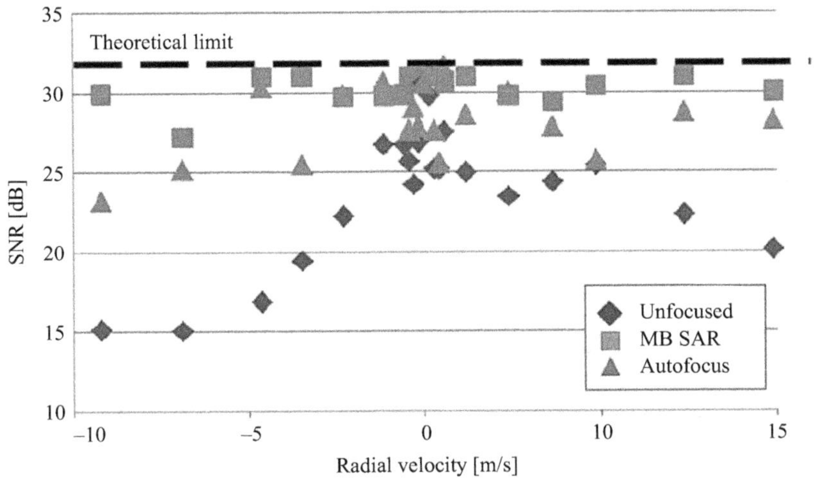

Figure 7.45 Improvement in SNR due to focus before ATI [21]

8. When the detection curves for sationary clutter is compared to moving clutter, a clear indication of the impact of ICM on moving target detection on the detection probability is seen, especially in SAR images. This verifies that focusing moving targets can improve the ATI detection performance [20].

Intuitive results, focusing on moving targets, improve coherent integration in back-projection and improve detection performance. However, there is a clear need for improvements in these algorithms for focus before detect. An important evolving technique is the use of ME autofocus on SAR images. Similar improvements

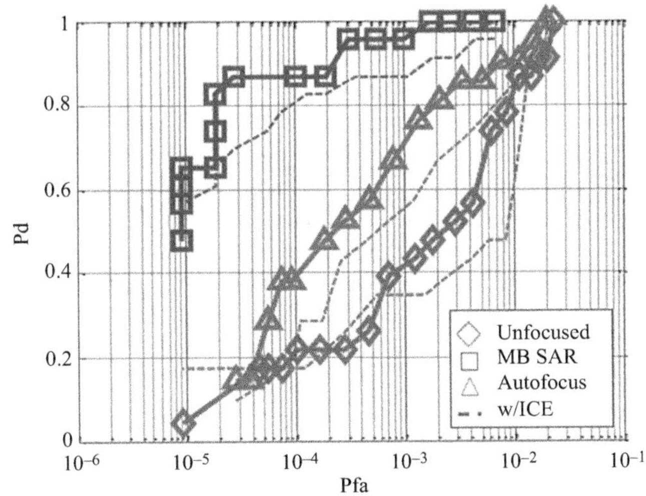

Figure 7.46 Receiver characteristics of moving target detection [20]

in SNR have been obtained, but with a strong maintenance of the ATI phase. This effort will be beneficial when it has been validated with multichannel radar data collections [24].

7.3.5 Minimum entropy focus

Entropy is a measure of the random spatial distribution of the objects in an image. For objects that have highly focused scattering centers, the entropy is minimized. This minimum entropy (ME) feature has been used for many applications, most recently in the quest for focus before detection of moving targets in SAR images [7, 26].

ME has been used to align range profiles and adjust phase functions. When there are both range and Doppler migration of the target, a two-dimensional algorithm is used. The two-dimensional entropy function $H(S_{mn})$ is defined as [27]

$$H\left(S_{mn}\right) = -\sum_{m,n} p\left(k,m\right)\ \log\left(p\left(k,m\right)\right) \tag{7.50}$$

where the probability distribution function is

$$p\left(m,n\right) = \frac{S_{mn}}{\sum\limits_{mn} S_{mn}} \tag{7.51}$$

The ME method is carried out iteratively to find $\hat{m}$, $\hat{n}$ that satisfy

$$\left(\hat{m},\hat{n}\right) = \arg\min\left\{H\left(S_{mn}\right)\right\} \tag{7.52}$$

The algorithm is used to calculate the ME conditions and is applied to both channels of the ATI images, as shown in Figure 7.36. The MATLAB code for ME was modified from [9] to apply to the two SAR images.

Figure 7.47 shows an area of mixed open land and sparse foliage [28]. Included in the scene are 21 targets moving between ±6 m/s in three directions, including within a sparse foliage. The objective of this analysis was to determine the importance of ME focusing of the targets for improvement in the ATI detection of moving targets. The process being used for applying ME to the two SAR channels was to select an area around a target (obtained by initial CFAR processing). The target-to-clutter ratio (TCR) for each target after SAR imaging is given in Figure 7.48, where the seven targets in the forested area are greatly reduced. The peak target return is indicated along with the mean target where there is range and Doppler migration.

The target area used for this effort was a 32 by 64 pixel chip centered on the centroid of individual target images. The steps in the ME processing include calculating entropy from (7.53) and estimating the actual time delay to each pixel in the chip from the platform kinematics. The next step is to center the target in the middle of the chip, similar to the quadtree process in Figure 7.42. The chips for each target are used to calculate the entropy along the range dimension. From this point, the minimization is carried out for range of velocity from –6 to 6 m/s, and acceleration from 1 to 4 m/s^2. A phase correction to the chips is then provided for each pixel in the image, where the phase correction aligns the pixels and focuses the images. Finally,

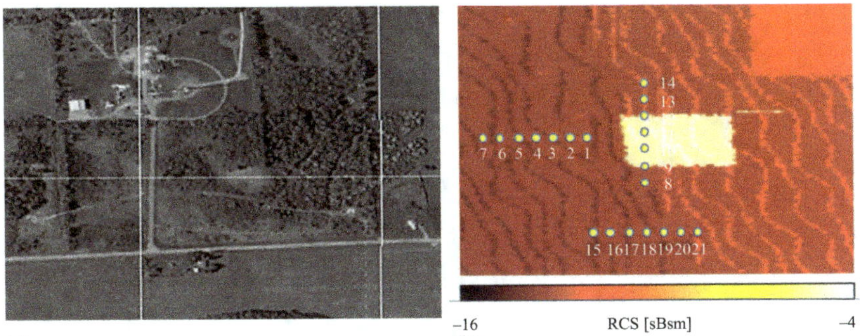

(a) (b)

Figure 7.47 *Simulation of ATI in sparse foliage: (a) Map image and (b) LULC and target scattering*

the entropy minimization for image phase adjustment from (7.54) is iteratively carried out before recreating each SAR image after phase adjustment. Figure 7.49 shows the SAR image after back-projection with the effects of both target motion and foliage attenuation on the peaks of the targets. These image chips for both channels were then placed into the SAR image frames. Figure 7.49 illustrates the images after a CFAR threshold, for each of the two channels. This was then processed with the ATI algorithms in (7.41). It should be noted that the residual clutter in the image is not shown. Figure 7.48 shows the TCR improvement after ME processing. There is approximately a 25 dB improvement in the peak target pixel TCR.

An important aspect of the ATI processing is that the target in each image has to be highly correlated. If the correlation factor is below 0.9, the estimation of target range rate has a high error. This in turn provides a geographic error in placement of the target. Figure 7.50 illustrates the correlation coefficient of each target in the scene. For the original ATI processing, the correlation value is significantly below 0.9. This is especially true for the targets 8–14. However, after ME processing, the estimated correlation coefficient is above 0.95 for targets in the open, and averages above 0.9 even for the targets in the forest area.

The ATI phase is not shown in this example. The ME algorithm used did not preserve the phase at each target location. This is due to the fact that each processing step is independent, and an arbitrary phase is produced. A recent result enables the phase to be preserved, as long as the same ME filter is applied to each image and target [26]. It is anticipated that further development in ME algorithms will provide the accuracy and restoration of moving target focus before detection to enhance this important new capability [8].

7.4 Summary

Moving target indication exploits target Doppler to separate the target return from main-beam clutter. A radar with a fast platform or low radar frequency complicates

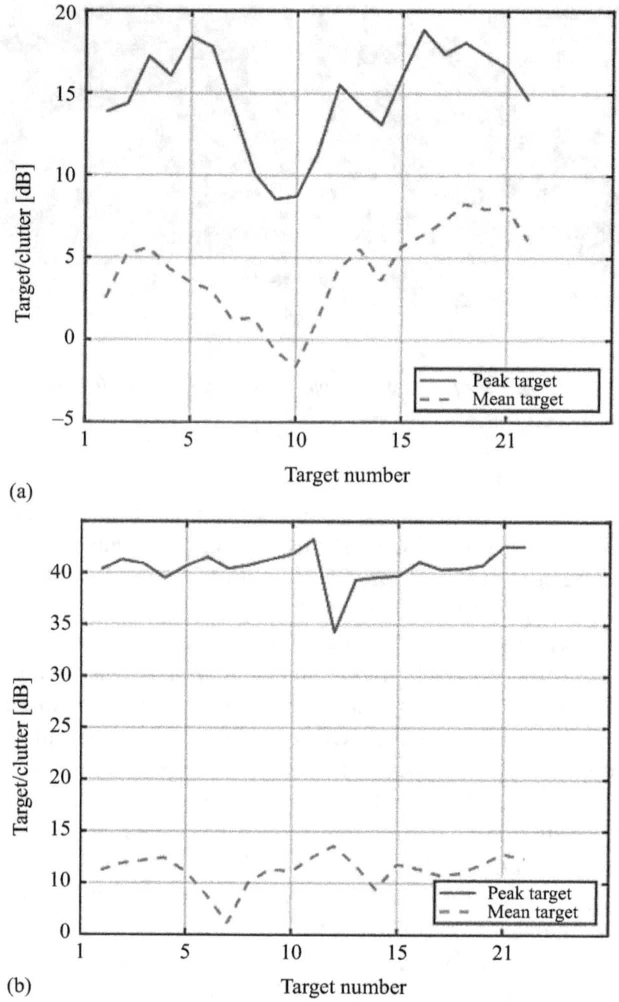

Figure 7.48 *SAR moving TCR. (a) Original SAR image and (b) after minimum entropy focus*

the GMTI detection. More importantly, the continuing need for fine range and cross-range resolution demands new radar architectures and improvements in coherent signal processing. This chapter has covered the effects of wavelength, long coherent integration times and ultra-wide bandwidth on GMTI design.

UWB modes are beneficial for characterizing the moving targets. However, as the resolution is made finer, the integration times will be significantly increased. New radar waveform and signal processing techniques have been developed to maintain the fine range and cross-range resolution of moving targets. But as expected, with the UWB waveforms, the signal processing has to compensate for movement through range and Doppler cells to maintain target coherence.

STAP is useful for slow to moderately fast targets, depending on the wavelength. However, if the bandwidth of the signal is increased to obtain resolution accuracy, the multiple spatial and temporal characteristics suffer from decorrelation. This effect is directly attributable to a loss in SINR and MDV. STAP has been applied to GMTI solutions from UHF to Ku-band and provides excellent reduction in the mainlobe clutter spread. It is necessary that additional preprocessing be applied to the multiple channel data prior to employing STAP. The most likely return on processing will be to use polyphase filter bands prior to STAP.

An alternative approach to UWB STAP is to use ATI for detecting the slow-moving targets in SAR images. The regime of target detection and localization is very slow-moving endo-clutter targets. ATI phase enables repositioning of moving

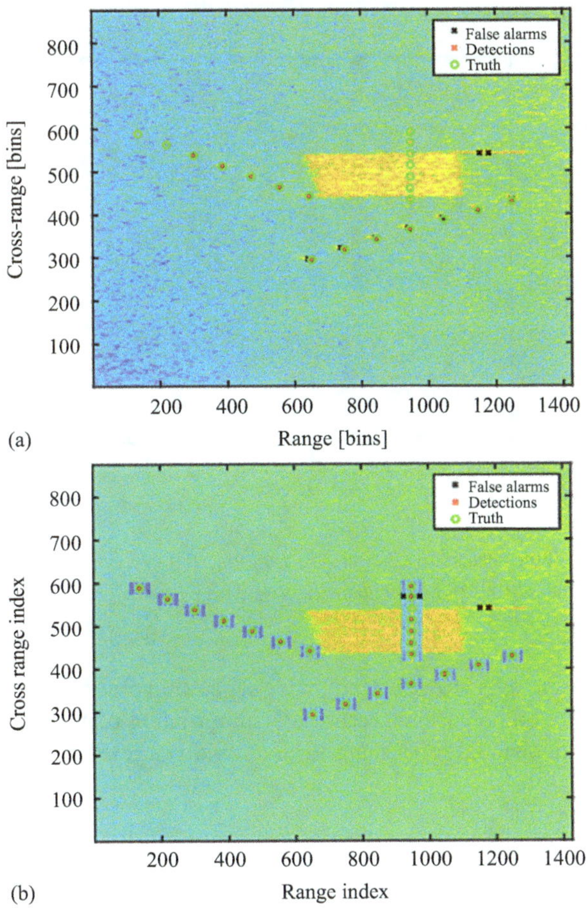

Figure 7.49 ATI detection of targets post-ME: (a) Original target detection and (b) target detection after ME focus

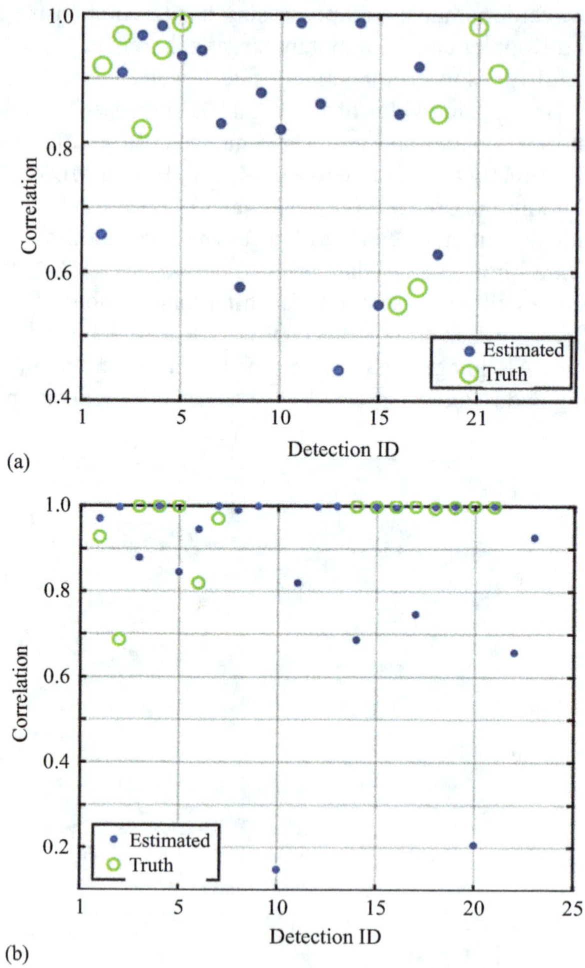

Figure 7.50 ATI correlation of two ATI targets: (a) Original target correlation and (b) target correlation post-ME focus

targets in the SAR image. However, any wide separation of the phase center baseline creates Doppler ambiguities. As a result, it is important to plan for multiple channel parallel imaging and signal processing to cover wide target surveillance coverage.

References

[1] Fowler C.A. 'The standoff observation of enemy ground forces from project peek to JointSTARS-A prolusion'. *IEEE Aerospace and Electronic Systems Magazine*. 1997;**12**(6):3–17.

[2] Bryant T.G., Morse G.B., Novak L.M., Henry J.C. 'Tactical radars for ground surveillance'. *Lincoln Laboratory Journal*. 2000;**12**(2):341–54.

[3] Ward J. *Space-Time Adaptive Processing for Airborne Radar*. Lincoln Laboratory Technical Report TR-1015; 1994.

[4] Ward J. 'Maximum likelihood angle and velocity estimation with space-time adaptive radar'. *Record of The Thirtieth Asilomar Conference on Signals, Systems and Computers*. Pacific Grove, CA; 1996. pp. 1265–7.

[5] Rosen P.A., Hensley S., Joughin I.R., *et al.* 'Synthetic aperture radar interferometry'. *Proceedings of the IEEE*. 2000;**88**(3):333–82.

[6] Davis M.E. 'Ground moving target indication, Chapter 6'. *Foliage Penetration Radar - Detection and Characterization of Objects Under Trees*. Raleigh, NC: Scitech Publishing; 2011.

[7] Xu J., Peng Y.-N., Xia X.-G., Farina A., Long T., Mao E.-K. 'Focus-before-detection radar signal processing: part i—challenges and methods'. *IEEE Aerospace and Electronic Systems Magazine*. 2017;**32**(9):48–59.

[8] Xu J., Peng Y.-N., Xia X.-G., Long T., Mao E.-K., Farina A. 'Focus-before-detection radar signal processing: part II–recent developments'. *IEEE Aerospace and Electronic Systems Magazine*. 2018;**33**(1):34–49.

[9] Chen C.W. 'Performance assessment of along-track interferometry for detecting ground moving targets'. *Proc. 2004 IEEE Radar Conference*. Philadelphia, PA; 2004.

[10] Wang H., Cai L. 'On adaptive spatial-temporal processing for airborne surveillance radar systems'. *IEEE Transactions on Aerospace and Electronic Systems*. 1994;**30**(3):660–70.

[11] Brennan L.E., Mallett J.D., Reed I.S. 'Adaptive arrays in airborne MTI radar'. *IEEE Transactions on Antennas and Propagation*. 1976;**24**(5):607–15.

[12] Billingsley J.B. *Low-Angle Radar Land Clutter – Measurements and Empirical Models*. Norwich NY: William Andrew Publishing; 2002. pp. 575–82.

[13] Billingsley J.B. 'Exponential decay in windblown radar ground Clutter Doppler spectra: multifrequency measurements and model'. *Technical Report 997*. Lexington, MA: MIT Lincoln Laboratory; July 29 1999.

[14] Techau P.M., Bergin J.S., Guerci J.R. 'Effects of internal clutter motion on STAP in a heterogeneous environment'. *Proceedings 2001 IEEE Radar Conference*. Atlanta, GA; 2001. pp. 2004–9.

[15] Guerci J.R., Bergin J.S. 'Principal components, covariance matrix tapers, and the subspace leakage problem'. *IEEE Transactions on Aerospace and Electronic Systems*. 2002;**38**(1):152–62.

[16] Nguyen T.Q. 'Near-perfect-reconstruction pseudo-QMF banks'. *IEEE Transactions on Signal Processing*. 1994;**42**(1):65–76.

[17] Rabinkin D., Nguyen T. 'Optimum subband Filterbank design for radar array signal processing with pulse compression'. *Proceedings 2000 IEEE Sensor Array and Multichannel Signal Processing Workshop*. Cambridge, MA; 2000. pp. 315–21.

[18] Hoffman A., Kogon S.M. 'Subband STAP in wideband radar systems'. *Proceedings 2000 IEEE International Radar Conference*. Arlington, VA; 2000. pp. 256–60.

[19] Himed B., Kim K., Zhand Y., Haggari A. 'A new approach to wideband space time adaptive processing (W-STAP)'. *Proceedings 2004 IEEE Sensor Array and Multichannel Signal Processing Workshop*. Barcelona, Spain; 2004. pp. 672–6.

[20] Kapfer R.M., Davis M.E. 'Along track interferometry for foliage penetration moving target indication'. *Proceedings 2008 IEEE RADAR Conference*. Rome, Italy; 2008.

[21] Kapfer R.M., Pohl M., Davis M.E. 'Detection improvement for UWB along track interferometry of moving targets under foliage'. *Proceedings 2013 IEEE Radar Conference*. Atlanta, GA; 2013.

[22] Fienup J.R. 'Detecting moving targets in SAR imagery by focusing'. *IEEE Transactions on Aerospace and Electronic Systems*. 2001;**37**(3):794–809.

[23] Davis M.E., Maher J., Hancock R., Theis S. 'High fidelity modeling of space-based RADAR'. *Proceedings 2003 IEEE RADAR Conference*. Huntsville, AL; 2003.

[24] Samczynski P., Kulpa K.S. 'Coherent MapDrift technique'. *IEEE Transactions on Geoscience and Remote Sensing*. 2010;**48**(3):1505–17.

[25] Wahl D.E., Eichel P.H., Ghiglia D.C., Jakowatz C.V. 'Phase gradient autofocus-a robust tool for high resolution SAR phase correction'. *IEEE Transactions on Aerospace and Electronic Systems*. 1994;**30**(3):827–35.

[26] Kantor J.M. 'Minimum entropy autofocus correction of residual range cell migration'. *Proceedings 2017 IEEE Radar Conference*. Philadelphia, PA; 2017.

[27] Chen V.C., Martorella M. *Inverse Synthetic Aperture Radar Imaging*. Edison, NJ: Scitech Publishing; 2014. pp. 91–8.

[28] Davis M.E. 'Moving target detection under sparse foliage'. *Proceedings 2018 IEEE Radar Conference*. Philadelphia, PA; 2018.

Chapter 8

Ultra-wideband spectrum compliance

A political and technical challenge for developing an operational ultra-wideband (UWB) surveillance radar is the ability to comply with the many international regulations on the radio frequency (RF) spectrum. It is clear that the worldwide regulations on the RF spectrum have made it more difficult to get licenses for radar operation. More importantly, the recent auction of RF spectrum for civilian personal communications in several countries has reduced the traditional bands for radar spectrum access [1]. This chapter will go over the beneficial uses of UWB surveillance radars, along with the attendant frequency allocation process. It is very difficult to get an RF allocation license for operating in any populated area. And the process is always changing, with the advent of personal communications, a widespread reliance on wideband communications, digital television and radio broadcast [2]. A short history of the RF frequency allocation in recent time will show a summary of the process, and the story will present several lessons learned from obtaining licenses for UWB radars [3].

The previous chapters have given examples of UWB surveillance radar systems and their spectral demands for waveforms. The driving factor for UWB surveillance is the ability to discriminate between man-made objects and cultural clutter. This requires spatial resolution as well as use of multichannel and polarization, for characterizing the objects and clutter in observed regions. Furthermore, there are two emerging radar technical areas to be considered. The first is cognitive radar where the knowledge of the local and temporal usage of RF spectrum can be employed for partially solving the technical hurdles [4]. A software-defined radar, similar to software-defined radio, has the potential for sensing the spectrum usage and adjusting transmissions to the "white spaces". However, the spectrum environment is rapidly changing, and the radar systems need to be adapting accordingly. The second is a very preliminary discussion of new RF waveform and processing needs to comply with the future RF regulation issues [3]. The communications world has created a spectrum of orthogonal waveforms that minimize the co-channel interference in a dense communications environment. The radar community has been exploring this approach to mitigate some of the regulation problems. It is important that these research areas continue beyond 2020 to preserve the abilities for operating UWB radar systems to meet increasing surveillance and humanitarian needs.

The original application of UWB long-wavelength radars was foliage and ground penetration [5]. There are many areas where the ability to detect objects under trees is needed for border monitoring and mitigation of poaching and illegal harvesting of wildlife and forests. There are several applications for civilian and humanitarian applications where UWB radar has unique applications. These uses can be either for all-weather surface surveillance or wide-area coverage for characterizing the terrain. Most recently, the UWB radar has been used for characterization of the biomass as a method to monitor global warming effects on forests and croplands [6]. The final application for UWB radar is for fine spatial resolution to monitor and discriminate objects close to buildings or other structures. This is important in surveillance of areas where objects are hidden or camouflaged against monitoring. Reduction of pixel size is important to detect these important objects, where coarse resolution would not be effective.

UWB radar systems have been developed over the past three decades for several military and geosciences applications. These applications include humanitarian and natural disaster monitoring operations and characterization of biomass for global warming research. Modern radar systems demand more bandwidth for discriminating obscured objects in close proximity. This spatial and spectral resolution enables improved resolution of targets and provide better obstacle detection and tracking. A critical impediment to realizing these new radar sensor capabilities is the international regulations on frequency allocation, which creates a "barrier for entry" for many new radar applications [7].

8.1 Motivation for UWB radar operation

Microwave frequencies have been the classical RF bands used for synthetic aperture radar (SAR) imaging of the Earth's surface. When the objects are in the open, there is no substitute for the shorter SAR imaging collection times. However, when the objects are camouflaged or are obscured under foliage, the need for operating at longer wavelengths is essential [5]. Figure 8.1 demonstrates the reasons for ultra-high-frequency (UHF) versus X-band for detection objects under trees. Figure 8.1a represents a 1-m resolution X-band image, where only the tops of the trees are evident. As a result, the X-band image is not considered to be UWB, since the bandwidth is under 10%, and this class of SAR does not have difficulty obtaining frequency authorization within that spectrum. On the other hand, Figure 8.1b is covering the same geographic image but collected at UHF. A 0.5-m resolution radar at UHF has a fractional bandwidth over 30%, and as a result is classified as UWB. This foliage penetration mode utilizes a long wavelength to minimize the attenuation of the foliage obscuring the man-made objects. Now, the ground and trunks are clearly seen without showing the tops of the trees. There are also several military tanks hidden under the foliage, along with a trihedral target in the bottom of the image. It is well known that a resolution of as fine as 0.5 m in range and cross-range is needed to discriminate the tanks from the forest. These image metrics require a UWB waveform along with wide-angle integration; as a result, the UHF image is definitely UWB. As a result of the international spectrum policy, it is difficult to get frequency

(a) (b)

Figure 8.1 *SAR images for detecting targets on battlefield [5]. (a) X-band microwave image and (b) UHF foliage penetration image*

authorization even with many stipulations. This is because the radar occupies the same frequency band as many communications and emergency beacons.

This section now goes over the spectrum compliance process for operating a UWB radar system. It is much more difficult at UHF and L-band to get the required bandwidth. This is primarily due to the number of RF communications and radar systems that already use these frequencies. However, there are also several communications bands at L-, S- and X-band that will also restrict radar systems to meet the UWB regulations imposed by all of the worldwide spectrum agencies.

Radio frequency interference (RFI) is important in any SAR application, where the data collection requires long dwells over a specific area and very wideband waveforms to obtain the spatial resolution for the collection objectives. At UHF and L-band, there are numerous RF transmissions and many more receivers that need to be considered for spectrum sharing. The two aspects of RFI impact on a UWB surveillance radar include tailoring the transmit waveform and interference on the radar receiver. The primary user in these bands is the worldwide satellite navigation capabilities of the global navigation and satellite system. The four main satellites systems are global positioning systems (GPS; USA), Galileo (EU), GLONASS (Russia) and Beidou (China) [8].

The UHF spectrum has two problems in terms of frequency allocation and interference from other RF sources. The image in Figure 8.2 illustrates the result of operating a UWB radar in this wide spectrum, showing the results of using its receiver as a spectrum analyzer to characterize all of the other transmissions within receiver reception. All of the peaks are transmitters in the region that are either other radars, television or communications sources. This sample was taken in the 1990–1995 timeframe, before the advent of digital TV and wideband mobile and fixed communications. As a result, the width of these interference sources is now

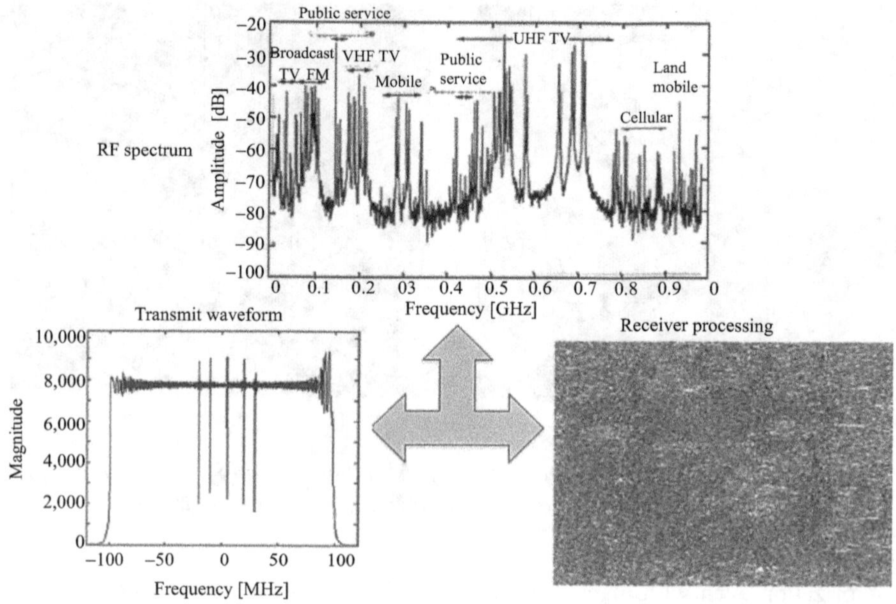

Figure 8.2 Impact of cluttered RF spectrum on UWB radar operation [9]

more pronounced due to the new digital communications systems. The image in the lower left illustrates the requirements for notching the transmission of the UWB UHF radar for approximately 10% of the bandwidth. And the lower right image is a UWB SAR image that has the effects of RFI on the SAR image. This image has a signal-to-interference plus noise ratio (SINR) of approximately –3 dB, which is useless for any reasonable applications. This chapter will go through the rationale for regulations on UWB radar operation, the need and approaches for notching on transmit, two techniques for adaptively removed RFI from the image and finally some potential systems and signal processing approaches to operate in the evolving RF environment.

There are two engineering problems to be covered. First, the frequencyauthorization process inherently requires notching on transmit to avoid interferingwith emergency and satellite navigation receivers. This notching of transmitwaveforms is essential under the National Telecommunications and InformationAdministration (NTIA) Part 15 regulations in the United States [10]. Similarregulations exist in Europe and Far East, where there exists a large populationemploying the massive increase in personal communications devices [11]. Andthe second engineering problem is a need for adaptive cancellation of the manyother RFI sources to repair the SAR image. As the number of RF communicationsdevices increase with more bandwidth utilization, the background RF interferencewill be more of a problem for high-quality radar operations. The following section willpresent the policy and practices needed to get a license for UWB to operate in apopulated region.

8.2 UWB spectrum regulations

UWB emissions are defined in the IEEE Standard 1675 and by frequency regulators as any signal over 25% fractional bandwidth or 500-MHz absolute bandwidth [12]. The US NTIA is the governing body for any system that transmits in North America, and there are similar definitions and regulations in Europe and Asia. Any system that operates above an equivalent isotropic radiated power threshold of –41.3 dB/MHz must be tested to obtain permission to transmit in a specified geographic area. However, the power spectral density interference levels vary with frequency, within specific bands historically allocated to narrow bands (e.g., aircraft communications, satellite communications and radar) [10]. More recently, the RF spectrum has been sold-off taking the classic radar and communications bands and allocating much of those spectra to WiFi commerce [13]. This selling off of radar spectrum has become a significant issue for UWB radars, since these radars fall under NTIA Part 15 and will potentially affect many receivers in any crowded geographic area.

The historic NTIA regulation (called Part 15) lists many narrow-band receiver frequencies, where a radar needs to avoid if the receivers were associated with life-safety or emergency communications. These systems were normally very narrow band, and the interference from UWB surveillance systems did not exceed the regulatory threshold. However, many of these communications systems are also wideband, and the NTIA protects them more closely. The short description of the NTIA regulation of spectrum compliance of any UWB transmitter implies that the power spectral density received at a victim receiver, as measured at 1 m from the receiver antenna, will not interfere with those protected systems [10]. It is important to recognize that the more recent applications of the regulations are even more stringent. Much of the radar band allocations have been sold off to telecommunication providers.

Figure 8.3 illustrates the NTIA spectrum mask used for compliance by UWB system, and it provides the threshold against which they will be tested. The dotted line is the requirement for intercept power limits as a function of frequency. This line is the maximum power spectral density level, since there are many receivers where their functions are classified as safety and require more stringent, reduced levels. The power per MHz of bandwidth incident on a receiver in the field of regard must be below that line. We can see that for the UHF band, that limit is –41.3 dBm/MHz, until the frequency gets to the satellite navigation bands between 900 MHz and 1.5 GHz. At that point, the limit is at –78 dBm/MHz. And there are several levels for power spectral density depending on the center frequency going out to over 3 GHz.

Figure 8.4 is the representative UWB signal for a UHF SAR with the waveform bandwidth, shown in the light gray shading. This wideband signal represents the most difficult part of the Part 15 where the wideband transmission may interfere with any receiver in the operating region that fall under the definition of a "sensitive receiver." The list of emergency and life-critical bands is given in Table 8.1, which have a much more stringent requirement on UWB radiation interference. The dark black lines indicate five of these bands for a particular region that the radar would

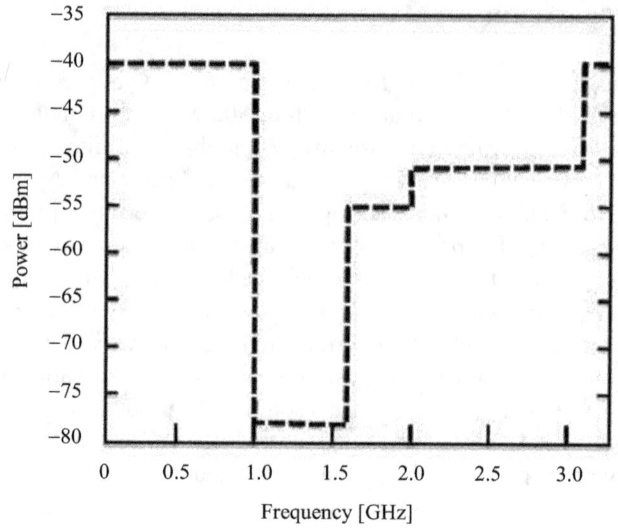

Figure 8.3 NTIA spectrum mask [9]

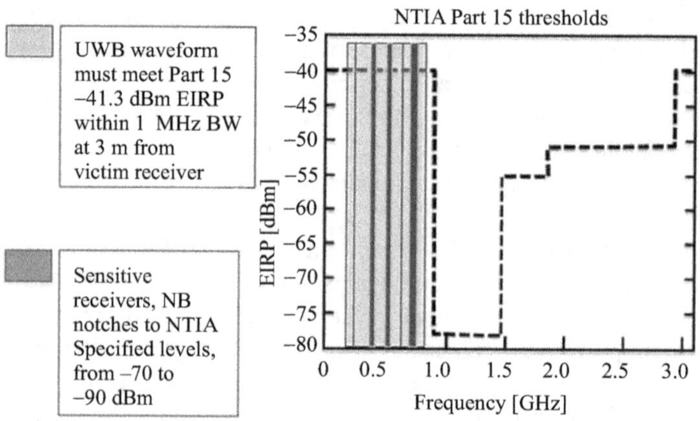

Figure 8.4 UHF UWB frequency utilization [9]

be operated. In these cases, the required notching on transmit is significant, and the radar needs to be tested to avoid these frequencies [5].

Table 8.1 lists the frequencies from low VHF to L-band that must be avoided in any UWB radar operation. The receivers are historically fixed locations, so that the route planning to avoid them was easier. However, mobile communications now make it more difficult to satisfy the frequency allocation regulations. We will cover this further in Section 8.2, because the requirements on frequency avoidance have become more stringent since 2008. Restricted bands are potential interference to

Table 8.1 Restricted band frequencies in the United States

Function	Frequency (MHz)	Function	Frequency (MHz)
Radio astronomy	37.5–38.25	AIS (automatic identification service)	162.0125–167.17
Radio astronomy	73–74.6	Fixed mobile, public safety, forest fighting	167.72–173.2
Aeronautical radionavigation	74.8–75.2	Fixed mobile, emergency	240–285
Aeronautical radionavigation	108–121.94	Fixed mobile, radio astronomy	322–335.4
Aeronautical mobile location	123–138	Radio navigation satellite	399.9–410
Mobile satellite (Earth to space)	149.9–150.05	Radio astronomy	608–614
Maritime mobile safety	156.52475–156.52525	Aeronautical radio navigation satellite	960–1,240
Maritime mobile distress	156.7–156.9	Radio navigation satellite	1,300–1,427

sensitive radio communications such as aircraft radio navigations, radio astronomy and search and rescue operations.

The last aspect of NTIA Part 15 is that any transmitter must have the out-of-band emissions below the NTIA requirements out to three times the center frequencies for the RF system bandwidth. In Figure 8.5, the shaded area represents the

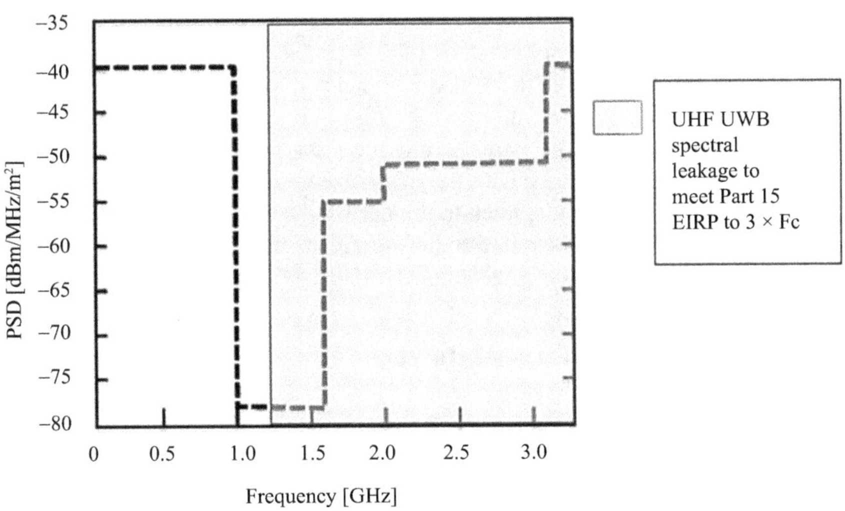

Figure 8.5 Requirement on three-times spectral leakage effect [9]

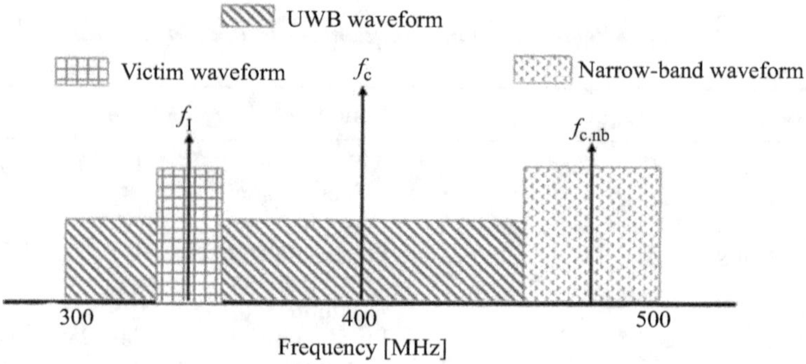

Figure 8.6 Conditions for frequency-dependent rejection [9]

region where spectral sidebands need to be controlled over the 3× bandwidth corresponding to the transmission RF bandwidth from Figure 8.4. This factor will be explored later in terms of controlling the transmitter sidebands and ensuring out-of-band spectral purity.

Figure 8.6 illustrates the NTIA Part 15 process for determining the frequency avoidance. Two cases are illustrated. The first is for on tune rejection (OTR), where the victim receiver is within the bandwidth of the UWB waveform, shown in light gray. In this case, the transmit waveform must be notched, if it cannot satisfy the maximum intercept power level. The second condition is for off frequency rejection (OFR), where there is no overlap between the narrow-band waveform and the victim receiver. In the OFR case, the major factor is the transmit waveform spectral sidebands, which may be overlapping the victim receiver bandwidth. All of these limits make it very important to design the radar system with a clear attention to spectral details [10].

The measurement process to obtain frequency authorization of a UWB RF emitter established by NTIA is very conservative. A process for instrumented testing is carried out by slowly sweeping the transmit signal across the system operating band, in order to determine whether the intercept power levels are contained below the applicable receiver limits. The issue here is the slow sweep of the transmit waveform. Many UWB transmitters have a very fast sweeping transmit waveform, which presents a time--bandwidth mismatch to the narrow-band victim receiver. This testing condition has been demonstrated to be significantly conservative, especially for a UWB waveform, where the signal changes rapidly within the spectrum [9].

8.3 Calculation of intercept power

Figure 8.7 illustrates the impact of slant range and bandwidth on the intercept signal at the victim receiver. The UHF NTIA Part 15 threshold is shown at –41.3 dBm/MHz, along with the safety system receiver limits of –70 to –90 dBm/MHz. The peak and average power are illustrated for the R_{RI}^{2} variation of power at the receiver plotted in kilometers. Part 15 allows a reduction of the intercept power limits by the

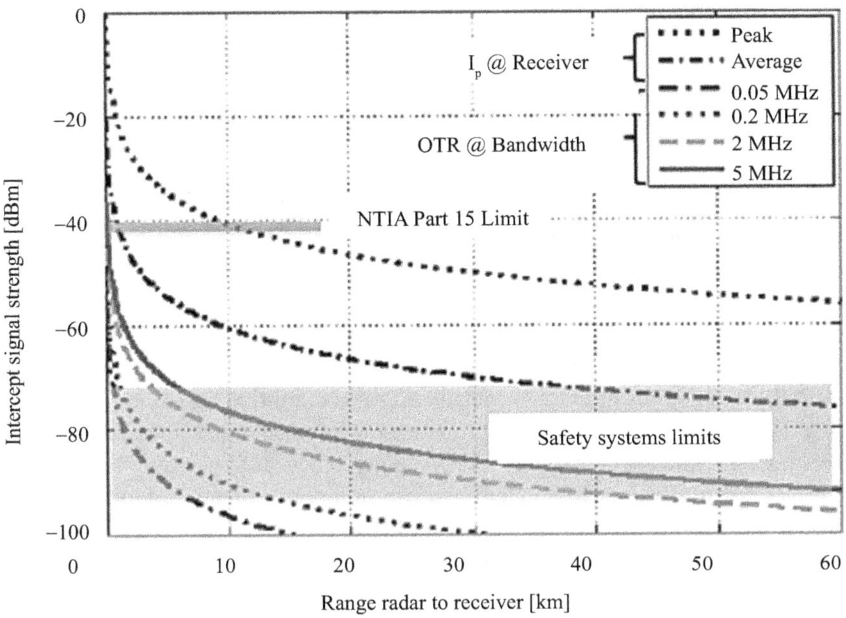

Figure 8.7 Intercept signal versus range to victim receiver [9]

ratio of the bandwidth between the transmit waveform and the victim receiver. The intercept power I_P is given by [10]

$$I_P = \frac{\widehat{P_T} G_{TI} G_I \eta_I}{L_b \left(R_{RI}^2\right) FDR\left(\Delta f\right) \eta_0 \left(\Delta f\right)} \tag{8.1}$$

where $\hat{P}_T$ is the peak transmit power, G_{TI} the antenna-on-antenna gain between the radar pattern (main lobe or sidelobe) toward the victim receiver, η_I the waveform duty factor and L_b the propagation loss that is proportional to the slant range distance R_{RI}^2. The frequency-dependent reduction FDR is defined by Part 15 to be

$$\text{FDR}\left(\Delta f\right) = \begin{cases} \text{OTR} & \left|f_c - f_{cI}\right| \leq \dfrac{B_T}{2} + B_I \\ \text{OFR}\left(\Delta f\right) & \text{otherwise} \end{cases} \tag{8.2}$$

where f_c is the waveform center frequency, B_T the transmit bandwidth, f_{cI} the center frequency of the intercept receiver tuning and B_I the intercept bandwidth. This reduction is calculated in Figure 8.7 for a 160-MHz bandwidth UWB SAR waveform, and the FDR loss is shown by the lines for 50 kHz to 5 MHz victim receiver bandwidths. If the intercepted receive power is higher than the Part 15 limits, the waveform needs to be notched at these frequencies and bandwidths to meet the NTIA requirements.

Figure 8.7 also illustrates the impact of slant range and bandwidth on the intercept signal at the victim receiver. The UWB NTIA Part 15 threshold is shown at –41.3 dBm/MHz within the UHF band, along with the safety limits of –70 to –90 dBm/MHz. The peak power and average power are illustrated for the R^2 variation of power at the receiver in kilometers. The Part 15 allows a reduction of the intercept power limits by the ratio of the bandwidth between the transmit waveform and the victim receiver. This reduction, calculated for a 160-MHz bandwidth UWB SAR waveform, is shown by the lines for 50-kHz to 5-MHz victim receiver bandwidths. If the intercepted receive power is higher than the Part 15 limits, the waveform needs to be notched at these frequencies to meet the NTIA requirements

The National Aeronautics and Space Administration (NASA) Jet Propulsion Laboratory carried out extensive testing on the geographic synthetic aperture radar (GeoSAR) radar in an attempt to get license for operation in Continental United States. The solid line in Figure 8.8 indicates the "attenuation" of the transmitted waveform using the NTIA slowly swept receiver. These numbers will be introduced into the intercept receiver loss calculation for compliance testing. However, it was observed that there is a further attenuation for narrow-band receivers, if the signal is swept rapidly. Most UWB SAR systems have a high time--bandwidth (BT) product, and the frequency is swept at many MHz per microsecond. JPL tested several receivers that had a phase-locked loop on the front end of the receiver to determine the signal received by the receiver. With the high time--bandwidth signals, and with a fast frequency sweep of 10 MHz/μs, the receiver PLL had a further attenuation of the signal as indicated by the dotted line. This attenuation should have been included in the intercept calculation. However, NTIA has not recognized this important distinction in their system test procedures and its subsequent impact on granting of UWB operations of airborne radars.

The history of notching requirements has been using the solid line for operation of GeoSAR, instead of following the dotted line in Figure 8.8. This resulted

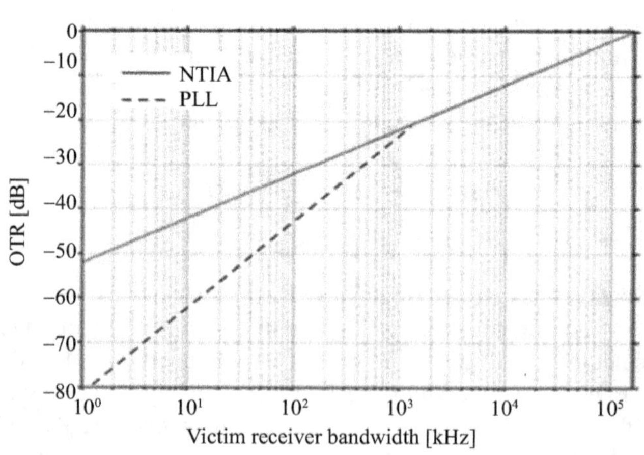

Figure 8.8 Example of OTR for UHF SAR waveform [14]

in excessive notching requirements for many of the very narrow-band receivers in the vicinity of radar operation. In addition, the current NTIA requirements call for the operator to file a flight collection plan at least 30 days before operation. This is so that the local users can be notified if there is any potential for interference with emergency or safety services in the area. The operation of UWB radars within the US borders have added a high logistics planning and operations cost.

8.4 RFI avoidance

In the previous section, it was stated that the process for obtaining a license for operating a UWB radar within a populated geographic area is very inefficient and logistically expensive. If these important radars are going to be useful in evaluating geophysical changes in the environment or helping with disaster remediation, a new approach is required. There are similar lessons to be learned from communications in congested environments, such as crowded cities or dynamic battlefields. The Defense Advanced Research Projects Agency (DARPA) Cognitive Radio program of the early 2000s [15] and the current Spectrum Collaboration Challenge [16] are adapting discovery of local, limited spectrum utilization and using artificial intelligence to reduce the interference within the environment in a highly dynamic interaction.

Figure 8.9 illustrates an approach for UWB radars to operate in an underutilized spectrum environment. The hypothesis is that communications spectrum bandwidth is sparse across a wide spectrum. A radar system could hypothetically sense the areas where there are no users at that instant, and the radar waveform scheduler can use that bandwidth until the instant that a communications user appears. There are two radar spectrum compliance approaches being considered. First, the idea of cognitive processing is being developed in several research areas, leveraging the

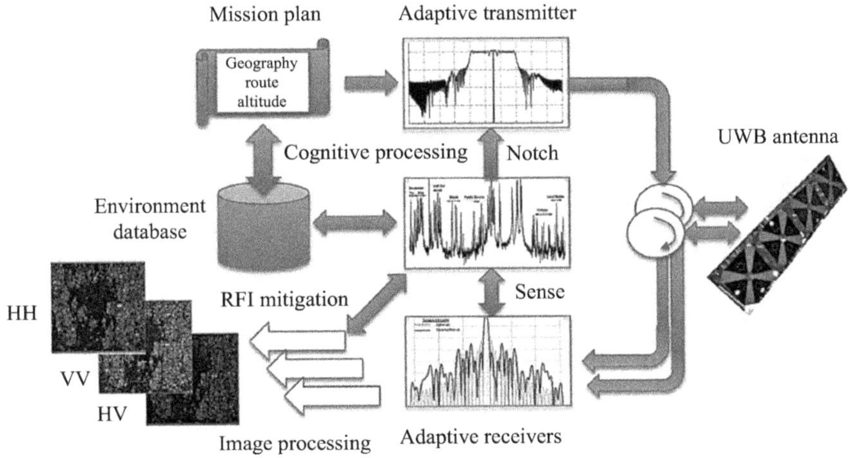

Figure 8.9 Processing architecture for cognitive UWB radar [14]

cognitive radio community. But the cognitive approach has not been adequately adopted by the radar community for improving compatibility with the communications environment. Second, the radar waveform needs to more efficiently adopt the transmitter notching approach to directly improve the power spectrum compatibility with the communications environment.

The concept of cognitive radar in a dense environment is the feedback of the knowledge of the environment on the transmission and reception of UWB RF energy [4]. The idea is to sense the environment to determine "white spaces" in the spectrum. These are parts of the frequency spectrum that have been allocated for either datalinks or RF communications, but which are not fully being utilized in the geographic region. The radar will periodically sense the spectrum and determine those areas where transmission is occurring.

8.4.1 Cognitive radar processing

The cognitive process uses the knowledge of the RF spectrum and any potential users for planning operations [17]. Under the current NTIA process, this requires 30-day notification before a UWB radar can operate. However, the location and characteristics of the primary emergency and life-safety RF systems can be readily understood for preflight and even within flight operation. This environmental database can be used, along with a mission plan, to determine the necessary modifications to either the waveform or the direction of collections.

Notching is the primary requirement for any UWB radar, whether it is at UHF or X-band. There are navigation and commutations receivers that are known a priori. The real challenges are the wideband communications on mobile radars that are not readily identified. The cognitive radio community has developed protocols to look for white spaces and operate selectively on their frequency spectrum schedule [15]. It is postulated that the radar community can and should have a similar approach.

Finally, the measurement of the RFI environment during the collections can be excised from the radar received energy in an attempt to provide the necessary radar capabilities. RFI adaptive processing in UWB radars have already been developed and operated on many radar systems. However, there is a need for more extensive RFI removal from the UWB SAR imaging.

The challenge for developing and operating a UWB surveillance radar is to balance the need for operating in a dense RF environment, restrictions on waveforms that can be used and the image quality needed for meeting the system objectives. Because of the interplay between the NTIA regulations, and significant RF signals in the environment, the process is ideal for utilizing several techniques adopted by the cognitive radar community. Figure 8.9 illustrates the processes that can be considered in future UWB radar operations and will be briefly covered in this section. The general system components are summarized as follows [7]:

- UWB mission plan: It is necessary to plan a radar mission in advance to be approved for operation. In today's regulations, this advance notice can be as much as 30 days.

- Environmental database: The proposed future process would include planning the mission within an approved environmental database and then calculating the necessary waveform controls to abide by the need to protect sensitive receivers in the vicinity.
- Cognitive processing: The proposed mode waveform control is to be implemented with cognitive processing technologies. By sensing and adapting to the RFI environment, the UWB surveillance system can operate without interfering with the local safety systems operation [4].
- Adaptive transmitter: The major system design attribute is utilizing knowledge of the transmitter and antenna characteristics and using preconditioning to satisfy the spectral purity of the system. There will always be sensitive frequencies that need to be notched, and those data will come from the environmental database. However, there may be new communications channels that need to be sensed and notched, if they are critical for noninterference operation.
- UWB antenna: The primary component that needs to be characterized for successful radar imaging characteristics is the UWB antenna. This antenna will generally have multiple polarimetric elements that must have channel balance and spatial--spectral match for both NTIA and SAR objectives.
- Adaptive receivers: There will be multiple receiver channels (e.g., polarization) that interface to the environment. These channels are used to sense the RFI and determine a need for subsequent notching, or are used for RFI mitigation to improve the image quality.
- RFI mitigation: The most critical signal processing is the RFI mitigation that adapts to strong, nonstationary communications signals in the environment. These processing steps have been implemented in current UWB radars, with varying degrees of success. The single most important measure of performance is the noise equivalent sigma-0 of the image.
- Image processing: The processing of multiple SAR channels needs to preserve the signal characteristics being used to detect targets or characterize the biomass. Depending on the degree of notching on transmit, these processes need to reconstruct the image point spread function as well as conduct radiometric calibration.

The rest of this section will present the proposed technologies that will enable obtaining future acceptance of UWB surveillance radar for worldwide operation.

8.4.2 Signal conditioning for spectrum compliance

The spectral purity of RF waveforms is imperative to meet recent and future NTIA requirements of operation in the RF environment. Without controlling the out-of-band emissions, there is limited possibility for a UWB radar to gain acceptance. The techniques of predistortion in communications waveforms are widely used to comply with the evolving constraints. Solid-state amplifiers introduce many of the nonlinearities that affect the system spectral purity. However, it is possible to predistort the signal to compensate for the final amplification and other nonlinear effects in

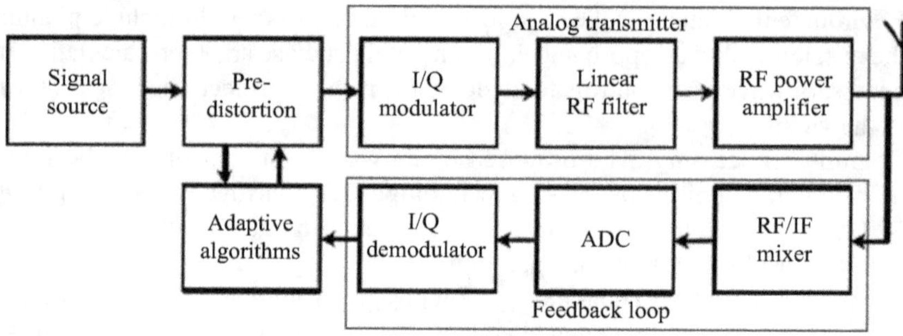

Figure 8.10 System model of transmitter for spectrum compliance [18]

the system. This section introduces the need for compensation on receive to match the channels and introduces the approach of predistortion to match multiple transmit channels and to satisfy the intermodulation and out-of-band signals [18].

The RF power amplifier has a prevalent problem of producing out-of-band signals due to the nonlinear nature of the operation. In particular, a solid-state transmitter biased into Class-C saturation, for maximizing the power added efficiency, has a problem with signal linearity because of the nonlinear operation of the solid-state amplifiers. This saturation is one of the major causes of broadening the spectral content of transmissions, which require remediation in the output to satisfy the NTIA regulation of the operation. Figure 8.10 shows the system model of a solid-state transmitter. In addition to the nonlinear operation of the transmitter, there are other components in the transmitter chain that require spectral matching. Not only is the matching required in-band, but the spectral power level has to be managed for up to three times the signal bandwidth. The figure illustrates several portions of the system model to meet spectral compliance.

The cause of nonlinearity is shown in the top of the system model, comprising the I/Q modulator, a filter bank for channel equalization and spectrum compliance, and the transmitter chain itself. The I/Q modulator will typically introduce a DC offset between the in-phase and quadrature phase parts of the signal. For the DC offset and I/Q imbalance in amplitude and phase of the channels, the system model takes the form of [18]

$$y(t) = \mathbf{M}x(t) + \mathbf{Ma} \tag{8.3}$$

where input signal is $x(t) = [x_I(t), x_Q(t)]^T$ and output signal is $y(t) = [y_I(t), y_Q(t)]^T$. The in-phase (I) and quadrature-phase (Q) components are modeled by thefollowing relation:

$$\mathrm{M} = \begin{pmatrix} \alpha\cos\left(\frac{\varphi}{2}\right) & \beta\sin\left(\frac{\varphi}{2}\right) \\ \alpha\sin\left(\frac{\varphi}{2}\right) & \beta\cos\left(\frac{\varphi}{2}\right) \end{pmatrix}, \ \mathbf{a} = \begin{pmatrix} a_1 \\ a_2 \end{pmatrix} \tag{8.4}$$

where α and β are the gains of the I and Q paths, φ the phase error between the components, and a_1 and a_2 are the DC offsets for both components. This DC offset of the

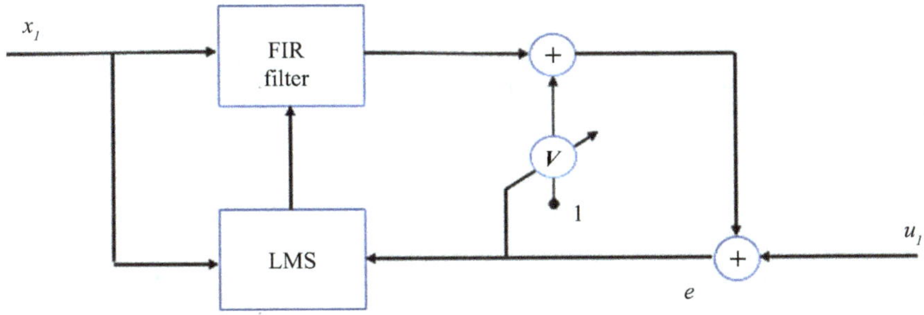

Figure 8.11 *LMS estimation of DC offset [18]*

baseband signal must be compensated in order to limit the unwanted multiplication of signal sidebands as well as the system noise. In this model, the DC offset is at a high level, and it must be minimized to avoid nonlinear distortion in the acquisition phase. While estimating the DC offset, we also identify the linear filter to define constraints for linear filter post-distortion training. An adaptive linear finite impulse response (FIR) system identification procedure extended with an averaging filter to estimate the DC offset is used to null the DC component. The model shown in Figure 8.11 is expressed as

$$y(i) = \sum_{k=0}^{K-1} h_k \left[x_1 (i-k) + d_c\right] \tag{8.5}$$

where h_k are the coefficients of the FIR filter, x_1 the input signal before any predistortions, i the time index of the discrete signal and d_c the DC offset. K is the number of coefficients in the FIR model. To compensate the I/Q imbalance, $y(i)$ is added to the input signal.

The coefficients h_k are estimated with a least means square (LMS) algorithm, as shown in Figure 8.11. A recursive averaging filter is used, given as

$$v(i+1) = v(i) + \mu\, e(i) \tag{8.6}$$

where μ is the adaptation step size and e is the error signal. Using $v(i+1)$ as the converged $v(i)$, the coarse DC offset is given by

$$d_c = -v \cdot \left(\sum_{k=0}^{K-1} h_k\right)^{-1} \tag{8.7}$$

The sum of the estimated coefficients h_k from (8.5) approximates the amplitude gain of the filter at zero frequency and the amplification of the DC offset at the output of the filter.

The development of the linear filter and residual DC offset post-distorters is shown in Figure 8.12, where the I/Q imbalance is active only during tracking. Coarse DC offset is compensated prior to this stage. The LMS algorithm is constrained by

$$w[i+1] = \mathbf{P}\left(\mathbf{w}[i] + \mu e[i]\, u_2[i]\right) + \mathbf{G} \tag{8.8}$$

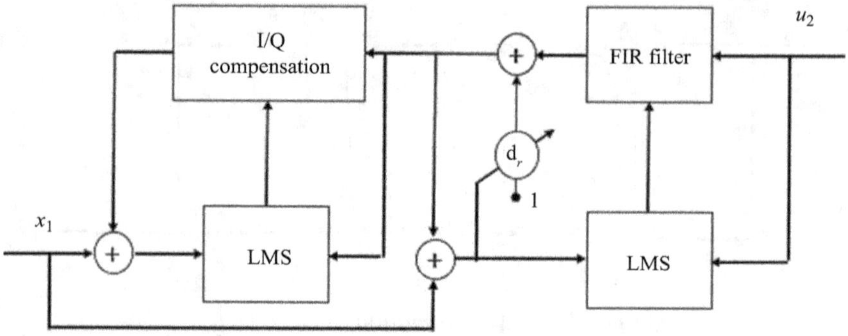

Figure 8.12 Training of FIR, DC offset and I/Q imbalance post-distorters [18]

where $w[i]$ is the column vector with N filter coefficients at time i, $\boldsymbol{P} = \boldsymbol{I} - \boldsymbol{C}^H(\boldsymbol{C}\boldsymbol{C}^H)^{-1}$ g is the $N{\times}N$ projection matrix, μ the step size parameter and $\boldsymbol{u}_2[i]$ the feedback signal vector consisting of N samples.

The adaptive predistortion employs constraints on the filtering with the matrix C. As the constraint matrix is

$$C = \left[c_{mn}\right]_{M\times N} = \left[\exp\left(-\mathrm{j}2\pi f_m n\right)\right]_{M\times N} \tag{8.9}$$

where $n = 0, \ldots, N-1$, and $f_m = F_m/F_s$ form a set of equally spaced normalized frequencies within the bandwidth of the feedback loop. The constraints are indexed with $m = 0, \ldots, M-1$. Next the gain vector is defined as

$$\boldsymbol{g} = \left[g_m\right]_{M\times 1} = \left[\frac{\exp\left(-\mathrm{j}2\pi f_m \tau\right)}{\sum_{k=0}^{k-1} h_k \exp\left(-\mathrm{j}2\pi\, k\, \mathrm{sgn}\left(f_m\right) f_e\right)}\right]_{M\times 1} \tag{8.10}$$

where $f_e = \pm BW/F_s$ are the normalized band corner frequencies of the acquisition training signal and τ is the desired group delay of the system.

The post-distortion is applied in the following sequence. First, the input signal is used to train the nonlinear post-distorter using a recursive least-squares algorithm. Next, the I/Q imbalance is trained as shown in Figure 8.12, with one coefficient for each element of M. Figure 8.13 shows the filter characteristics giving the compensated RF filter and the predistorter filter shape. It is clearly illustrated that the predistorter covers the out-of-band performance, and the RF filter has attenuated the spectrum to fall within the desired bandwidth. Finally, Figure 8.14 shows the total spectrum of the waveform requiring mitigation of the nonlinear performance. The input signal experiencing the nonlinear distortion during the transmitter chain is seen with its wideband interference spectrum. After the post-distortion filtering, the spectrum has been constrained within the system bandwidth (in this example ±36 MHz). As a result of this filtering, the interference spectrum has been reduced to –90 dB outside of the desired ±25-MHz signal bandwidth. These performance levels are comparable with the expectations for satisfying NTIA requirements for UWB transmitters.

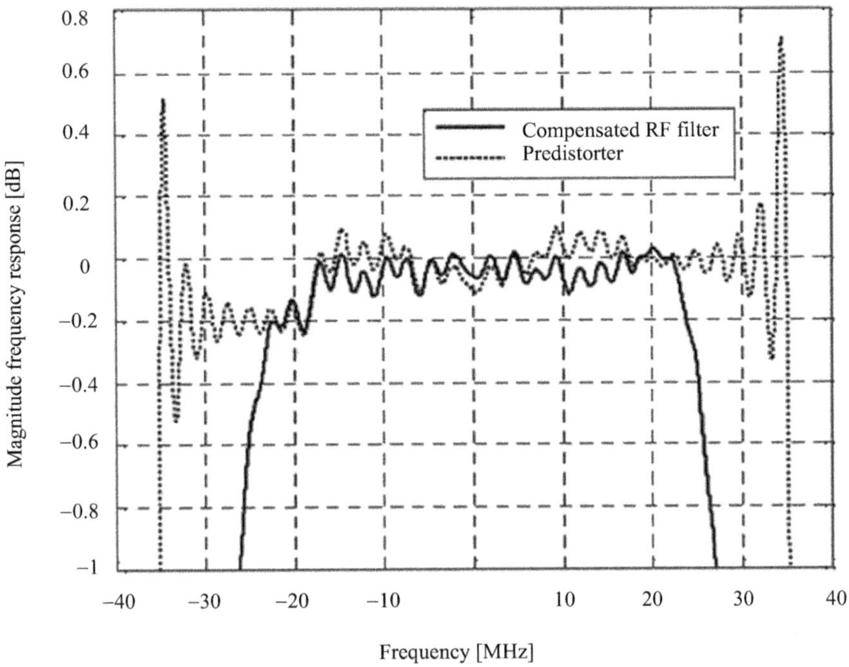

Figure 8.13 Magnitude response of compensated RF filter and predistorter [18]

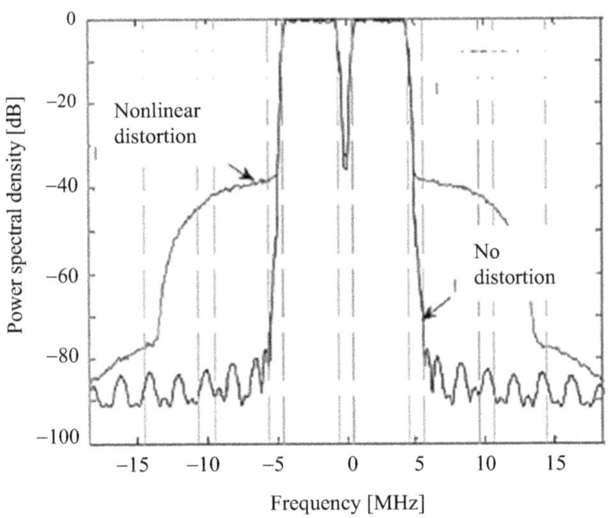

Figure 8.14 Pre- and post-distortion processing for UWB waveform [18]

8.4.3 Waveform notching

When one system transmits in the vicinity of and at the same time as another operating receiver, the radar energy is considered to be RFI. RFI also includes the effects of other emitters on the radar operation. It is important to note that no RF part of the spectrum is sufficiently allocated for UWB radar. As a result, it is imperative to plan on notching the radar transmit pulse to avoid interference of the other RF receivers within the geographical area and RF band. However, the efficient notching of a transmit waveform by using only phase has been a difficult process.

Solid-state transmitters running in saturation do not provide the capabilities for amplitude modulation of the transmit waveform. The system design problem is to provide single or multiple sequences with constant modulus in the time domain, such that their respective matched filter outputs ideally resemble delta functions while the cross-matched filter outputs are zeros. Since the matched filter outputs represent auto- and cross-correlations of the original sequences, the problem is to design constant modulus time series whose autocorrelations approximate the delta function and the cross-correlations approximate the all-zero function. Therefore, the use of a unimodal or phase-only waveform was developed to enable multiple nulls in the transmit waveform [19].

A technique is also needed for restoring the range sidelobes to the required low levels, because of the loss of waveform uniformity. A similar technique has been developed to compensate for notching on transmit and provide the required low time sidelobes. A linear least-squares iteration on the waveform in the time domain has been shown to reduce the sidelobes [14].

Table 8.2 lists the specifications for the full band LFM signal and the desired frequency nulls. Figure 8.15a illustrates the full band chirp spectrum, and Figure 8.15b illustrates the desired notched chirp spectrum. These specifications will be used in the waveform design examples that follow. Constant envelope finite duration waveforms with prescribed notches are designed using approaches similar to convex projection type operations in the time-frequency domain. Notches shape the spectrum and prescribed notches give a spectrum with given characteristics. Hence, the problem is to design a constant modulus waveform with the prescribed spectrum. In addition, the resolution properties of the resulting waveform should be comparable to a standard LFM signal.

Table 8.2 Waveform and null specifications [14]

Component	Center frequency	Bandwidth (MHz)	F_{low} (MHz)	F_{high} (MHz)
Fullband LFM	400	200	300	500
1	335	2	334	336
2	348	5	345.5	350.5
3	372	2	371	373
4	425	8	421	429
5	458	5	454.5	460.5

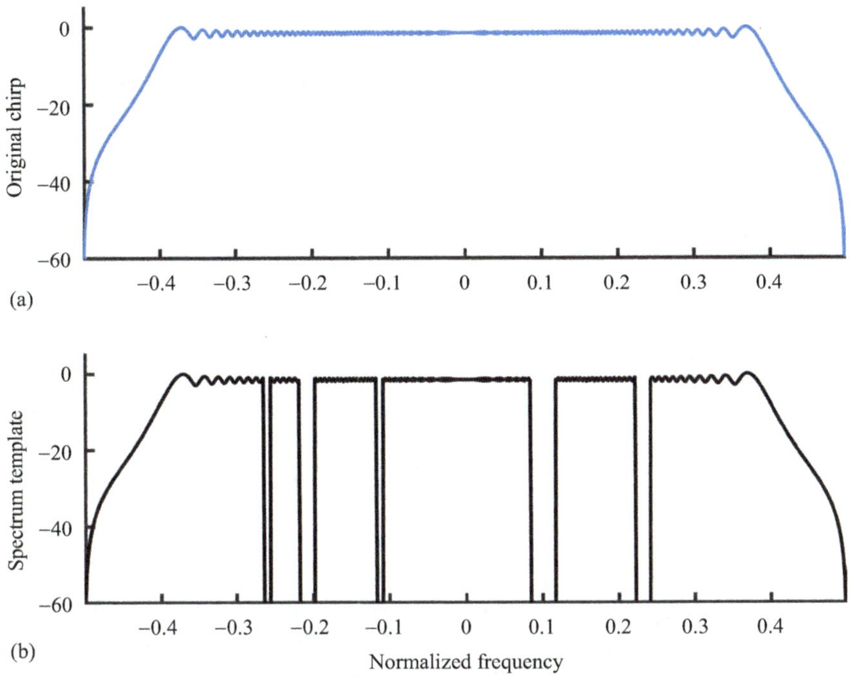

Figure 8.15 LFM waveform with five-band spectrum avoidance [14]. (a) Original LFM spectrum and (b) notched LFM templae

If these notches are to be applied as part of a cognitive radar, the implementation technique needs to be iterative and predictable. The location and spectral bandwidth of the notches can be calculated from the environmental database or the spectrum sensing functions. For fixed emitters such as airfields or civil operations, the depth of these notches must be determined to meet the frequency avoidance criteria.

The goal addressed here is the construction of a time domain constant modulus waveform of $y(t)$ specified duration $[0, T]$, whose frequency spectrum $Y(f)$ has nulls at k specified frequencies f_k for $1 \le k \le K$. Several nulls can be placed near each other to form a stop-band in the spectrum. At each step of the algorithm, we will project the signal y onto the subspace of signals having nulls at the specified frequencies.

To do so, the first projection is onto the complex exponentials having those frequencies to be notched and followed by subtracting the resulting signal from y. Let $\boldsymbol{Q}$ be an orthogonalization of the columns of the $N \times K$ constraint matrix $\boldsymbol{C}$ where

$$C_{n,k} = \exp\left(\mathrm{j}2\pi f_k n\right),\ 0 \le n \le N-1,\ 1 \le k \le K. \tag{8.11}$$

That is, $\boldsymbol{Q}^*\boldsymbol{Q} = \boldsymbol{I}$ and the columns of $\boldsymbol{Q}$ span the same space as the columns of $\boldsymbol{C}$. Then the required constraint can be written as

$$\boldsymbol{Q}^*y = 0 \tag{8.12}$$

where $y = [y(0), y(2), \ldots, y(N-1)]^{\mathrm{T}}$. And the projection matrix P can be expressed as

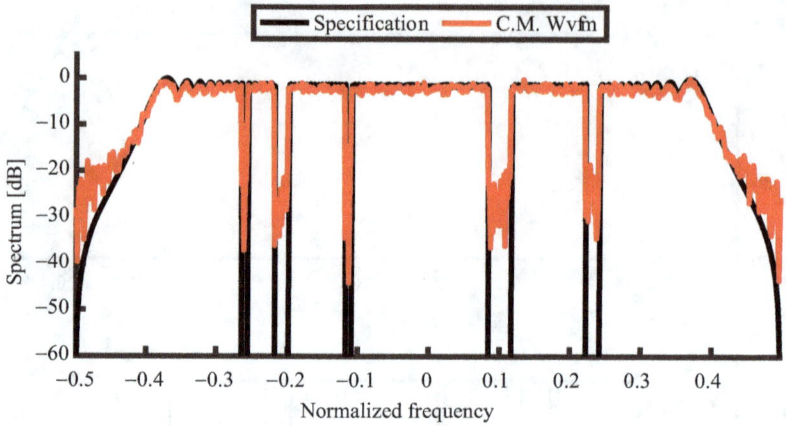

Figure 8.16 Spectra for alternating projection method [14]

$$P = \boldsymbol{I} - \boldsymbol{Q}^{*}\boldsymbol{Q} \tag{8.13}$$

By defining the projection in this way, the iterative algorithm to obtain a chirp-like constant modulus signal $y(t)$ having frequency domain nulls at K distinct frequencies, f_k. The function is given by first initializing the waveform using the fullband chirp

$$y(n) \leftarrow A(n)\exp\left[\mathrm{j}\left(2\pi f_0 n + \beta n^2\right)\right] \tag{8.14}$$

The notch constraint is applied in the time domain using the projection matrix from the following process:

$$\text{set } \mathbf{Q}^{*}y = 0, \text{ and } y \leftarrow y - \mathbf{Q}\,\mathbf{Q}^{*}y \tag{8.15}$$

Finally, the constant modulus constraint is applied before iterating, by repeating from (8.15):

$$y(n) \leftarrow A(n)\exp\left[\mathrm{j} \measuredangle y(n)\right],\ 0 \le n \le N-1 \tag{8.16}$$

Figure 8.16 illustrates the waveform designed with the null specifications given in Table 8.2. In this case, each null has a multiplicity of 2.

Match filter with sidelobe minimization

The major problem with notching waveforms, no matter how efficient, is that they introduce higher sidelobes that obscure object detection and discrimination. The objective is to design an input unimodular sequence x, where$|x| = 1$ of length N. When a receiver is filtered with the following spectral components [19]:

$$h = \left[h_0, h_0, \ldots, h_K\right] \tag{8.17}$$

it can be designed to satisfy the "desired" output property. Thus, when the transform $z(n)$:

$$z(n) = \sum_{k=1}^{K} x(k)\, h(n-k+1),\ n = 1 \to N+K-1 \tag{8.18}$$

represents the output at time instant, it is ideally required that

$$z(n) = \delta(n - n_0) \tag{8.19}$$

where n_0 corresponds to the desired peak term at the output. In general, the receiver length K is a free parameter left to other design considerations. One approach in this context is to use a criterion similar to the merit factor. Thus, we define

$$\eta = \frac{|z(n_0)|^2}{\sum_{n \neq n_0} |z(n|^2} \tag{8.20}$$

to represent the ratio of the receiver output peak value power to the total undesired sidelobe power at each receiver output, then the receiver design problem is to maximize η. Observe that the criterion in (8.20) is well known and it is the same as the approach behind the classical matched filter. As a result, it can be expressed in matrix form as

$$\eta = \frac{|\boldsymbol{b}_{n_0}\boldsymbol{h}|^2}{\boldsymbol{h}^{*}\boldsymbol{A}\boldsymbol{h}} \tag{8.21}$$

where

$$\boldsymbol{A} = \boldsymbol{F}_0^{*}\boldsymbol{F}_0 \tag{8.22}$$

Represents a $K \times K$ positive-definite matrix and $\mathbf{F}$ represents the $(N + K - 1) \times K$ convolution matrix generated from the input sequence $\boldsymbol{x}$. In (8.22), $\boldsymbol{F}_0$ represents the $(N{+}K{-}2) \times K$ matrix generated from $\mathbf{F}$ with its n_0 row deleted. Finally, in (8.21), $\boldsymbol{b}_{n_0}$ represents the row vector corresponding to the n_0 row of $\boldsymbol{F}$.

$$\boldsymbol{b}_{n_0} = [0, \ldots, x(N), x(N-1), \ldots, x(1), 0, \ldots, 0] \tag{8.23}$$

for $n_0 > N$. The number of zeros in (8.23) depends on the value n_0 with respect to N. For example, with $n_0 = N$ and $K = N$, the vector $\boldsymbol{b}_{n_0}$ is the time reverse of $\boldsymbol{x}$. Applying the Schwarz inequality to (8.23) yields the optimum receiver to be

$$\boldsymbol{h} = \boldsymbol{A}^{-1}\boldsymbol{b}_{n_0} \tag{8.24}$$

Figure 8.17 illustrates the waveform designed using Alternating Projection method with the null specifications given in Table 8.2. The time domain waveform corresponding to the red spectrum in Figure 8.16 has a constant amplitude, and the nulls are approximately 30--35 dB lower than the passbands. Adding the nulls while maintaining a constant modulus generally has the effect of perturbing the remaining passbands of the waveform; however, there is minimal perturbation here. Figure 8.17 shows the outputs of applying both the matched filter (MF) and sidelobe-minimizing receiver (SLR) to the notched waveform. Here we see that using the MF results in undesirably high sidelobes. But by applying the SLR technique, the mainlobe width can be preserved, while simultaneously minimizing the sidelobe level. The MF output of the fullband LFM is provided for reference.

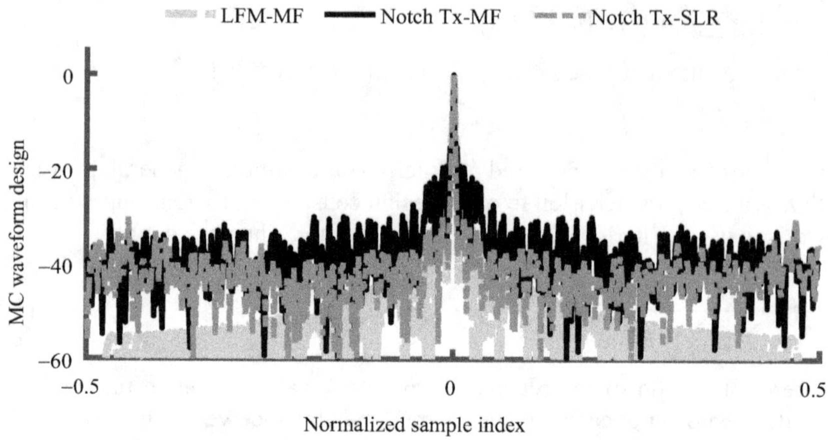

Figure 8.17 Receiver output using alternating projection [14]

GeoSAR notching results

In order to meet NTIA regulations, multiple notches for up to 30% of bandwidth are often required and up to 60% in high-density populations in the United States. These requirements need a technique that is compatible with solid-state, saturated amplifiers, hence the development of unimodular sequences. If a cognitive radar approach is accepted to reduce the total instantaneous notch bandwidth, then the approach requires fast calculation of notches that minimize waveform sidelobes to adapting to the environment. In the first 6 years of worldwide GeoSAR mapping, the image quality was satisfactory with spectral notching less than 20% of signal bandwidth. Furthermore, in the GeoSAR history from 2003 to 2015, no GeoSAR interference was reported to spectrum regulators, when signal notching was used for ranges less than 20 km from the nadir track. After 2008, the new policy within the continental United States was changed to notch all potential receivers within 150 nmi of the radar. And the prior notification of flights operations was extended to 90 days. These changes severely affected the economics of flying a UWB UHF radar for commercial and Earth resource characterization. The majority of the flights for GeoSAR after this NTIA policy change have to be in remote geographic areas, such as equatorial South America and Malaysia. While GeoSAR was operating in remote geographic areas, the full bandwidth of the radar could be used. Figure 8.18 shows mappining in Columbia, where the local frequency regulations did not require notching. The image quality was very good, and the land features were such that the interferometric synthetic aperture radar (InSAR) mode would provide excellent measurement of the terrain below the tops of the dense forests as shown in Section 6.1.7.

Notching transmit waveforms have been employed in the GeoSAR radar for interferometric SAR terrain height applications. However, the notching depends on the geographic region of operation. Figure 8.19 shows the waveform used in Canada, where only two notches were required: the instrument landing system and an emergency personnel recovery beacon. The MF response maintained IPR sidelobes out to

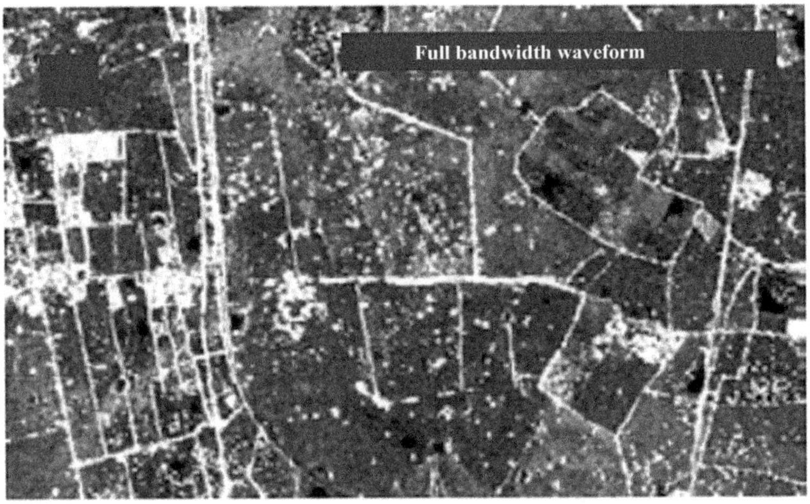

Figure 8.18 GeoSAR map – Columbia, South America [7]

30 m from the object being measured. This was sufficient for meeting the multiplicative noise response for InSAR operation [7].

The notching of a transmit waveform adds an additional step to the image formation processing. There are two types of notches. One is very narrow band, typically between 25 and 200 kHz. This notch can be applied within the pulse width, and the phase across the notch is small compared to the phase slope. The second notch is many MHz in width or over 2% of the bandwidth. If this notch were to be applied by simply notching the whole transmitted waveform, then both the signal to noise level and waveform sidelobes will be severely affected. In this case, the phase is continuous across the waveform, and the frequency jumps over the portion of the waveform that is notched [20].

While GeoSAR was operating in sparsely populated areas, the full bandwidth could be utilized for characterizing the InSAR return. With the 160 MHz of bandwidth and no notching on transmit, the system demonstrated 0.83 m of resolution, and an integrated sidelobe ration (ISLR) that rolled of at –12 dB per decade. Even in lightly populated areas, such as southern Canada, the waveform was only notched by 11% and the resolution was approximately 1 m. This was sufficient to obtain the full digital terrain elevation measurements that were desired. And the ISLR was sufficient that the multiplicative noise was not seriously degraded in a mixture of strong clutter and urban building. However, due to the changes in NTIA approvals on operation in continental United States, the waveform needed to be notched by over 35%. In this case, the IPR sidelobes only decay by –4.2 dB per octave. This is unusable for digital terrain elevation data (DTED) generation or calibration of this commercial mapping system.

However, when the change in NTIA policy occurred after 2008, the notching percentage was raised to over 50%. And the range resolution was degraded to 3.75 m. The spectrun notching in densely populated are%as, such as Southern Maryland, up to 50 oercent of the band nothcing was required, as shown in Figure 8.20. The

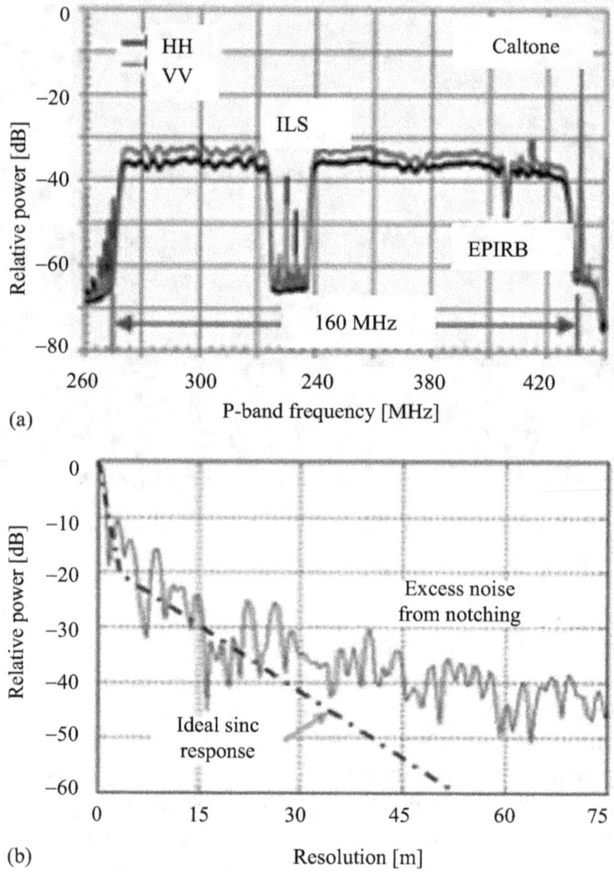

Figure 8.19 *GeoSAR waveform for Southern Canada [7]. (a) Transmit spectrum and (b) P-band impulse response*

sidelobes of the waveform are so seriously degraded that strong clutter would mask any cultural features in the image. This makes the viability of a P-band InSAR platform questionable for operation in any populated region where the NTIA regulations were conservatively employed. An SAR image from Southern Maryland is shown in Figure 8.21 with visibly strong artifacts in the image. The inset within the figure illustrates the effects of very high sidelobes in the radar impulse function, with layover of strong scatterers away from the imaging area. These artifacts destroy the utility of this class of mapping radar.

Table 8.3 summarizes the impact of notching percentage from three geographic regions. The results of notching provide distinct impacts on the SAR image quality. Without notching the GeoSAR transmit waveform has a resoluton of 0.83 m and the ISLR will fall off at −12 dB per decade in spatial distance. When the notching is only 11%, the resolution is modestly degraded to 1 m, and the ISLR has a −10 dB

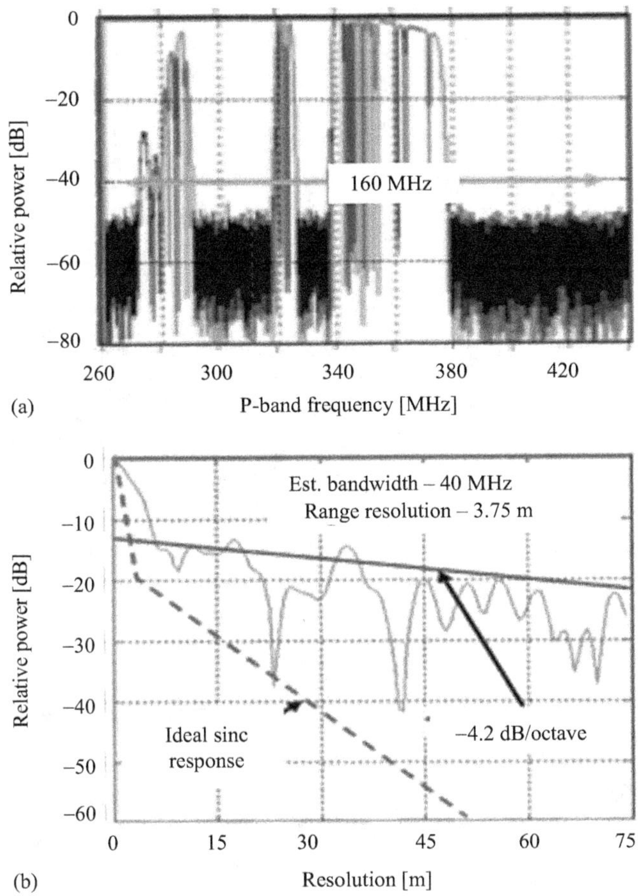

Figure 8.20 GeoSAR waveform for Southern Maryland [7]. (a) Transmit spectrum and (b) P-band impulse response

per decade. Both these image measurements are adequate for the intended application. However, with the very conservative frequency allocation in populated areas, the image quality is destroyed. It is important that the radar community develop a means to reduce the impact of UWB radars in wide-area surveilance applications.

8.4.4 Potential for spectrally compliant radar waveforms

The development of digital RF communication systems was able to achieve higher data rates and throughput, and at the same time to use low-cost consumer devices. This objective resulted in the increase of the interference levels in the physical layer. Frequency division multiple access (FDMA) is used in the fourth-generation LTE mobile wireless communications in orthogonal frequency division multiple access

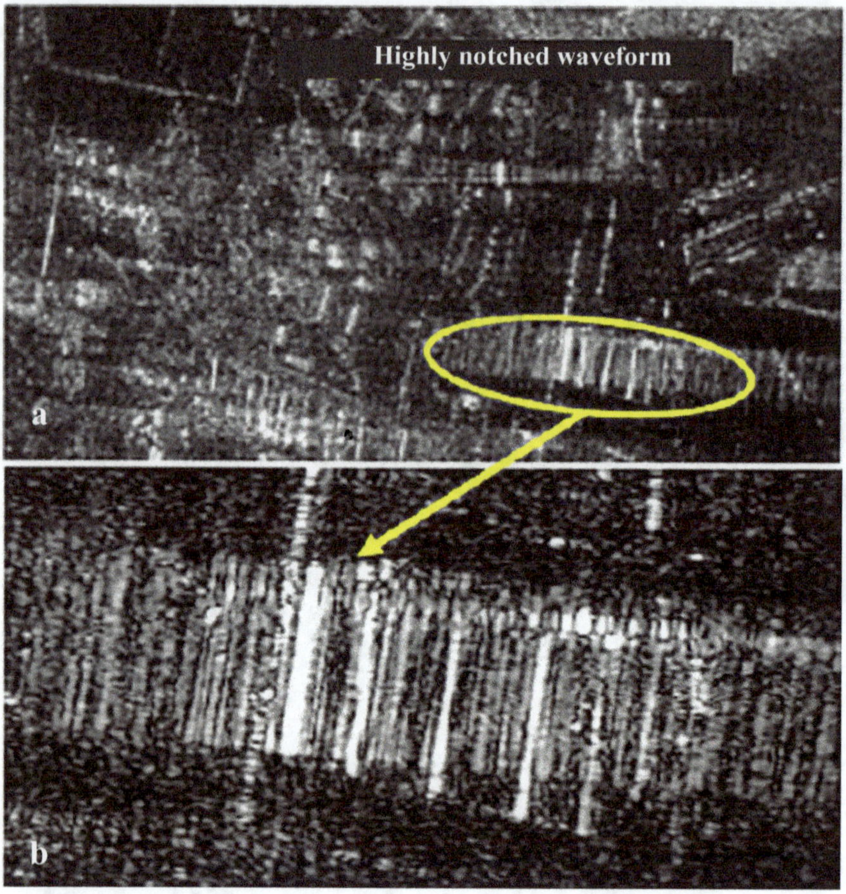

Figure 8.21 *GeoSAR map – Western MD, USA [7]*

(OFDMA), and the signal bandwidth is split into *N* orthogonal subcarriers, resulting in a high spectral efficiency, as shown in Figure 8.22. The left panel shows eight frequencies in adjacent code channels, while the right panel shows only the even code channels. The design of these channels is such that the nulls of the receiver channel reduces the cross-coupling of the energy in the two channels [21].

Table 8.3 *GeoSAR waveform performance under varied notching [7]*

Data set	Notch (%)	Range resolution (m)	ISLR dB/decade
Papua New Guinea	0	0.83	−12
Southern Canada	11	1.04	−10
Western MD	60	3.75	−4.5

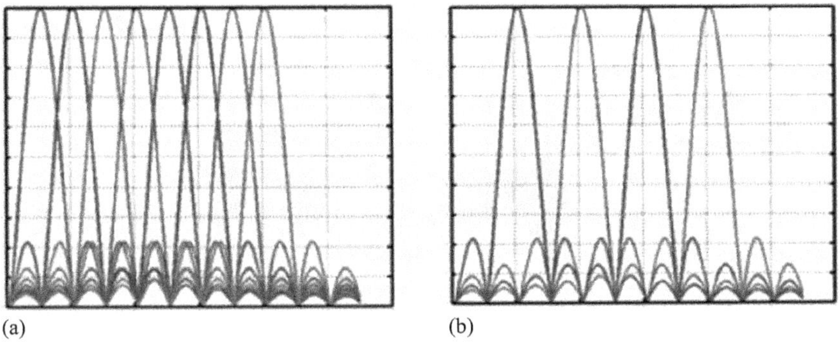

Figure 8.22 OFDM communications waveform [21]. (a) Eight adjacent carriers and (b) even filter

In contrast to the OFDM waveform, a frequency jump burst (FJB) in radar applications is analogous to FDMA and has the potential for being orthogonal to communications waveforms. However, the FJB waveform shown in Figure 8.23 is pulsed and the spectrum spill-over may cause problems with the frequency hopping filters in radar receivers [22]. This section introduces the benefits of OFDM for possible radar applications [23].

OFDM signals are relatively new waveforms in the radar community. The main difference with other common radar waveforms (chirp, step frequency, etc.) lies in the simultaneous transmission of the frequencies. For that reason, OFDM waveforms have been developed for short time measurements over a wide frequency band (e.g., micro-Doppler analysis). It is a challenge to the radar community to develop the OFDM for a long-range surveillance system and maintain the coherencies over long time epochs and motion of the platform.

The frequencies are orthogonal over the symbol duration, T_0, which means that the frequency difference between two adjacent carriers is equal (or proportional) to the inverse of T_0. The signal has thus a two-dimensional structure, over N carriers (in frequency) and M chips (in time), as shown in Figure 8.22. The complex envelope of the baseband signal can be expressed by [21]

$$s(t) = \sum_{n=1}^{N} \sum_{m=1}^{M} a_{n,m} h\left[t - (m-1)T_0\right] e^{\left[j2\pi\left(n - \frac{N+1}{2}\right)\frac{t}{T_0}\right]} \tag{8.25}$$

where T_0 is the symbol period (time of a chip) and $a_{n,m}$ the complex code associated with the nth carrier and the mth chip. The function h is defined by $h(t) = 1$ for $0 \le t < T_0$, and zero elsewhere.

Unlike linear frequency-modulated chirp or stepped frequency waveforms, OFDM signals are digital waveforms by nature and can be easily generated by arbitrary waveform generators. An important and rather negative aspect of OFDM waveforms is their high peak-to-mean envelope power ratio. A noncoded signal will exhibit constructive interference between the different sine waves over very short time.

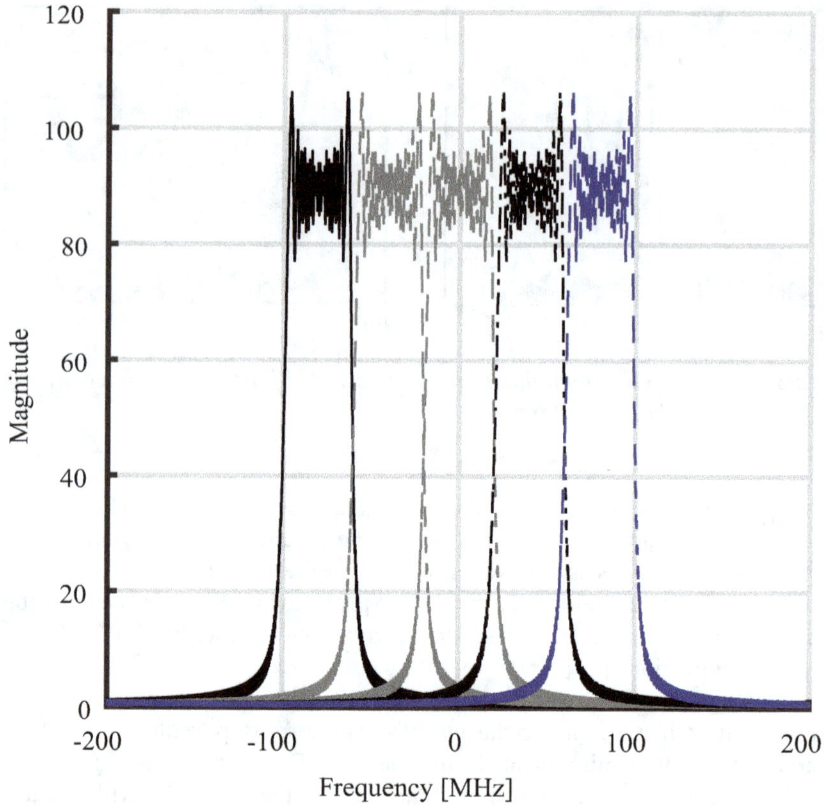

Figure 8.23 FJB – radar waveform [9]

The idea of applying OFDM signals in radar systems has been a hot point recently. Since the waveforms are generated digitally with pulse-to-pulse shape variation, it is easy to resist narrow-band jamming by simply turning off certain sub-bands. This feature also brings the advantages such as simultaneous communication in parallel to radar sensing, increased the low probability of intercept/detection, transmitting large amounts data in real time, etc. Second, OFDM signals are reported to be advantageous regarding Doppler shift.

The ambiguity function of an SAR with an LFM signal can be expressed as [24]

$$\chi(\Delta x, \Delta y) = \delta_1 \delta_2 \frac{\sin\left(\dfrac{2K_r T_r}{c r_0} y_0 \Delta y\right)}{\dfrac{2K_r}{c r_0} y_0 \Delta y} \frac{\sin\left(\dfrac{2\pi v_a f_c T_s}{c r_0} \Delta x\right)}{\dfrac{2\pi v_a f_c T_s}{c r_0} \Delta x} \tag{8.26}$$

where K_r denotes the range frequency modulation (FM) rate, T_r represents the duration of time pulse in the range and f_c the center frequency of SAR.

The OFDM signal can be considered as a group of multifrequency systems or a monostatic system equipped with one transmitter and one receiver. The OFDM time domain signal with N orthogonal sub-bands can be expressed as

$$u(t) = \frac{1}{\sqrt{N}} \sum_{i=1}^{N} d_i(t) \exp(j2\pi f_i t), \ 0 \le t \le T \tag{8.27}$$

with $d_i(t)$ denoting the envelope of the ith sub-band, f_i being the individual subcarrier frequencies and T being the envelope duration. The sub-band signals will be transmitted simultaneously. Usually $d_i(t)$ stands for one OFDM symbol and $\{d_i(t)\}$ represents an arbitrary information series consisting of complex modulation symbols. The forms of $d_i(t)$ include a large selection of signals, such as a rectangle, phrase modulation, random modulation, LFM signal, etc. The OFDM signal must be orthogonal between each single subcarrier to avoid the interference, which is fulfilled in the case of

$$\Delta f = \frac{k}{T}, \ k = 1,2,\ldots \tag{8.28}$$

The two-dimensional OFDM signal in SAR can be expressed as

$$u(t-\tau) = \frac{1}{\sqrt{N}} \sum_{i=1}^{N} d_i(t-\tau) \exp(j2\pi f_i(t-\tau)) \tag{8.29}$$

where $\tau = 2R(t)/c$ is the time delay of the target. Assuming that the SAR is operating at broadside, and $T\Delta f = k$ to obtain the ambiguity function:

$$\begin{aligned} x(\Delta x, \Delta y) &= \frac{\delta_1 \delta_2}{N} \sum_{i=0}^{N-1} \sum_{m=0}^{N-1} \exp(j\pi\Delta\tau(i-m)\Delta f) \frac{\sin(\pi\Delta\tau(i-m)\Delta f)}{\pi(i-m)\Delta f} \\ &\cdot \sum_{m=0}^{N-1} \exp\left(-j\frac{2\pi \upsilon_a f_m \Delta_x}{cr_0} T_s\right) \frac{\sin\left(\frac{2\pi \upsilon_a f_m T_s}{cr_0}\Delta_x\right)}{\frac{2\pi\ \upsilon_a f_m \Delta_x}{cr_0}} \end{aligned} \tag{8.30}$$

From (8.30), the ambiguity function of an SAR based on an OFDM signal only changes with the relative position of targets and not the absolute positions. The ambiguity function in range depends on the sum of the sub-bands returns from both targets. This result embodies the feature of the OFDM signal. These sub-band signals also determine the range resolution. If the center carriers between two sub-bands are farther and the bandwidth of the sub-band is wider, the range resolution will be higher. When all sub-bands work, this ambiguity function shows the equivalent range resolution Δr as conventional SAR:

$$\Delta r = \frac{c}{2B} = \frac{c}{2N\Delta f} \tag{8.31}$$

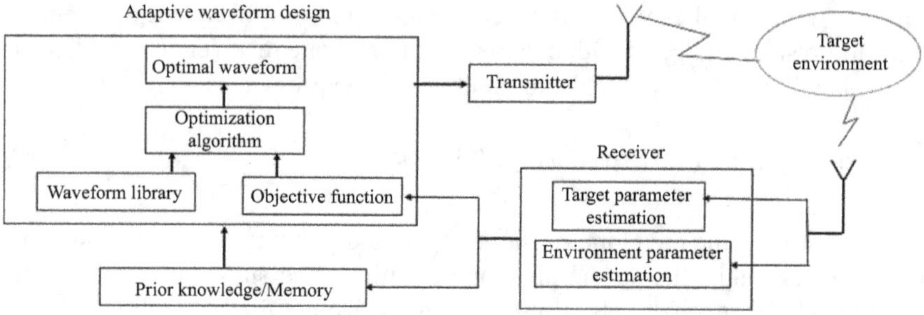

Figure 8.24 Block diagram of the OFDM cognitive radar [21]

The ambiguity function in azimuth or cross-range depends on the sum of the sub-bands returns only from one target, which is the same as the conventional SAR. The longer the length of the synthetic aperture is, the higher the azimuth resolution is.

The next-generation surveillance radars, namely cognitive radar, has the feedback function from the receiver to the transmitter. As such, the radar can not only process data adaptively in receiver, but also emit adaptive waveforms from the transmitter. Therefore, the whole radar is a dynamic closed-loop system. The idea of cognitive radar is to improve the performance of the next-generation radar and fit the fast changing environment of the modern battlefield, which leads to a bright future both in civilian and military area [25].

As shown in Figure 8.24, the block diagram of the proposed OFDM cognitive radar describes a closed-loop system. First, the transmitter emits OFDM signal to the target and environment. The echoes are processed by two functional blocks in the receiver, which are the target parameter estimation block and the environment parameter estimation block. Their aim is to obtain information of the target and environment. Then, the information is sent to the prior knowledge storage/memory block and the adaptive waveform design block. The prior knowledge of the RF environment is stored in the system memory in order to transforms the information into "knowledge" and makes these data useful for match filtering the received waveform for SAR imaging.

The OFDM radar single pulse $\boldsymbol{s(t)}$ is expressed by [21]

$$s\left(t\right) = \sum_{n=0}^{N} w_n \operatorname{rect}\left(\frac{t - T_p/2}{T_p}\right) e^{j2\pi f_n t} \tag{8.32}$$

where N is the number of subcarriers, T_p duration of a pulse and w_n the weight of the nth subcarrier.

$$\operatorname{rect}(t) = \begin{cases} 1 & -\frac{1}{2} \le t \le 1/2 \\ 0 & \text{elsewhere} \end{cases}, \quad f_n = f_0 + n\Delta f. \tag{8.33}$$

and f_0 is the center frequency, $\Delta f = 1/T_p$.

The technical hypothesis is that a waveform can be developed for radar applications that remains orthogonal to the communications OFDM waveforms. This has been demonstrated only in small applications and in radar system analyses. Therefore, the challenge is whether these orthogonal, communications waveforms be applied to radar to reduce the mutual interference. Can the waveform be transmitted from a radar active electronic scanned antenna without excessive spectral spreading? Are there techniques, such as the reference tones in IFSAR, that will maintain coherence for long integration? And can the waveforms enable sufficient subclutter visibility for ground moving target indication applications?

8.5 RF interference removal

Image formation processing was a significant development challenge for UWB SAR. This is especially true for foliage penetration radars operating in the UHF or L-band. Because of the long wavelength, the signals do not allow lend themselves to linear approximations in terms of integration angle and SAR processing. The collection angles are very wide compared to microwave SAR. And the range migration is significant within any desirable swath, requiring large data collection volumes. But the most challenging part of UWB SAR is the presence of other RF receivers in the vicinity. It is also true for UWB X- and Ku-band radars, because of other sources in the same band. The sources can be datalinks, aviation landing radars and other radars. As a result, the process of removal of these sources is important for maintaining the image characteristics of the SAR systems. And when the application is InSAR, it is as important to make sure the two RF interference removal filters are the same for the two channels.

We have noted several times that RFI will directly affect the background noise in the image, affecting the ability to discern features of the SAR image and any target detection. Figure 8.25 illustrates the recorded spectrum from an experimental P3 UWB

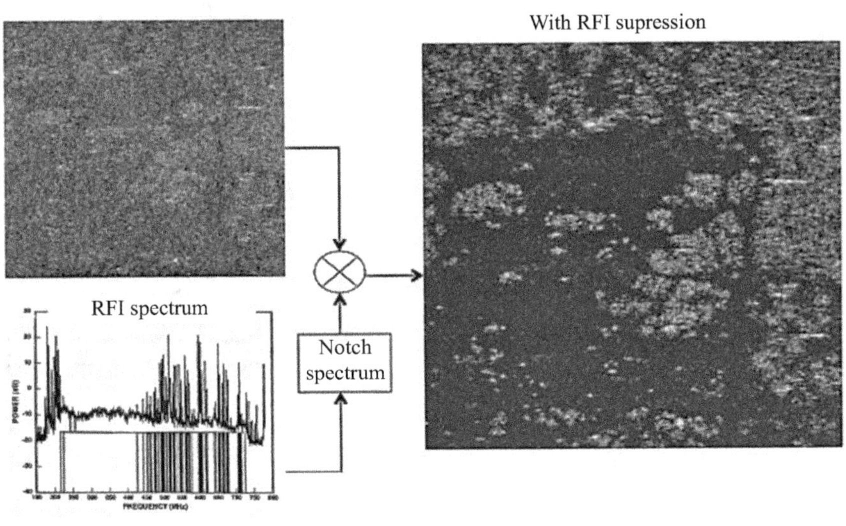

Figure 8.25 RF interference impact on SAR images [9]

collection in Grayling Michigan. The image in the upper left was the result without subtracting the RFI interference from the image processing. When the spectrum was notched at the appropriate frequencies, the image at the right was obtained. In this example, the noise equivalent sigma 0 (NES0) was improved from –6 to –18 dB.

These were very satisfactory results in the 1995 time period. However, important improvements in RFI mitigation have been developed. The next section will illustrate two different approaches to mitigate RFI suppression. In addition, the changes in the RFI sources from narrow-band TV to wideband digital communications and TV. The former exploited the stationarity of the RFI. However, the later made the RFI very nonstationary and has required more significant adaptive processing to counter the changing spectral nature of the interference.

An adaptive transversal filter has been developed to handle the stationary high-power RFI due to analog TV and radio stations. The large number of emitters in the environment were successfully cancelled for the UWB radar on the DARPA/Navy P3 test platform. However, the limitation to this approach is the proliferation of mobile communications and digital television, where the signal bandwidth is increased to 5 MHz.

8.5.1 Adaptive transversal filter

The adaptive transversal filter, as shown in Figure 8.26, starts with collecting a background RF spectrum from the platform receiver. Under the assumption that the RF interferences are stationary, the taps and weights can be calculated offline and applied to the fast-time data on receive. Offline implies that an UWB receiver channel can collect sufficient data when the transmitter signal is being received, and

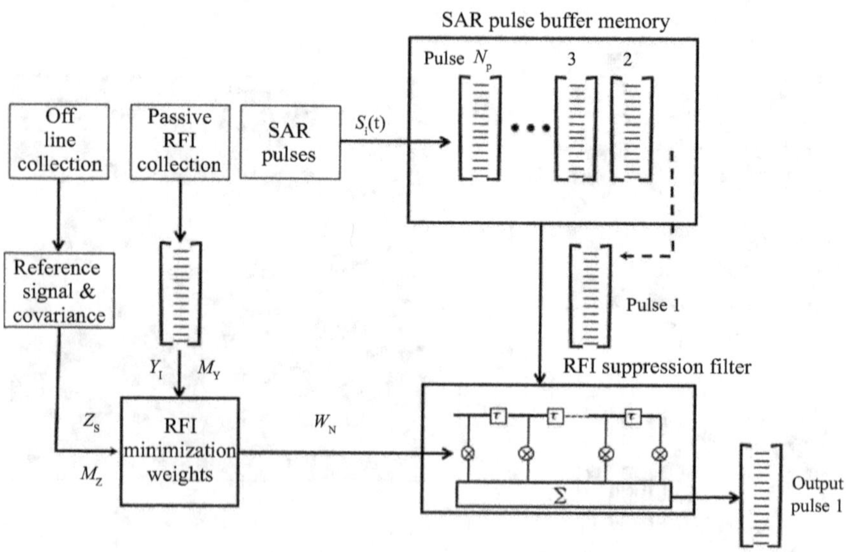

Figure 8.26 *Adaptive transversal filter for RFI removal [26]*

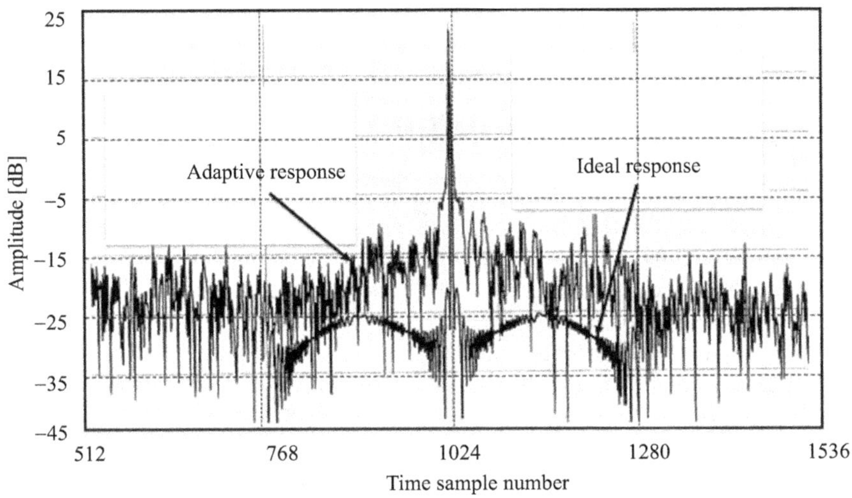

Figure 8.27 Transversal filter reduction of RF interference [26]

the weights are calculated prior to SAR image collection during sub-apertures of the SAR imaging. When the aircraft flies, each fast-time line of received data will pass through the FIR filter. These high-power but narrow-band interferences are cancelled to lower the noise background and improve the NES0 of the radar image.

The reference signal from convolving weights and receiver model [26]:

$$Z_S = w_Q^T s_R \otimes x_S \tag{8.34}$$

Weights from SINR minimization processing

$$W_N = [M_Z + M_Y]^{-1} r_D \tag{8.35}$$

With the weights determined, each record of the SAR image is processed serially through the transversal filter. For example, the P3 UWB radar used a 515-MHz bandwidth LFM pulse, which exhibited a 7.6-dB input signal to the interference ratio. A transversal filter with 251 taps was obtained after using 1,500 adaptive iterations to obtain suitable weights. A comparison of the ideal LFM signal with the adaptive output with RFI removed is shown in Figure 8.27. The quantitative signal response is summarized in Table 8.4.

Table 8.4 Comparison of RFI removal with ideal UWB waveform [26]

	Ideal response (dB)	Adaptive response (dB)
Peak signal amplitude	22.7	20.0
Peak sidelobe level	43.8	22.9
RMS sidelobe level	56.5	38.9

8.5.2 Least–squares adaptive RFI removal

The LMS adaptation process has been widely published in the adaptive processing literature [27]. The performance of a multichannel radar relies on optimum weights derived in terms of the correlations between the antenna channels. However, optimum weights are never achieved due to uncertainties in the antenna gain over the bandwidth and propagation differences in the environment. The implementation of an adaptive sidelobe canceller using correlation feedback loops is developed by Applebaum and Howells [28] and was utilized in this RFI algorithm. An algorithm that attempted to directly estimate the covariance matrix was the Reed, Mallett and Brennan adaptive processor [29]. These algorithms started the space-time adaptive processing research for battlefield surveillance radar in the early 1990s.

An adaptive processing technique was developed on the GeoSAR system to enable RFI removal in the frequency domain rather than the time domain. The bandwidth of GeoSAR was smaller than the UWB SAR testbed (UWB P3), and direct digital sampling at the intermediate frequency was used [30].

A technique that resembles adaptive array processing was developed. By using a reference channel, the RFI environment was sampled and transformed into the SAR image domain. An LMS adaptive algorithm then correlated the reference signal to the primary input and subtracted the weighted signal. Several techniques were explored using both temporal and frequency domains for the interference subtraction.

The dual requirement of a low radar frequency for foliage and/or ground penetration and an UWB signal for high range resolution leads to radars operating in frequency bands occupied by other radio systems, such as radio communications, navigation, police and emergency rescue. The LMS adaptive filter is applied to suppress narrow-band RFI in wideband SAR signals, as illustrated in Figure 8.28. The input signal from the radar fast-time signal is given by $d(n)$. $\mathbf{x}(n)$ is the L-dimensional reference vector given by [30]

$$\mathbf{x}(n) = [x(n), x(n-1), \ldots, x(n-L+1)]^{\mathrm{T}} \tag{8.36}$$

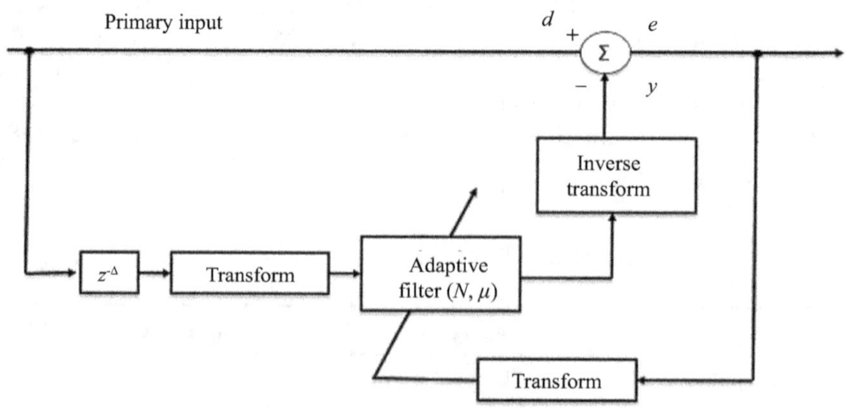

Figure 8.28 Frequency domain adaptive RFI removal [30]

and $x(n)$ is the reference signal with a constant decorrelation parameter Δ:

$$x(n) = d(n - \Delta) \tag{8.37}$$

The delay Δ causes decorrelation between the wideband components (radar signal) of the primary input *d* and the reference input ***x***. The adaptive filter tries to estimate the narrow-band component *y* (RFI signal), and in doing so, effectively forms an equivalent transfer function, which is similar to that of narrow-band filters centered at the frequencies of the narrow-band components of the input signal. The wideband component of the delayed input is rejected, while the phase difference of the narrow-band components is readjusted so that they cancel each other at the summing junction, producing a minimum error signal consisting mainly of the wideband component.

For each iteration and when the next sample is available, the weights of the transversal filter are adapted according to the equation

$$\boldsymbol{w}(n+1) = \boldsymbol{w}(n) + \mu \boldsymbol{x}(n)\, e^*(n) \tag{8.38}$$

where $\mathbf{w}(n)$ is the complex filter weight on the *n*th iteration

$$\boldsymbol{w}(n) = [w_o(n), w_1(n), \ldots w_{L-1}(n)]^{\mathrm{T}} \tag{8.39}$$

where L is the filter length, and μ is the adaptation constant or step size parameter. This parameter controls the tradeoff between convergence speed and final misadjustment. Large values of μ lead to faster convergence, at the expense of large final misadjustment. The step size is preselected based on the desired performance characteristics that are the radar performance parameters.

The output of the filter is a linear combination of these delayed past input values weighted by the filter weight vector $y(u)$:

$$y(n) = \boldsymbol{w^T} \cdot \boldsymbol{x}(n) = w^T(n) \cdot d(n - \Delta) \tag{8.40}$$

This gives an estimate of the RFI signal. The prediction error, which is the radar signal of interest, is obtained by subtracting the RFI estimate $y(n)$ from the received signal $d(n)$:

$$e(n) = d(n) - y(n) \tag{8.41}$$

The value of Δ is chosen to remove the correlation between the wideband components of the input signal $d(n)$ and the (predicted) filter output $y(n)$. The filter weights $w(n)$ are selected so as to minimize the mean square error $E[e^2(n)]$, where $E(.)$ is the expectation operator.

A summary of the main features and important results is as follows. Let λ_n be the eigenvalues associated with the input correlation matrix are [31]:

$$R_{xx} = \langle x(n)\, x^H(n) \rangle \tag{8.42}$$

Then, the learning time of the adaptive filter for each mode is

$$\tau_n = \frac{1}{4\mu\lambda_n} \tag{8.43}$$

With one sinusoid in white Gaussian noise, the convergence time τ can be expressed by

$$\tau = \frac{1}{2\mu\sigma_n^2\left(1 + \frac{L}{2}\mathrm{SNR}\right)} \tag{8.44}$$

where σ_n is the average noise power and SNR is the signal-to-noise ratio. We note that in the GeoSAR RFI problem, σ_n is the power of the radar signal, i.e., the radar signal is treated as noise as far as the adaptive filter is concerned.

Figure 8.29 illustrates LMS adaptive interference subtraction in the frequency domain, by applying the LMS algorithm. The top signal spectrum shows the received signal including the RFI and the transmitted waveform. The output of the adaptive RFI suppression shows the interference cancelled to within 3 dB of the ideas signal. In this example, almost full system sensitivity was achieved. The research evaluated time domain, frequency domain and filter bank architectures to exploit wideband radar and narrow-band RFI sources. The long time delays decorrelate the wideband SAR and reference inputs. By using an adaptive filter, the RFI is predicted and subtracted as narrow-band notches.

The LMS RFI approach was an efficient processing algorithm, but it was only feasible for non-real-time operation. However, with increase in signal processing efficiency and advent of very high-speed integrated circuits, these techniques are being evaluated for real-time operation. These algorithms may suffer in

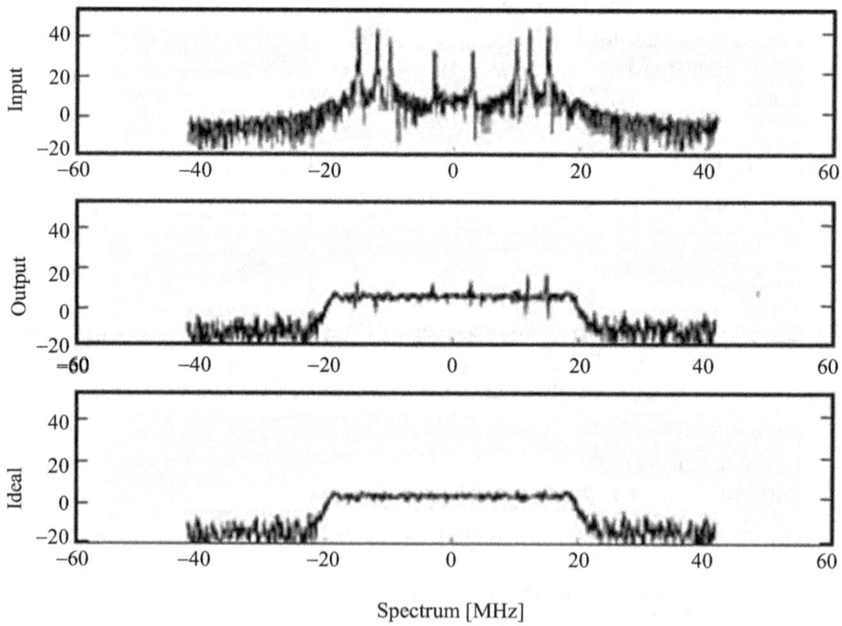

Figure 8.29 Frequency domain RFI removal results [30]

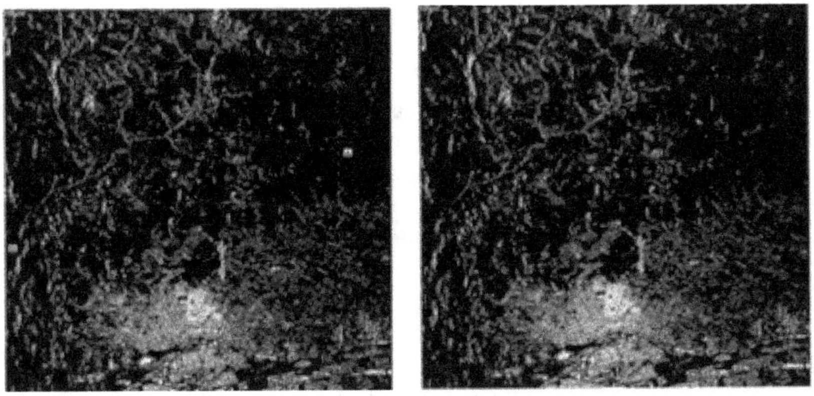

(a) (b)

Figure 8.30 SAR image removal of RFI artifacts [31]. (a) Before RFI removal and (b) After RFI removal

non-stationary interference due to the block nature of the filter. Figure 8.30 provides an example of a high-quality image in Southern Californial remote collection, with the RFI removal from high-power television and radio towers within the line of sight to the radar.

8.6 Summary

The benefits of UWB radars are many. They have direct applications for military, civilian and geospatial applications around the world. The most compelling applications of a UWB interferometric radar system are for disaster relief and remediation after major events such as earthquakes, volcano eruptions and major floods. The ISR platforms have been used sparingly to monitor flooding and widespread destruction of infrastructure. However, the need to obtain frequency allocation before these events are most difficult.

The NTIA had set up a process for providing license for UWB radars. It has been a difficult process to get a license for operating in many geographic areas, especially in highly populated regions. However, it has been possible to get these permissions on a case-by-case basis for military applications. For civilian applications, it has become much more difficult. This resistance by the regulatory community has been increasing due to the rapid expansion of civilian and personal communications.

The UWB radar community needs to start to adopt many of the techniques from cellular communications and digital telephone implementation. The rapid development of wideband digitization technology and integrated signal processing for adapting the transmit and receive waveforms to the environment are important for future applications.

UHF SAR for foliage and ground penetration requires UWB waveform for spatial resolution to detect objects under dense forest or in areas where there is local volumetric clutter obscuring the targets. It is necessary but difficult to obtain permission for operation of these UWB transmission systems that need to comply with the regulations of NTIA and CEPT. These regulatory organizations use very conservative signal interference metric that assumes a fully coherent receiver and does not consider fast RF waveform sweep or transition effects on the receiver. The NTIA Part 15 dictates that the transmission provides multiple notches on transmit that cover a very high percentage of the RF spectrum, regardless of whether there are any receivers in the line-of-sight environment. A digital matched filter can alleviate some sparse spectral aperture effects on the ISLR.

However, the evolution of digital communications may allow a rational approach to licensing UWB systems. On the front end, adaptive selection of RF waveform frequency and bandwidth can be used. Cognitive radios have demonstrated the approach of sensing “white spaces” and providing rapid selection of RF sub-bands for transmission. This requires a fast electronic sensing and characterization receiver that can sense the environment and first predict open spectra availability, and sensing changes in the occupied spectrum for adaptation of the radar waveform to avoid interference.

This will still present a sparse spectral waveform for long integration times. As a result, the requirement for compressive sensing algorithms need to be developed to control fidelity of the transmit waveforms and improve the ISLR on the received energy.

A key enabler for cognitive UWB operation is fast digital waveform synthesis and high-speed control logic on selecting the transmit waveform on a pulse-to-pulse basis and enable matched filtering on reception. The dynamic range and bandwidth of analog-to-digital converers are improving. This advance in RF waveform implementation provides opportunities for better match for range waveform coherence and significant reduction in cross-channel interference between the radar and digital communications systems.

References

[1] Wikipedia. *United States Wireless Spectrum Auction*. 2008. Available from https://en.wikipedia.org/w/index.php?title=2008_United_States_wireless_spectrum_auction&oldid=913221545 [Accessed 30 August 2019].

[2] Griffiths H., Cohen L., Watts S., *et al*. ‘Radar spectrum engineering and management: technical and regulatory issues’. *Proc. IEEE*. 2015;**103**(1):85–102.

[3] Griffiths H.D. ‘The spectrum crunch – a radar perspective’ in Blunt S.D., Perrins E.S. (eds.). *Chapter 2 Radar and Communication Spectrum Sharing*. Stevenage: IET; 2018.

[4] Haykin S. ‘Cognitive radar: a way of the future’. *IEEE Signal Process. Mag*. 2006;**23**(1):30–40.

[5] Davis M.E. ‘Foliage Penetration Radar - Detection and Characterization of Objects Under Trees’. *Chapter 6 Radio Frequency Interference*. Raleigh, NC: Scitech Publishing; 2011.

[6] Hensley S., Van Zyl J., Lavalle M., Neuman M., Moghaddam M. 'L-Band and P-Band Studies of Vegetation at JPL'. *Proceedings 2015 IEEE Radar Conference*. Pretoria, South Africa; 2015. pp. 515–20.

[7] Davis M.E., Reis J.J. *Operation of GeoSAR foliage penetration modes in spectrum regulated environment*. Pretoria, South Africa: Proceedings of 2015 IEEE Radar Conference; 2015.

[8] European Space Agency. 'ESA Global Navigation Satellite System Agency'. *GSA GNSS Market Report – Number 6*. Luxembourg; 2019. Available from https://www.gsa.europa.eu/system/files/ reports/market_report_issue_6_v2.pdf.

[9] Davis M.E. 'Frequency allocation challenges for ultra-wideband radars'. *IEEE Aerosp. Electron. Syst. Mag.*. 2013;**28**(7):12–18.

[10] Federal Communications Commission. *Understanding the FCC Regulations for Low Power, Non-Licensed Transmitters, OET Bulletin No. 63*. Columbia, MD: FCC Office of Telecommunications Technology; 1993.

[11] DECAWAVE. *UWB Regulations - A Summary of Worldwide Telecommunications Regulations Governing the Use of Ultra-Wideband Radio. Application Note APR01*. Dublin, Ireland; 2015. Available from https://www.decawave.com/sites/default/files/apr001_uwb_worldwide_regulations_summaryrev1.2.pdf.

[12] Bruder J.A. 'IEEE radar standards and the radar systems panel'. *IEEE Aerosp. Electron. Syst. Mag.* 2013;**28**(7):19–22.

[13] Auction73. *Factsheetatfcc.gov*. Available from http://wireless.fcc.gov/auctions/default.htm?job= auction_factsheet&id=73)Auction 73 Factsheet from FCC.gov.

[14] Davis M.E., Pillai S.U. 'Waveform diversity for ultra-wide band surveillance radars'. *IET Radar, Sonar & Navigation*. 2014;**8**(9):1226–33.

[15] McHenry M., Livsics E., Nguyen T., Majunmdar N. 'XG dynamic spectrum sharing field test results'. *Proceedings of the IEEE International Symposium on New Frontiers in Dynamic Spectrum Access Networks*. 2nd. Dublin, Ireland; 2007.

[16] Tilghman P. *Spectrum Collaboration Challenge, DARPA*. 2019. Available from https://www.darpa.mil/program/spectrum-collaboration-challenge.

[17] Guerci J.R. *Cognitive Radar – The Knowledge Aided Fully Adaptive Approach*. Norwood, MA: Artech House; 2010.

[18] Lasanen M., Kotelba A., Hekkala A., Jarvensivu P., Mammela A. 'Adaptive Predistortion Architecture for Nonideal Radio Transmitter'. *Proceedings 2008 IEEE Vehicular Technology Conference*. Calgary, Canada; 2008. pp. 1256–60.

[19] Pillai S.U., Li K.Y., Zheng R., Himed B. 'Design of Unimodular Sequences Using Generalized Receiver'. *Proceedings of 2010 IEEE International Radar Conference*. Arlington, VA; 2010. pp. 729–34.

[20] Hensley S., Madsen S.N. '*Interferometric radar waveform design and the effective interferometric wavelength*'. *Proceedings 2007 Waveform Diversity and Design Workshop*. Kuai, HI; 2007. pp. 287–91.

[21] Paichard Y. 'OFDM waveforms for multistatic radars'. *Proceedings 2010 IEEE Radar Conference*. Washington, DC; 2010.
[22] Lord R.T., Inggs M.R. 'High resolution SAR processing using stepped frequencies'. *Proceedings 1997 IEEE International Geoscience and Remote Sensing Symposium*. Singapore; 1997. pp. 490–7.
[23] Dimitrov S., Privitera N., Suffritti R., Boccolini G., Awoseyila A., Evans B.G. 'Spectrally efficient waveforms for the return link in satellite communication systems'. *IEEE Transactions On Communications*. 2015;**63**(6).
[24] Fan C., Huang X., Jn T., Wenge C. 'Ambiguity function of SAR based on OFDM waveform'. *Proceedings of 2010 IEEE International Radar Conference*. Washington, DC; 2010. pp. 397–401.
[25] Huo K., Qiu Z., Liu Y., Jiang W. 'An adaptive waveform design method for OFDM cognitive radar'. *Proceedings 2014 International Radar Conference*. Lille, France; 2014.
[26] Koutsoudis T., Lovas L. 'RF Interference Suppression in Ultra Wideband Radar Receivers'. *Proceedings SPIE*. 2487. Orlando, FL; 1995. pp. 107–18.
[27] Nitzberg R. *Adaptive signal processing for radar*. Boston, MA: Artech House; 1992. pp. 81–106.
[28] Applebaum S.P. *Steady State and Transient Performance of the Sidelobe Canceller*. Syracuse, NY: Special Projects Laboratory Report, Syracuse University Research Corporation; 1966.
[29] Reed I.S., Mallett J.D., Brennan L.E. 'Rapid convergence rate in adaptive arrays'. *IEEE Trans. Aerosp. Electron. Syst*. 1974;**AES-10**(6):853–.
[30] Le C.T.C., Hensley S. 'RFI rejection in wideband SAR signals using LMS adaptive filters'. *Adaptive Sensor Array Processing Workshop, MIT Lincoln Labs*. 1; 1998. p. *295*.
[31] Le C.T.C., Hensley S., Chapin E. *Adaptive Filtering of RFI In Wideband Radars Using the LMS Algorithms. Part I: The TDLMS Adaptive Filter*. Pasadena, CA: NASA Jet Propulsion Laboratory, Technical Memo JPL00-1309; 2000.

Chapter 9

Ultra-wideband multimode operation

This chapter presents a history and some future directions for multimode ultra-wideband (UWB) surveillance radars. Many radar applications are now multimode operation, including moving target detection and track, synthetic aperture radar (SAR), interferometric SAR and adaptive processing to cancel radio frequency interference (RFI) and clutter. For analog radars, the components were limited by lack of correlation or coherence between the channels, waveforms and target characteristics. The solution to this approach was to create high-resolution modes and post-process the signal returns to eliminate spatial and spectral differences between the hardware signal paths. As these efforts were repeatedly reported in the literature, there are significant improvements in performance as a result of paralleling the transmit, receiver and signal processing paths in the system. The goal of many UWB multimode radar is to maintain a high area coverage rate (ACR) in search and selectively improve the resolution of the imaging or detection by additional signal processing.

An early radar concept was proposed by Merrill Skolnik in 2002. He termed it as "ubiquitous," i.e., "a radar that looks everywhere, all the time." The idea was to have an illumination source that could cover a large fraction of a hemisphere, coupled with multiple simultaneous receive beams to provide the signal processing for multiple, simultaneous radar modes. The principal focus of the concept was to employ digital beam forming and adaptive processing to enable a wide range of radar modes for detection and tracking of entities in the surveillance volume [1]. The concept traded a reduction of antenna gain on transmit for an increased coherent integration gain on receive. The staring receive beams give better Doppler resolution during the long coherent processing intervals (CPIs). However, there is an increased expense of having to operate several receivers along with a significant increase in computational resources. This chapter covers several of the evolving radar architecture and signal processing algorithms as a basis for further development.

Section 9.1 will start with a more detailed description of the Multibeam Modular Survivable Radar (MB/MSR) – a system that attempted to develop significant capabilities for wide-area surveillance using the emerging RF and digital signal processing technologies of its time. Chapter 5 presented the development of the multibeam antenna in the mid-1980s. The initial concept of a multiple beam antenna came about to improve ACR of these systems, without having to radically increase the power aperture product. The key objective of the MB/MSR project was the development of a novel antenna with real-time multimode signal processing, accompanied by airborne demonstrations of target detection and characterization. These approaches of a novel concept development were conducted in the 1980s: before the subsequent developments in monolithic microwave integrated circuits and active electronic scanned arrays (AESAs).

Section 9.2 covers the emerging radar technologies that would enable an order of magnitude improvement of the radar's surveillance effectiveness. The assumption is that AESA antennas coupled with high-performance signal processing will become practical and affordable with the continued advance of Moore's Law. By developing simultaneous SAR and ground moving target indication (GMTI) modes, coupled with multiple beams and parallel signal processing on receive, an evolving, conceptual radar architecture will demonstrate a significant ACR improvement.

9.1 Modular survivable radar

The concept of an MB/MSR came about during the 1980s to leverage the rapid development of electronically scanned antennas (ESAs) for battlefield surveillance. GE Aerospace provided a major departure from the single-beam, electronically scanned surveillance radar, as depicted in Figure 9.1. An advanced concept radar was developed in 1983, cooperatively between GE Aerospace, the Defense Advanced Research Projects Agency (DARPA) Tactical Technology Office and the US Air Force Wright Avionics Laboratory to evaluate emerging radar architectures.

The project's objective was to demonstrate improved surveillance ACR, with nearly simultaneous SAR and GMTI operation. The wide-area transmitter illumination would provide a persistent coverage of the battlefield. And the multiple beams enabled the radar to selectively process the data for detecting and tracking multiple ground targets. The challenge in the system architecture was to provide the

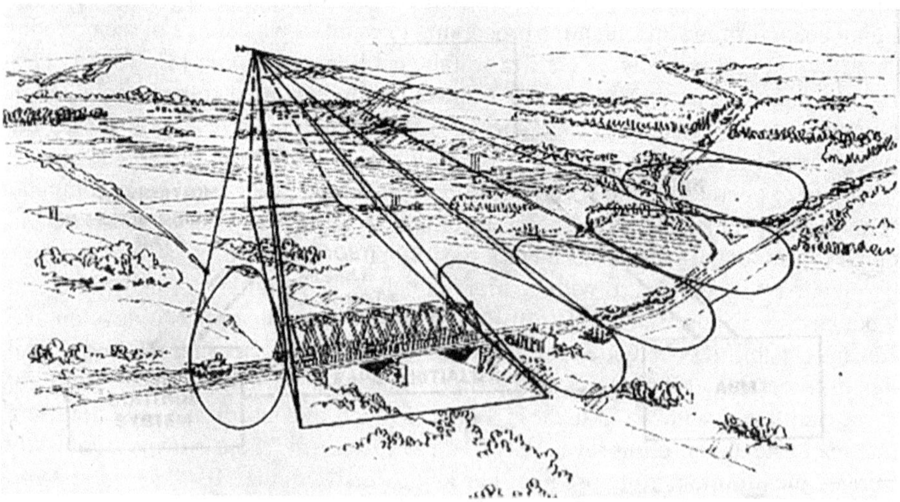

Figure 9.1 Multibeam multimode radar [2]

long dwell coherency necessary to obtain improved signal-to-noise and clutter levels. The multibeam antenna increased the ACR by integrating over several receive beams and channels to detect and register any targets or terrain features.

Very High-Speed Integrated Circuits (VHSICs) were emerging to provide more capability in avionics system, at smaller weight and volume and increased reliability. The MB/MSR was designed to demonstrate multiple channel, real-time signal processing in a small form factor. The real-time processing was developed to evaluate an improved ACR, while demonstrating improved battlefield surveillance detection and tracking of fixed and moving targets. The long-term objective was nearly simultaneous SAR and GMTI operation, consistent with low probability of detection operation. This section presents the system concepts, prototype radar mode characteristics and results of some limited flight demonstrations.

The MB/MSR radar system was developed as a flying testbed and flown on a commercial Douglas DC-4 aircraft, as shown in Figure 9.2. The multiengine aircraft had sufficient mission time for conducting research on new radar modes and space within the aircraft to demonstrate these modes to the radar community. This aircraft was operated out of a local airfield in Upstate New York and conducted trials and demonstrations in Virginia, Arizona and California.

Figure 9.2 Multibeam MSR aircraft over Southern California [3]

The ESA antenna was mounted in a pod under the aircraft, and the radar equipment was housed within the fuselage of the aircraft, for ease of operation and analysis of the data. An aircraft operating between 10,000 and 12,000 feet, with a speed of 105–200 knots was consistent with propeller-driven manned and unmanned aircraft, with mission times of up to 4 h. The project was not providing an operational capability, since there was insufficient schedule and funding. But the advanced radar demonstrator was used for future planning by the DARPA and US Air Force system development communities [2].

A conventional battlefield surveillance radar had a narrow beam that can scan electronically over a wide angle, as illustrated in Figure 9.3. The single-beam operation is similar to the Hostile Weapons Locator System (HOWLS) radar presented in Chapter 2 [5]. The ACR was defined by the time the tasked beam needed to dwell on a designated area, divided by the necessary CPI for target detection or characterization. The major challenge was to revisit a search area often enough that moving targets could be maintained in track. These single-beam systems could also employ a small number of receive channels to provide improved clutter rejection with the displaced phase center array technology. A single-beam system has been very useful for wide-area surveillance with rapid revisit of targets to track their movement. However, there was a need for higher ACR in a dynamic operation and improvements in detecting and characterizing both fixed and moving targets in a complex scenario.

The multibeam MSR antenna consisted of a corporate-fed, electronically scanned architecture on transmit, but with a large number of subarrays to enable formation of multiple receive beams. In the multibeam mode, the illumination was spread in angle using a quadratic phase across the array. And the multiple beams were formed using an intermediate frequency (IF) Butler matrix, with the input to the beamformer at the subarray level. Aperture weighting provided very low

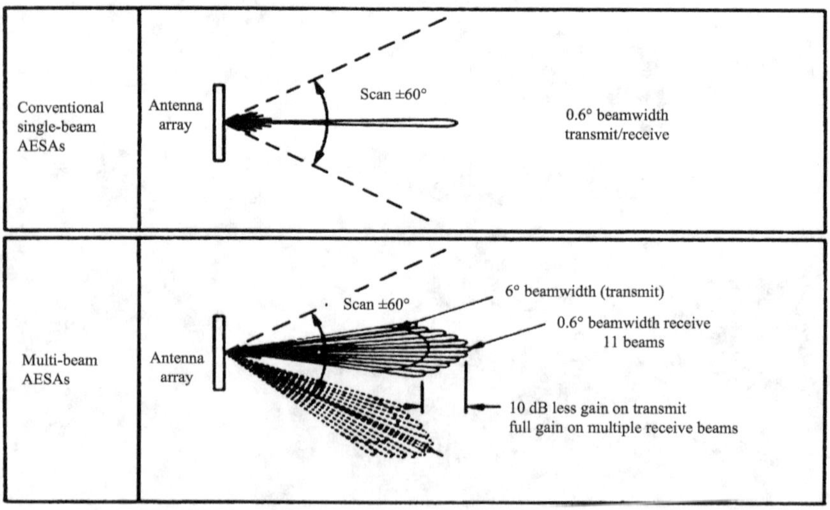

Figure 9.3 Multibeam surveillance operation [4]

sidelobes for both sum and difference patterns on each of 11 simultaneous receive beams. And weighting for low sidelobes was implemented to reduce the interference levels by both clutter and RFI in a cluttered environment. Furthermore, the advanced concept antenna was designed to provide the same angular scan volume as previous ESA developmental surveillance radars.

An electronic beam control provided for rapid switching between single beam for tracking and multiple beam for surveillance, as illustrated in Figure 9.3. The single-beam ESA provided narrow beam on both transmit and receiver for the best target detection and high signal-to-clutter operation for sub-clutter visibility. The multibeam ESA mode was obtained by spoiling the transmit beam and receiving multiple coherent aligned within the transmit beam, along with frequency agile returns over a wide-angle search region. These multiple, receive beams demonstrated wide-angle surveillance with detections accurately aligned for both GMTI and Doppler beam sharpening (DBS). The VHSIC signal processor provided real-time, onboard radar processing to demonstrate up to five simultaneous beams [4].

9.1.1 Multibeam antenna

There were several challenges in multichannel radar system implementation. The most important was the high cost of RF components to provide amplitude and phase control, along with channel match. The analog radar designs were limited by an availability of reciprocal phase and time delay control over wide bandwidths. And as mentioned in the previous section, there continues to be a basic requirement for amplitude weighting on transmit to achieve reduction of clutter and RF interference. However, there are difficulties with efficient amplitude weighting on AESAs due to the practical need for a Class-C solid-state amplifier. This technical challenge with solid-state T/R modules remains as an impediment to reducing transmit feed amplitude weighting to lower sidelobes.

DARPA and USAF Wright Avionics Labs funded GE Aerospace for the development of a novel antenna, as shown in Figure 9.4, for demonstrating multibeam surveillance radar for multimode operation, as well as increasing the ACR. The primary benefit was greatly limiting the radar RF emission characteristics in the battlefield. The concept was to have one broad transmit beam illuminating an area along with multiple receive beams to detect and characterize the tactical objects on the surface. The transmit gain of the MB/MSR was reduced by spoiling the beam by up to 10 dB over a single-beam system. ACR was maintained by forming up to 10 search beams collimated within the transmit beam. For low probability of intercept, the antenna had very low sidelobes, and the waveform was changed rapidly to make it more difficult to intercept. Many new radar modes were implemented by the frequency agility over 500-MHz tunable bandwidth and an instantaneous bandwidth of 20 MHz within a waveform [6].

The measured patterns, shown in Figure 9.5, demonstrated the ability to spoil on transmit for 10:1 beamwidth, using a quadratic phase across the azimuth face. This quadratic phase spoiling maintained the low sidelobes over the full azimuth scan pattern, previously covered in Section 4.1.2. And the independent receive beams

Figure 9.4 Ku-band multibeam antenna [2]

were collimated to the transmit beam. Since the azimuth phase scan was applied to each of the column radiators, both transmit and receive beams scanned as a group. More importantly, the receive beams were aligned within the transmit illumination pattern and with respect to each other. This alignment was essential to demonstrate the geolocation accuracy of features and targets within the individual beams.

An analog beamformer was developed at an L-band IF frequency using a Butler matrix amplitude and phase weighting, designed to achieve a 50 dB Taylor weighting. The Taylor weighting was approximated by 32 points on the 50-dB Taylor curve, as shown in Section 4.1.1. A separate beamformer for difference pattern using 40-dB Bayliss weights is given in Section 4.1.2. By controlling the amplitude and phase calibration at each subarray, both the peak and root mean squared (rms)

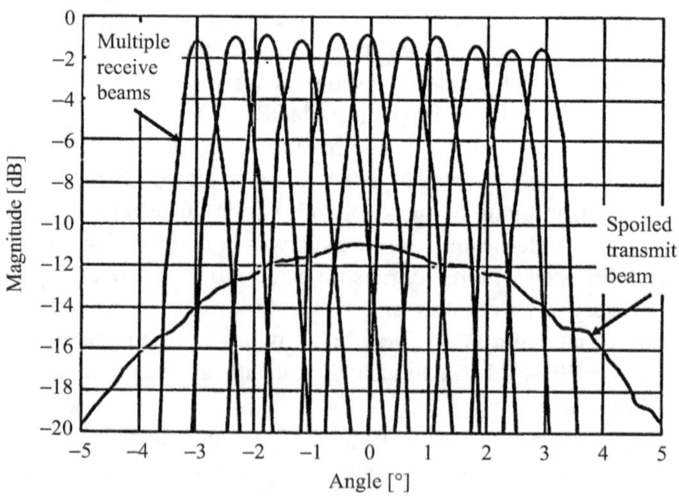

Figure 9.5 Multibeam antenna patterns [4]

sidelobes were maintained, where the pointing accuracy maintained one-tenth of a beamwidth. Moreover, the Ku-band receivers were calibrated for match across the array bandwidth. Finally, the gain variation was maintained within 0.5 dB between the independent beams, due to the low loss at L-band in the IF, and equalization in signal processing from a calibration table. These factors were instrumental in achieving the very-low-sidelobe levels across the wide electronic scan volume [2].

A simultaneous grouping of 10 Δ beams were formed in a separate Butler matrix. These Δ patterns enabled accurate cross-range tracking of ground moving targets. The output of the beam former went into IF amplification and followed by 50 Mbit/s analog-to-digital conversion. For onboard processing, a custom very large-scale integrated circuit real-time signal processor was developed to carry out up to five multiple simultaneous channels. These beams were then displayed on the flight monitors. A wideband digital recording subsystem captured the received data for each of the Σ and Δ data streams for offline analysis.

9.1.2 Multibeam modes

The multibeam radar system was flown to collect data for performance verification, along with real-time signal processing to evaluate the capabilities for operational feedback to the users. These modes included the GMTI modes using Doppler signal processing, DBS to improve the cross-range resolution of surface surveillance, strip map SAR for improved resolution and multiple GMTI detections for target tracking, with the real time 5-channel processing, as shown in Figure 9.6. The following sections will illustrate the benefits of a multiple beam system for wide-area surveillance.

DBS is a dwell time efficient method to get increased cross-range resolution, without the long integration times required by SAR. The conventional DBS mode uses the main beam to obtain cross-range resolution due to the Doppler spread across the broader receiver beam [7]. For the MB/MSR, the area of the collection can be increased by using all of the receive beams on the Earth's surface and combined with the spread of the main transmit beam for illumination, as illustrated in Figure 9.7.

The cross-range resolution can be increased by Doppler processing within the ground beam. The DBS mode starts with considering the Doppler of ground range within a beam [7]:

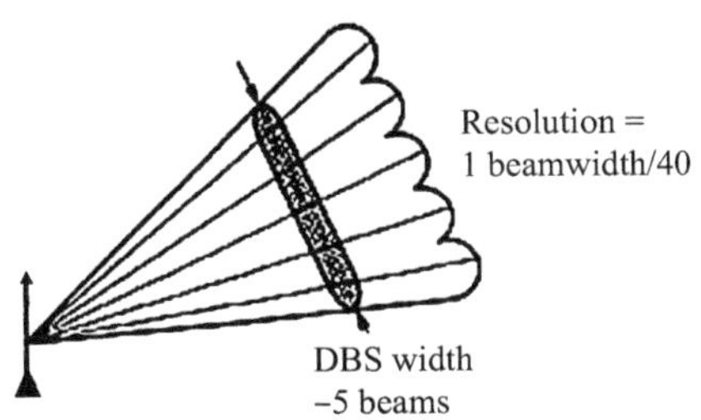

Figure 9.6 Multibeam antenna pattern for DBS [4]

$$f_d = \frac{2V_{plat}}{\lambda} \cos\alpha \cos\phi \tag{9.1}$$

where α is the azimuth angle and ϕ the elevation angle with respect to the velocity vector. The ground returns at a different azimuth angles have different Doppler shifts. Since cross-range is a function of azimuth angle, Doppler filtering allows partitioning the returns across the illumination within the scene.

A DBS mode then uses the range resolution and Doppler resolution to create a range-Doppler map in the scene coordinates. The cross-range resolution is simply the width of the beam at range, divided by the number of Doppler filters across the beam. To maintain high clutter returns for mapping, only the central point of the mainbeam is used. Typically, the usable Doppler filters are less than the number of pulses integrated. For example, in Figure 9.7, 40 Fourier filters out of a 64-pulse CPI gives the Doppler resolution within the individual receive beamwidths and maintains the clutter return to less than 3 dB degradation of the clutter-to-noise ratio.

To examine the DBS image formation process, a point spread function was derived in Section 3.2.2. If the quadratic phase in (3.12) is less than $\pi/2$, a DBS image limit is given by.

$$D_{SAR} < \sqrt{r\,\lambda_c} \tag{9.2}$$

When this criterion is satisfied, then the DBS phase function in (3.12) is simplified to.

$$\Delta\psi \approx -\frac{4\pi}{\lambda_c} u \cdot \left(\frac{x}{r}\right) \tag{9.3}$$

This linear relationship between phase and cross-range position enables a fast Fourier transform (FFT) to be used for image processing [7]. The DBS criterion in (9.2) was met with the MB/MSR demonstration system due to its limited bandwidth of 50 MHz and wavelength of 1.71 cm.

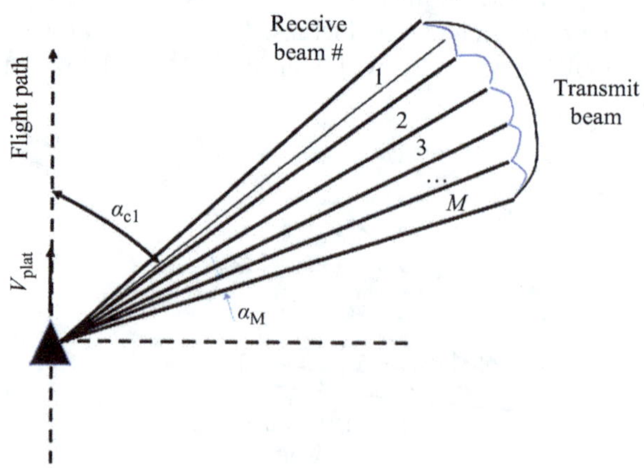

Figure 9.7 Multibeam DBS receive geometry

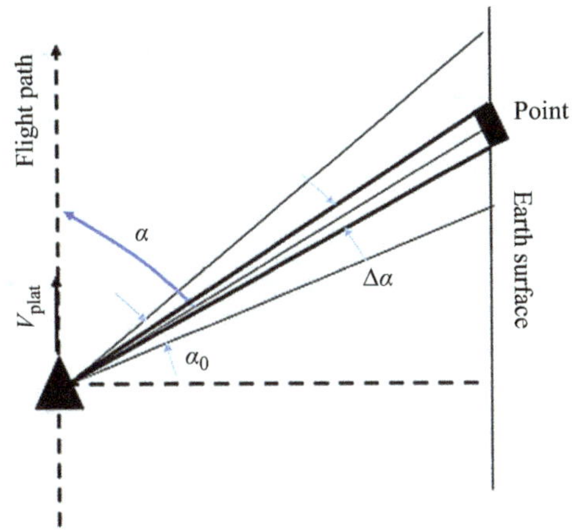

Figure 9.8 *DBS collection geometry [6]*

The geometry for a DBS collection is shown in Figure 9.8. The total beamwidth between the 3 dB points is given by α_0, and the Doppler width of a resolution cell is given by $\Delta\alpha$. It should be noted that the angular width of the beam is given by the electronic scan angle. And α_k is the *k*th angular beam within the physical beamwidth. The Doppler bandwidth of the sharpened beam is given by [6]

$$\Delta f = \frac{2\,V_{\text{plat}}}{\lambda}\left[\cos\left(\alpha_k + \frac{\Delta\alpha}{2}\right) - \cos\left(\alpha_k - \frac{\Delta\alpha}{2}\right)\right] = \frac{2\,V_{\text{plat}}}{\lambda}\,\Delta\alpha\,\sin(\alpha_k) \tag{9.4}$$

Based on (9.4), the cross-range resolution of the DBS collection can be estimated as

$$\delta_{\text{CR}} = R\,\Delta\alpha = \frac{R\,\lambda\,\Omega}{2V_{\text{plat}}\,\alpha_0\,|\sin(\alpha_k)|} \tag{9.5}$$

where Ω is the angular velocity of the platform motion around the imaging point. This angular velocity can be approximated by the DBS aperture time T_a and the angle from the velocity vector α. It should be noted that each point in the beamwidth has a different value of α:

$$\Omega \approx \frac{\alpha}{T_{\text{a}}} \tag{9.6}$$

The most efficient processing to obtain the DBS image characteristics is to use an FFT on the input data. The individual filters across the beam will provide the estimation of the Doppler offset from α. To achieve high-frequency resolution, the collected data should be as large as the real antenna beamwidth

$$N = T_{\text{a}}\,f_{\text{PRF}} \tag{9.7}$$

where f_{PRF} is the pulse repetition frequency. The spectrum of each FFT data record is centered around f_{d0} corresponding to the azimuth in the center of the beam. The spectrum of each block is computed in N points, and the K internal filters are used for

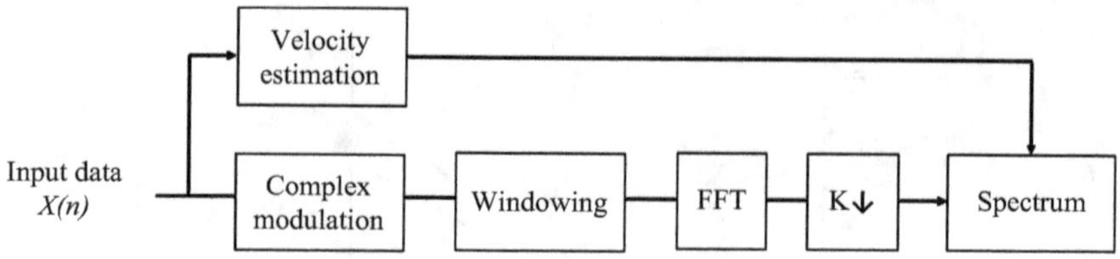

Figure 9.9 Motion compensation for DBS beam center [6]

mapping. K is chosen to minimize the scalloping of the image magnitude, normally chosen at the 3 dB point of the beam. As a result, each filter has a Doppler frequency of

$$f_n = \frac{2V_{plat}}{\lambda}\left[\cos\left(\alpha_n - \alpha_c\right)\right] \tag{9.8}$$

where α_c is the angle to the center of the beam.

For the multibeam DBS approach, the transmit beam covers a beamwidth of α_B, and there are M receive beams within the transmit beam coverage, as shown in Figure 9.7. For an ESA, the beamwidth varies by $1/\cos(\theta_s)$, where θ_s is the scan angle away from the broadside of the antenna. As such, there will be a slight increase in the cross-range resolution within each scanned or squinted beam. However, it is normal to process each receive beam with the same FFT size. The motion compensation for each beam is given in Figure 9.9. The velocity estimation is made for the center of each beam in the multibeam illumination. And each beam will use the same linear frequency modulation (LFM) waveform, where the range compression is applied in the fast time domain. If desired, a windowing function can be applied in the slow time followed by an FFT filtering, to lower the Doppler sidelobes at the penalty of increasing the cross-range resolution. After decimation of the data in each filter, the spectrum is magnitude converted and then displayed [6].

The real-time multibeam processing to demonstrate DBS included a transmit beam spoiled 5:1 and scanned in azimuth to cover an area, as shown in Figure 9.10. Multiple simultaneous receive beams were formed to cover 2.8° cross-range. Each beam was Doppler processed to form DBS images in each receive channel, with a 64-pulse FFT and motion compensated to middle of beam, with the azimuth prefilter for using center 40 contiguous filters for DBS. This processing provided image azimuth resolution of 17 m at 14-km range.

Figure 9.11 shows the DBS results from one of the test scenarios over Oneida Lake, NY. Using the multiple beams, a wider area was collected and demonstrated a faster image integration time compared to a single-beam radar. The MB/MSR platform had five real-time processing channels, so that these images could be viewed during the flight. This example verified that DBS is more time-efficient than SAR, but does not give the resolution needed for image feature recognition applications. However, it is easy to pick out the break water and canal within the multiple parallel images. As another DBS example, Figure 9.12 shows a short-range collection over the canals in Figure 9.11, with improved cross-range resolution.

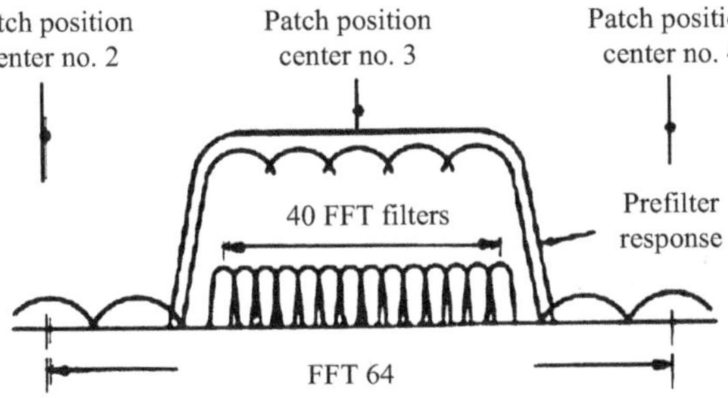

Figure 9.10 Formation of multiple patch DBS images [2]

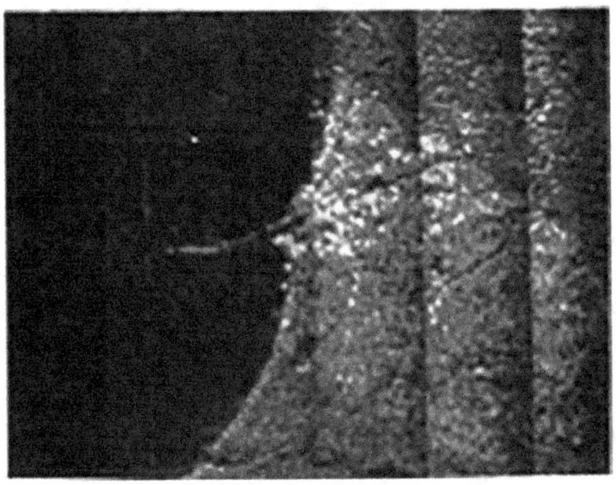

Figure 9.11 Oneida lake breakwater [4]

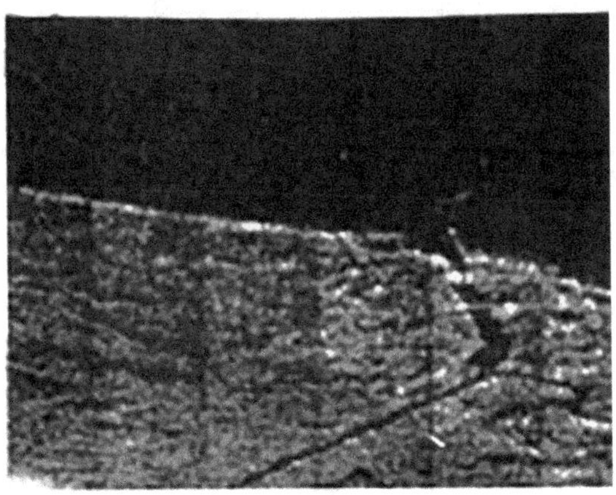

Figure 9.12 Oneida lake canals [4]

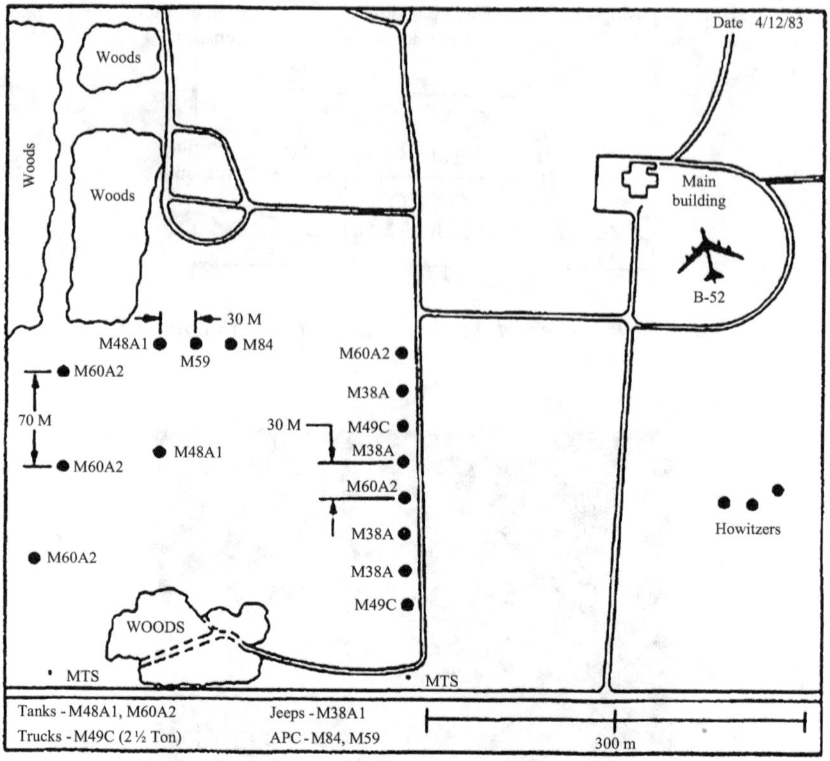

Figure 9.13 Stockbridge military test area [2]

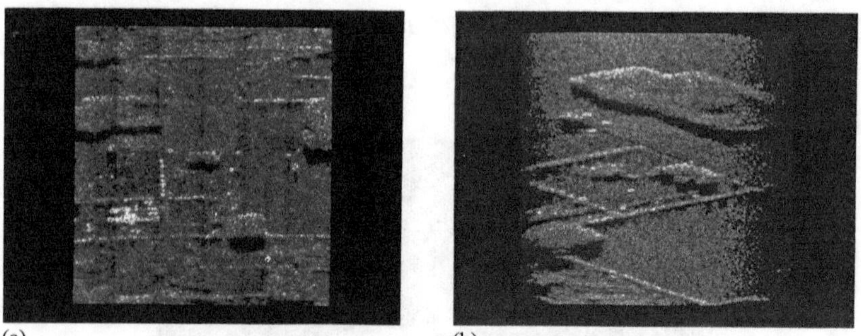

Figure 9.14 Multibeam radar images of Stockbridge test site [2]. (a) DBS image and (b) SAR image

A second experiment was conducted to compare DBS with SAR imaging. Figure 9.13 illustrates the location of tactical targets for MB/MSR tests over the Stockbridge, NY test site. This site near Rome, NY, covers a 1.8 by 1.5 km test area, with multiple tactical vehicles arranged in fields and woods. Figure 9.14 shows the

Figure 9.15 Military exercises in Southern California [3]

comparison of DBS with SAR in real-time images from VLSI onboard processor using a 15-MHz LFM waveform. The SAR imaging was a strip map collection as covered in Section 3.2. The DBS image gave 7.5-m range and 6-m cross-range resolution with approximately 64-ms collection time. By comparison, the SAR image yielded 7.5-m range and 2.5-m cross-range resolution, with an SAR collection time of 0.5 s. This resolution is not very impressive compared with that of modern radars. But it gave an indication of the emerging tactical SAR real-time processing. The DBS gave a modest image in one-eighth of the time with only a degradation of cross-range resolution of 2.5–1.

The MB/MSR was taken to California to collect moving and fixed target operations at Ft. Ord. The inset in Figure 9.15 shows a single look GMTI with the vehicles moving down the road, along with a 60-s scan of the area. Onboard the aircraft, the GMTI could be played back to give the operator a time sequence of the movement. In more modern radars, this is done with a real-time tracker. Once the GMTI indicated that the targets had stopped, a real-time SAR image was tasked. The use of these SAR images provided more details on the rally area, where a number of tanks were clearly shown [3].

During the exercises, the sequence of tasked modes were demonstrated on the aircraft. Figure 9.16 shows a single GMTI look with several moving targets on the road. The MB/MSR development radar was operated to illustrate multimode battlefield operation during army exercises and tests. The onboard display had a cartographic display of several details such as roads, latitude and longitude grids and major planned areas. In the upper left image, GMTI was overlaid on carto map for situation awareness. The second image shows a multiple scan of the moving targets

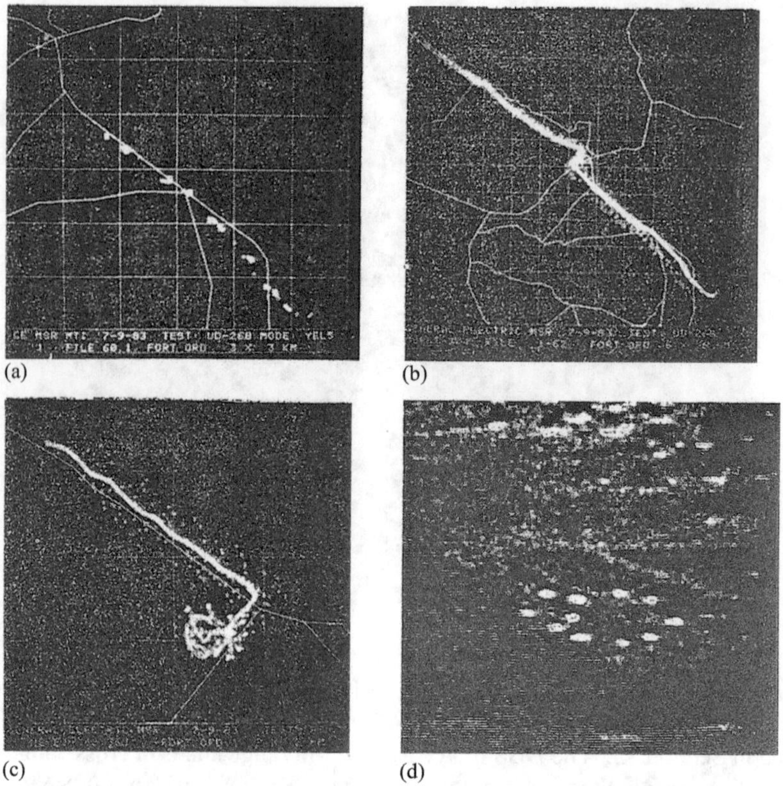

Figure 9.16 *Multimode data from test of multibeam MSR [3]. (a) Single GMTI look, (b) 60-s scans, (c) movement area and (d) SAR image*

on the road, where the density of detections can be readily seen. The radar displayed the speed of the targets moving up and down the road. And in a playback mode, the operator could determine when the targets would slow down and stop at a staging area. The immediate playback of moving targets was used to identify movement areas. Once these areas were observed, SAR images were tasked to provide target characterization.

9.1.3 Multibeam radar summary

The multibeam surveillance radar demonstration was successful in developing techniques for multiple, independent radar channels that had sufficient match to enable both fixed and moving target detection. The concept development of an early multibeam array demonstrated SAR and GMTI modes with flight test results. Most importantly, the multiple beam formation provided an improvement in ACR by multiple receiver beams. Finally, the real-time VHSIC processor demonstrated wide-area DBS and GMTI operation. However, the radar component technology at that time was limited to a passive phased array and analog beam forming. As such, the size, weight and power achieved made it impractical for deploying on small unmanned air vehicles. And the techniques for space-time adaptive processing (STAP) were not available at that time to fully exploit the number of subarrays. Furthermore, the losses in IF beamformer affected system noise figure due to the two-way losses in the analog components in the elevation radiators and the ferrite phase shifters in the azimuth beam steering.

With the development of active ESAs and digital wideband beam forming, it is now feasible to build a surveillance radar with greatly improved characteristics. Having the active array at the face of the antenna, the two-way losses are reduced, and the noise figure can be very low. Thus, the practical detection of small targets in difficult clutter environments is possible. More importantly, digital waveform synthesis and accurate motion measurement in the systems enable better geolocation accuracy of the radar image formation for commercial applications.

9.2 Future multimode UWB surveillance radar

Solid-state AESAs and digital formation of multiple beams are now feasible and are becoming more affordable due to commercial developments in component technology developed commercially for microwave digital datalinks and civilian communications. It has also become possible to use a common waveform for both SAR and GMTI, given the advancements in waveform diversity technology. However, the ACR of fielded surveillance radars still has limitations by switching from one mode to the other, e.g., GMTI to SAR. This is particularly difficult for the longer wavelength radars for penetration or characterization of objects under foliage and camouflage. Radar technology continues to develop waveforms that combine fixed and moving target detection. And increasing the bandwidth was needed to provide target characterization for reducing false alarms or discriminating between classes of targets.

Two radar architecture developments have been demonstrated and can now be made practical with improvements in the component technology. First, the frequency jump burst (FJB) waveform has been studied as one of those waveform diverse technologies [8]. It provides a high bandwidth signal for SAR and improved detection for ground moving targets. The next section will provide an analysis of these emerging capabilities, along with new algorithms based on the earlier mode designs from a medium-bandwidth surveillance radar. Second, digital beamforming has become feasible with wideband high dynamic range analog-to-digital converters (ADCs) and solid-state transmit receive (TR) modules at each element or subarrays. With a low noise figure, high gain receiver behind each element or subarray, multiple independent beams are practical to overcome the earlier analog multibeam prototype. By matching the amplitude and phase across the wide bandwidth for excellent channel match, independent characterization of the distributed clutter or point targets can be exploited.

When a surveillance radar interrupts GMTI for high-resolution SAR imaging modes, the ACR is seriously affected. SAR images are typically an order of magnitude smaller than the GMTI imaging area, and the coherent SAR dwells are a couple of orders of magnitude longer. This is especially critical in longer wavelength radars, where wide-angle integration and long coherent dwells are required for achieving the SAR image resolution. It was shown in Section 2.4 that frequency agility enables detection of small targets by converting Swerling-2 targets to Swerling-0 (non fluctuating) targets, thereby giving an improvement of detection probability by ~4 dB. The use of UWB waveforms with sub-banding can provide the benefits of frequency agility. With the FJB waveform, the benefits of frequency agility can be obtained within a dwell, as opposed to dwell-to-dwell. Another key enabling technology is the high dynamic range ADC combined with high-speed digital beam forming, which provide improved radar processing.

The objective of these studies was to overcome classic single SAR and GMTI mode limitations by innovative wideband waveform implementation. These limitations include high time—bandwidth waveform and low pulse repetition frequency (PRF) for SAR and low resolution for longer integration times for GMTI. The hypothesis is a utilization of FJB as a useful, interesting point of departure for a future UWB radar concept [6].

9.2.1 *Multibeam multichannel architecture*

Airborne radars have been developed for detecting moving targets in a large number of diverse geographic and clutter environments. Within these environments, targets can traverse the Doppler spectrum from endo- to exo-clutter, limiting their detectability by traditional radar. Very large phased array antennas and STAP are required for reliable detection of moving targets. For a stationary target detection, traditional systems deploy SAR modes to discriminate stationary targets or cultural features for improved situational information. Because of the long integration times, SAR competes for the system's valuable radar timeline, reducing the overall revisit and ACR of the surveillance capability. For early battlefield surveillance systems at microwave frequencies, the antennas can be designed and flown on medium-to-large aircraft and on unpiloted air vehicles. The large antenna provides excellent target location accuracy in both SAR and GMTI, as well as cancelling mainbeam clutter for a low minimum detectable velocity (MDV).

This section explores a combined or simultaneous SAR and GMTI waveform in a multiple beam antenna architecture. A novel system architecture is explored that attempts to synthesize a common waveform for simultaneous stationary and moving target detection and to provide for complete endo- to exo-clutter target detection and characterization. The issue of clutter spread is illustrated in Figure 9.17, with three clutter segments. First, exo-clutter detection of fast-moving targets is the simplest task, since only the sidelobes of the antenna or Doppler filters are competing with clutter. Second, as was illustrated in Chapter 6, SAR and along-track interferometry (ATI) detection of fixed and slow-moving targets can be obtained within the main beam of the radar. These are high-bandwidth modes, providing the opportunity to

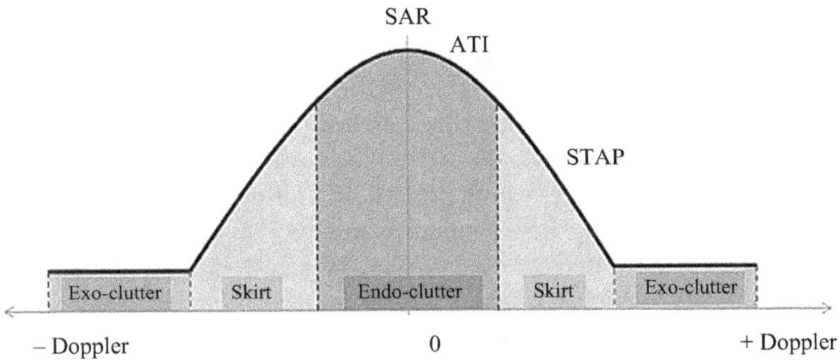

Figure 9.17 Fixed and moving target regimes within radar beam

characterize the targets. The effective Doppler bandwidth of these image features needs to be controlled by either presumming or by choosing a low PRF. Third, the issue of detecting moderately fast-moving targets that are still endo-clutter requires the multiple spatial and temporal processing of STAP. The clutter can be trained for adaptive cancellation, and the targets can be detected with modest signal-to-interference plus noise ratio (SINR) loss.

The main beam clutter spread in Doppler is the major obstacle in detecting targets. This factor leads to exploiting SAR for stationary target identification, ATI for very slow-moving target detection and STAP for moderately moving target detection, as shown in Figure 9.17. Multiple beam antenna designs will be investigated to enable satisfactory ACR and revisit rate for GMTI surveillance, as well as STAP for isolating moving targets from main beam clutter. A common or simultaneous SAR and GMTI waveform has several benefits. The high ACR of GMTI makes it more attractive for wide-area search of ground features and objects. Multiple passes are not required, if the instantaneous area is large enough. And the spatial diversity of discriminating the targets from clutter demands finer resolution in range and angle. The combination of SAR, GMTI and ATI has been examined for detection feasibility, using an advanced modeling and simulation tool.

The approach in this section is to explore a waveform and antenna architecture for simultaneous detection and identification for targets in a complex environment, which is achievable with a common waveform, i.e., the FJB waveform. FJB provides multiple, frequency diverse pulses for M-of-N moving target detection. And it can be reprocessed for ultra-wide bandwidth for stationary target detection, background mapping and feature collection (SAR, HRR and ISAR). A multibeam architecture with multiple phase centers and wideband characteristics allows for very slow-moving target detection, i.e., ATI.

Several advantages of a common waveform include the following: high ACR for wide-area surveillance, sub-clutter target detection with higher resolution to discriminate spatial features from clutter, target feature collection with wide bandwidth for target ID and long-term tracking with persistent dwells. Moreover, simultaneous waveforms have been demonstrated for exo-clutter and stationary target detection.

Moreover, ATI is a viable option for slow-moving target detection with the possiblity of high target resolution for characterization through feature discrimination. But what about exo-clutter, fast target detection? The key to these questions is the synthesis and analysis of a waveform that can lead the way for future system prototyping and data collection. This challenge will be covered in the following sections.

Ultra-high frequency (UHF) is the most difficult challenge for ACR because there is a need for switching between single modes. This is especially true when there is a desire for examining the context of targets that have transitioned from moving to stationary, which requires an SAR image. This UHF SAR image will take significant time, as shown in Figure 9.18, of the order of tens of seconds. For microwave frequencies, these numbers are significantly lower. But the need for both fixed and moving targets remains a system technology objective regardless of the frequency of operation.

GMTI performance as a surveillance mode is characterized by ACR, how wide and long of a surveillance area a can be surveyed as the platform traverses an area of interest. When moving targets stop or are hidden by obstacles, the platform needs to schedule an SAR mode, which severely impacts ACR. At UHF frequencies, the problem is greatly complicated by a need for wide integration angles for SAR image cross-range resolution and very large antennas for GMTI MDV.

As shown in Figure 9.19, the antenna architecture is different for the three modes. SAR has a wide beamwidth to image the area for cross-range resolution. And GMTI requires multiple spatial channels for STAP and should have a broad transmit beam and multiple receive beams for meeting the ACR requirements. Finally, the ATI mode needs to have multiple subarrays for forming SAR images, and interferometric processing for discriminating the moving objects in the scene.

The evolving RF technologies can be used to develop such a system. And the rapid development of high-performance embedded computing will provide multiple teraflops of processing to handle the data flow of this system. These new technologies will allow significant improvements over the MB/MSR discussed in Section 9.1 [9].

When multiple channel architectures are required for these future multimode capabilities, the emerging technologies will enable new modes. In particular, it is now feasible for wideband, coherent digital waveform synthesis, wide instantaneous bandwidth operation, high dynamic range ADC and high-performance embedded computing (multiple trillion floating point operations/sec (TFLOPS)). The following sections will explore the design, simulate a challenging architecture and finally assess the demands on real-time signal processing. The final analysis will be an update to the ACR from Chapter 6 with this simultaneous multimode waveform [8].

9.2.2 Common GMTI and SAR waveform

A potential common waveform for both SAR and GMTI is illustrated in Figure 9.20 using an FJB with short LFM waveforms of moderate bandwidth and switching the center frequency between sub-pulses to gain the UWB objectives. For SAR,

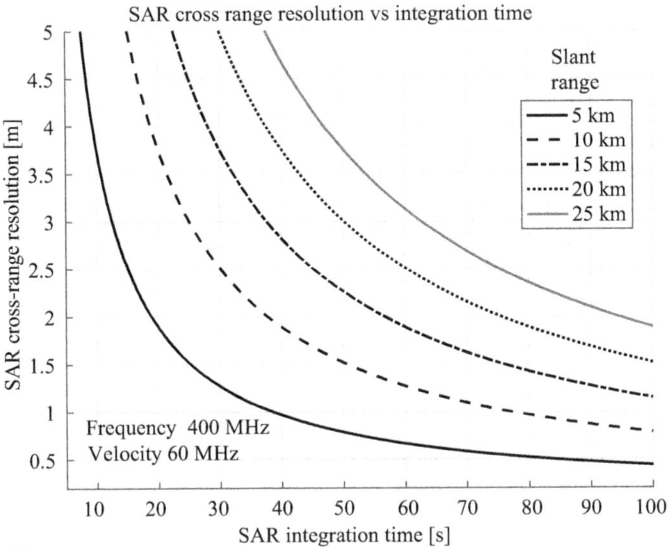

(a)

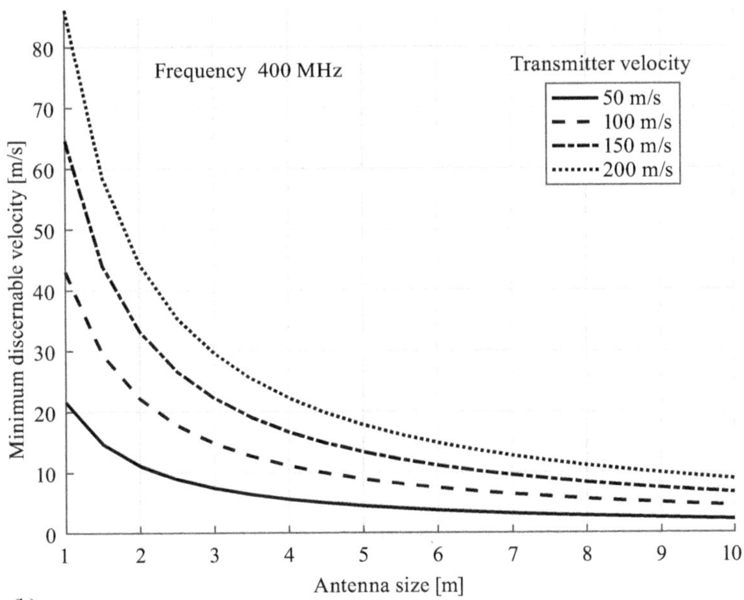

(b)

Figure 9.18 Radar design consideration for UHF SAR and GMTI operation. (a) SAR image collection time and (b) MDV each sub-pulse is combined by delaying the waveforms into a wideband signal over the scene to be imaged. For GMTI, each sub-pulse is coherently added between the pulse repetition intervals (PRIs) to get a modest bandwidth return. There exist multiple, narrow-band GMTI dwells, with frequency agility between the dwells to improve the signal-to-noise requirements. A common GMTI/SAR waveform has been proposed, based on the FJB waveform shown in Figure 9.20 to enable narrow bandwidth synthesis of an SAR signal. The pulse width T_P is divided into M sub-pulses of width $T_{PM} = T_P/M$ and transmitted at a fraction of the bandwidth $B_M = B/M$ per sub-pulse. The return from each sub-pulse is received during fractional PRI $T_M = T/M$ [9].

The composite signal s_{FJB} is given by

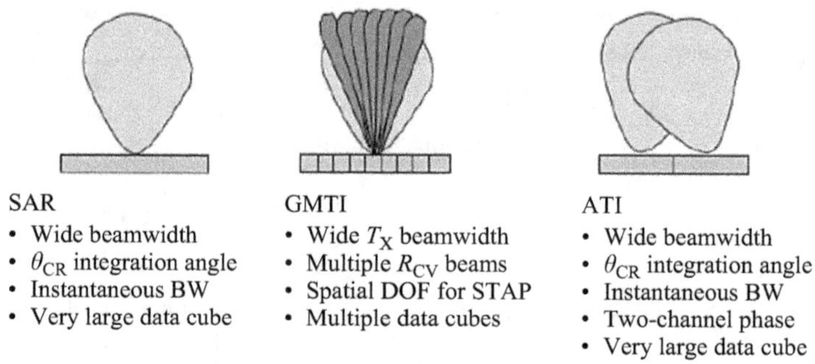

Figure 9.19 Multibeam/multichannel architectures [9]

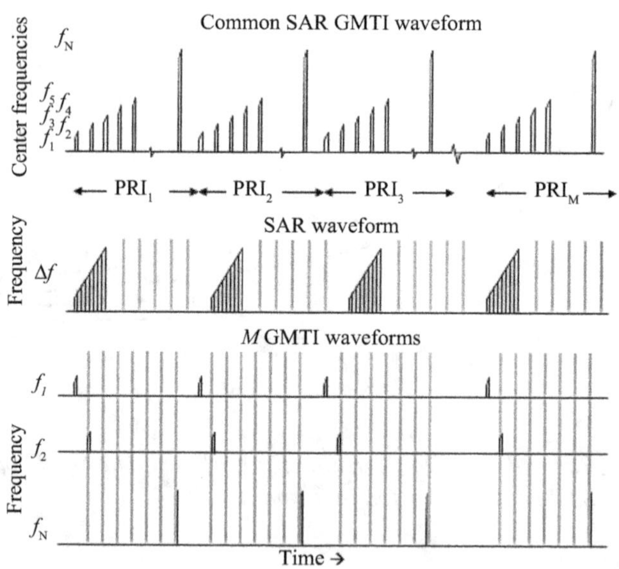

Figure 9.20 Waveform for simultaneous SAR and GMTI operation [10]

$$S_{\mathrm{FJB}}(t,m) = \mathrm{rect}\left[\frac{t}{t_{\mathrm{PM}}}\right] \exp\left[\mathrm{j}\,\omega_{\mathrm{c}}(m)\,t + \mathrm{j}\frac{2\pi\,B_{\mathrm{M}}}{T_{\mathrm{PM}}}\,t^2\right] \tag{9.9}$$

The center angular frequency of the mth sub-pulse is given by $\omega_c(m) = 2\pi f_c(m)$, where $m = [1{:}M]$. For SAR, these sub-pulses must be aligned to provide the full bandwidth required for range resolution. It is further assumed for this exercise that each segment of the composite waveform will have the same chirp slope as the baseline LFM waveform [11]:

$$\gamma_{\mathrm{M}} = \frac{B_{\mathrm{M}}}{T_{\mathrm{PM}}} = \frac{B}{T} = \gamma \tag{9.10}$$

In order to align the sub-pulses for match filtered compression, each sub-pulse must be time delayed by a progressive amount proportional to the sub-pulse width T_{PM}:

$$\Delta t(m) = T_{\mathrm{PM}}\left(m - \frac{M+1}{2}\right) \tag{9.11}$$

For GMTI operation, the combination of sub-pulses is not carried out. Instead, each pulse at center frequency $\omega_{\mathrm{CM}}(m)$ is coherently added to the corresponding pulse frequency at a PRI delay of T. In this way, M simultaneous CPIs can be constructed. The advantage of this approach is that M independent, frequency agile looks at the moving target are provided for reduction of target scintillation effects and improvement in target location accuracy [9]. A disadvantage is that there is range eclipsing by blanking the receiver at each sub-pulse PRI. The challenge is the ability to sample in the fast-time domain and reconstruct the single-frequency returns.

The common surveillance mode analysis will consider three different system applications: UHF, X-band and Ku-band of operation. The first is for foliage penetration applications such as UHF SAR and GMTI, and the second two match microwave applications for surveillance of fixed and moving targets in the open. Table 9.1 summarizes the top-level parameters to be used in investigating the surface Doppler returns around a platform. In all three cases, the platform altitude and speed is maintained the same. For the tree frequencies, the bandwidth of the waveform and PRF are different, in order to examine the fundamental differences. In a future system design, these parameters need to be examined for the specific application.

Table 9.1 Parameters for comparing FJB at three frequencies [10]

Altitude	**7.5**	**km**	**Frequency tradeoff**			
Speed	150	m/s	Frequency	f_0 (MHz)	PRF (MHz)	R_{amb} (km)
MFJB	8	Sub-pulses	UHF	350	300	500
BW (SAR)	240	MHz	X-band	9,500	1,200	125
BW (GMTI)	30	MHz	Ku-band	16,000	2,000	75
GRCA	300 × 300	km				

Range ambiguities

Range ambiguities are of two types in simultaneous SAR/GMTI analysis, with differing impact on the performance. First, the multiple time around returns in the basic PRF, such that range returns are folded into the detection. Second, the FJB has a higher effective PRF due to the sub-pulses. Because the transmitter subpulses are present during the receiver sampling, the receiver needs to be blanked on each sub-pulse. As a result, the range returns for each waveform frequency has an apparent ambiguity, affecting both the range coverage for GMTI and the sample support for SAR.

The basic equation for the first kind of range ambiguity is given by

$$R_{\mathrm{amb}} = \frac{c\,\mathrm{PRI}}{2} \tag{9.12}$$

where c is the speed of light and PRI is the overall pulse repetition interval. This factor is independent of frequency. Figure 9.21 shows the ambiguous range R_{amb} for PRFs between 300 and 2,400 superimposed on the range interval for analysis from 10 to 500 km. The minimum range is indicative of the transmit pulse width and maximum range is the limit of the surveillance window. At the UHF 300 Hz PRF, the ambiguous range is 500 km, well beyond the horizon, as shown in Figure 9.21. At the X- and Ku-band PRFs, the range interval from (9.12) is a decreasing fraction of the ground reference coverage area round surveillance radars generally operate in the low PRF regime, where there are no range ambiguities. However, they do need to consider the Doppler or velocity ambiguities in terms of the velocity spectrum of targets or the surface clutter with respect to the PRF for SAR.

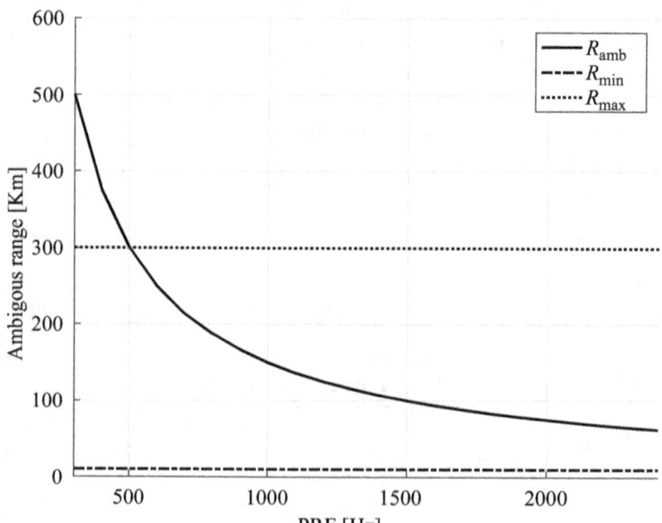

Figure 9.21   Range ambiguity versus waveform PRF

Doppler ambiguities

Equally important in examining the common waveform is the impact of Doppler ambiguities on the ability to detect targets. Based on the pulse Doppler waveform, there exist Doppler ambiguities f_{Damb} at the PRF rate. When the Doppler is compensated to have zero in the middle of the Doppler region, the ambiguities will occur at

$$f_{Damb} = \pm(2\,N_{\text{amb}} + 1)\frac{\text{PRF}}{2} \tag{9.13}$$

where N_{amb} is the number of the ambiguity. There is a comparable velocity ambiguity, v_{amb}, for radar returns in the scene that can be calculated as

$$v_{\text{amb}} = \frac{2\,v_{\text{t}}}{\lambda}\frac{1}{\text{PRF}} \tag{9.14}$$

These returns are either from ground clutter or from moving targets. For ground clutter, this is the phenomenon that will be exploited for SAR imaging. For the UHF case, the velocity ambiguities for platform velocity of 150 m/s can be calculated as a function of frequency as shown in Figure 9.22. The three lines are for the center frequencies at the low, mid and high pulses in the various sub-pulses. The high instantaneous bandwidth of the UWB FOPEN pulses presents a different velocity ambiguity that needs to be accommodated in the processing. This is accommodated in UWB SAR image formation. But it must also be considered when the GMTI or STAP processing is done at the different frequencies of $\omega_c(m)$. This effect of Doppler ambiguites is important in all modes as was demonstrated with the STAP results, previously shown in Section 7.2.4.

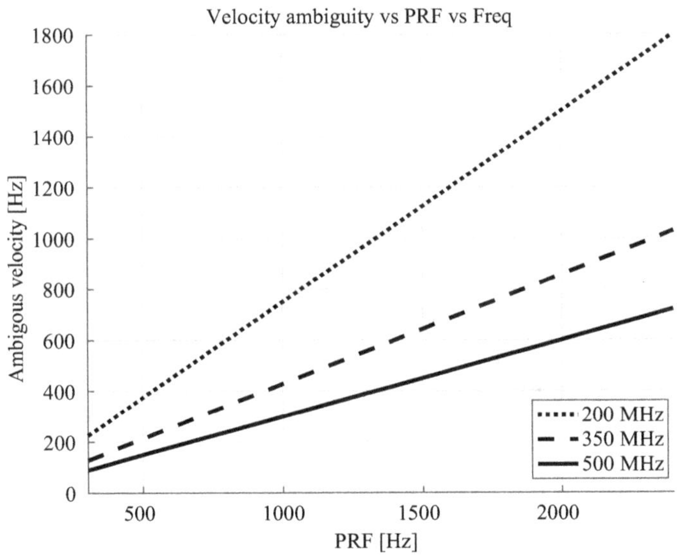

Figure 9.22 Doppler ambiguities of targets at UHF frequency

Comparable ambiguous velocity performance over the signal bandwidth is obtained for X-band and Ku-band operation, as shown in Figures 9.23 and 9.24, respectively. It should be noticed that the ambiguous velocity is significantly reduced by the wavelength. These Doppler ambiguity factors will be seen as a challenge

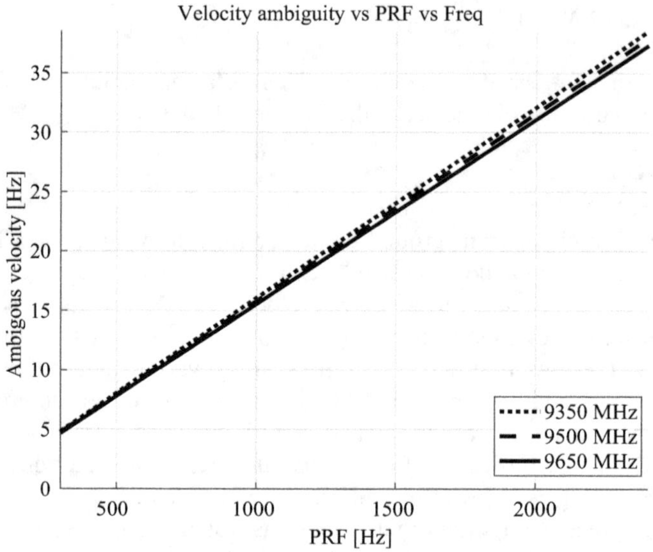

Figure 9.23 X-band velocity ambiguity versus PRF

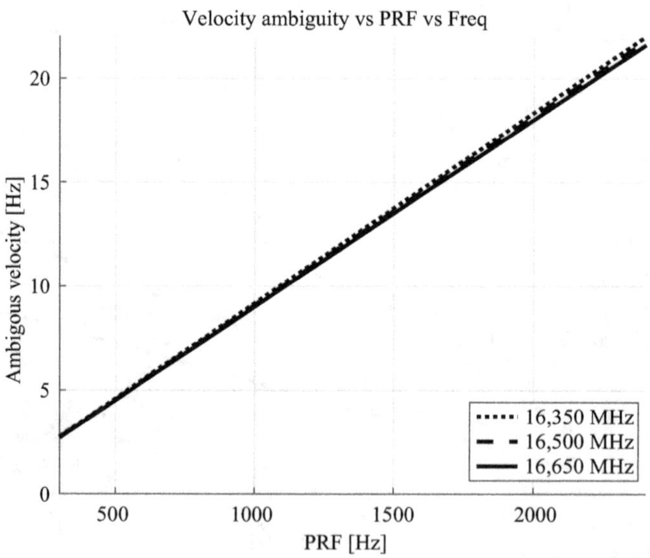

Figure 9.24 Ku-band velocity ambiguity versus PRF

for the ambiguous target velocity and range for GMTI operation. And because the instantaneous bandwidths of the X- and Ku-band waveforms are under a few percent, the divergence of the ambiguous velocities versus PRF is reduced from the UHF example.

Surface clutter Doppler spectrum

The velocity of the platform and the wavelength determine the surface clutter spectrum by the relationship:

$$f_{\mathrm{dop}} = -\frac{2V_{\mathrm{plat}}}{\lambda_0}\sin\phi_{\mathrm{t}}\cos\gamma_{\mathrm{g}} \tag{9.15}$$

The angles are the azimuth angle off of the velocity vector and the incident angle γ_g with respect to the local terrain. Figure 9.25a shows the surface clutter spectrum at UHF, for the GRCA and scenario parameters given in Table 9.1. Figure 9.25b only plots the iso-range and iso-Doppler contours for the range and Doppler ambiguities, respectively, along with the iso-range contours of the receiver blanking regions for the FJB waveform. As a result, the boundaries of FJB sub-pulses and Doppler ambiguities are more clearly understood.

Several important points of this scenario are that the Doppler spectrum is ±300 Hz, corresponding to the aircraft velocity from (9.15). As a result, the range ambiguity is beyond the horizon, an ambiguity is not shown in Figure 9.25. The receiver blanking contours are the closely spaced circles, which are separated by the subpulse width in the slant plane. The dark, solid lines illustrate the Doppler ambiguity boundary at PRF/2. The region between the two lines is Doppler or surface clutter unambiguous. So, there will be no folding of any returns within the wedge. The antenna beamwidth needs to be narrower than the PRF/2 boundary or cross-range Doppler will be folded into the imaging area.

Similar clutter spectra are shown for X-band at a PRF of 1200 Hz and Ku-band operation at 2000 Hz PRF are shown in Figures 9.26 and 9.27, respectively. Only nine iso-Doppler contours (corresponding to $N_{\mathrm{amb}} = 4$) are plotted, in order to illustrate the Doppler ambiguity issue. These figures illustrate an important point concerning the potential for simultaneous SAR and GMTI waveform for a microwave frequency. The primary PRI for X-band and Ku-band are shown as the heavy black lines. And the FJB sub-PRI is given as the dotted line. This illustrates the importance on timing for receiving and processing the multiple subpulses. It is also clear that the Doppler ambiguities are separated by very small angles θ_{amb}. This separation can be determined by

$$\theta_{\mathrm{amb}} = \mathrm{as\ in}\left(\mathrm{PRF}\,\frac{\lambda}{2\,v_{\mathrm{P}}}\right) \tag{9.16}$$

The angular separation between iso-Doppler ambiguities will require a large antenna to provide the attenuation of Doppler ambiguities within either GMTI or SAR modes.

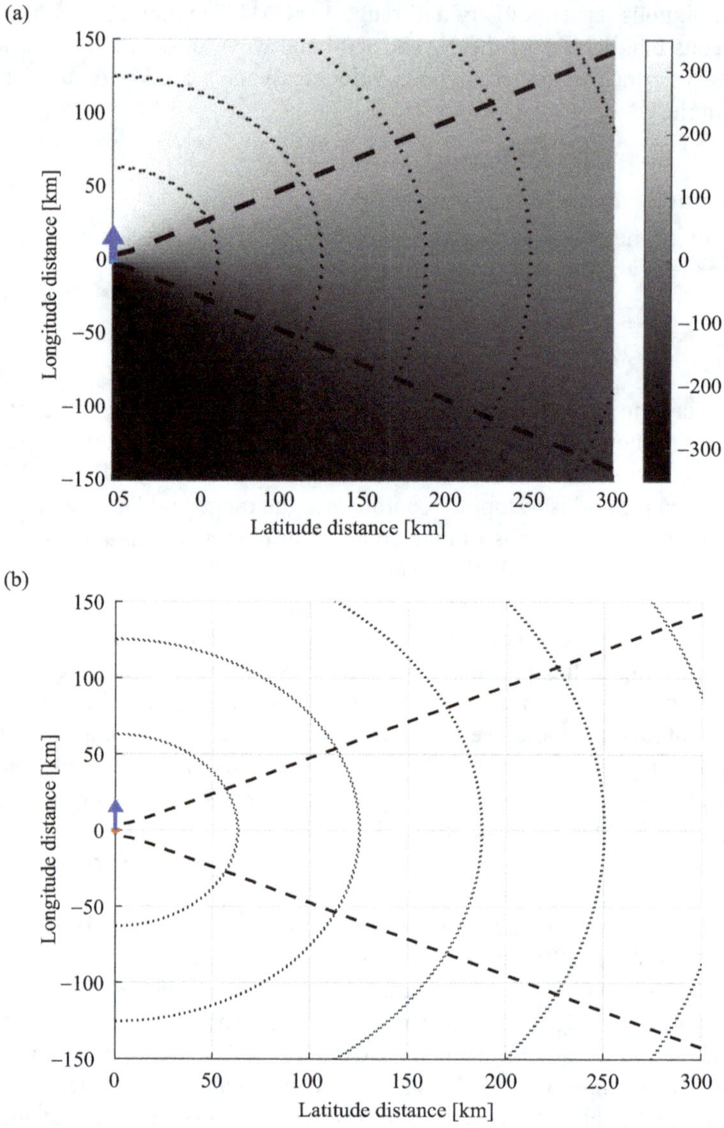

Figure 9.25 *UHF ground clutter Doppler spectrum (PRF 300 Hz). (a) Surface Doppler spectrum on GRCA and (b) iso-range and iso-Doppler contours*

The problem can be mitigated in part by raising the base PRF for the waveform. When a base PRF of 1,200 Hz is used, the X-band surface Doppler spectra are given in Figure 9.26. And for Ku-band, the ground clutter is given in Figure 9.27. It is clear in the figures that the ambiguous clutter lines are spread by 4:1 over the 300-Hz PRF. This gives a wider angular region to detect targets in either GMTI or

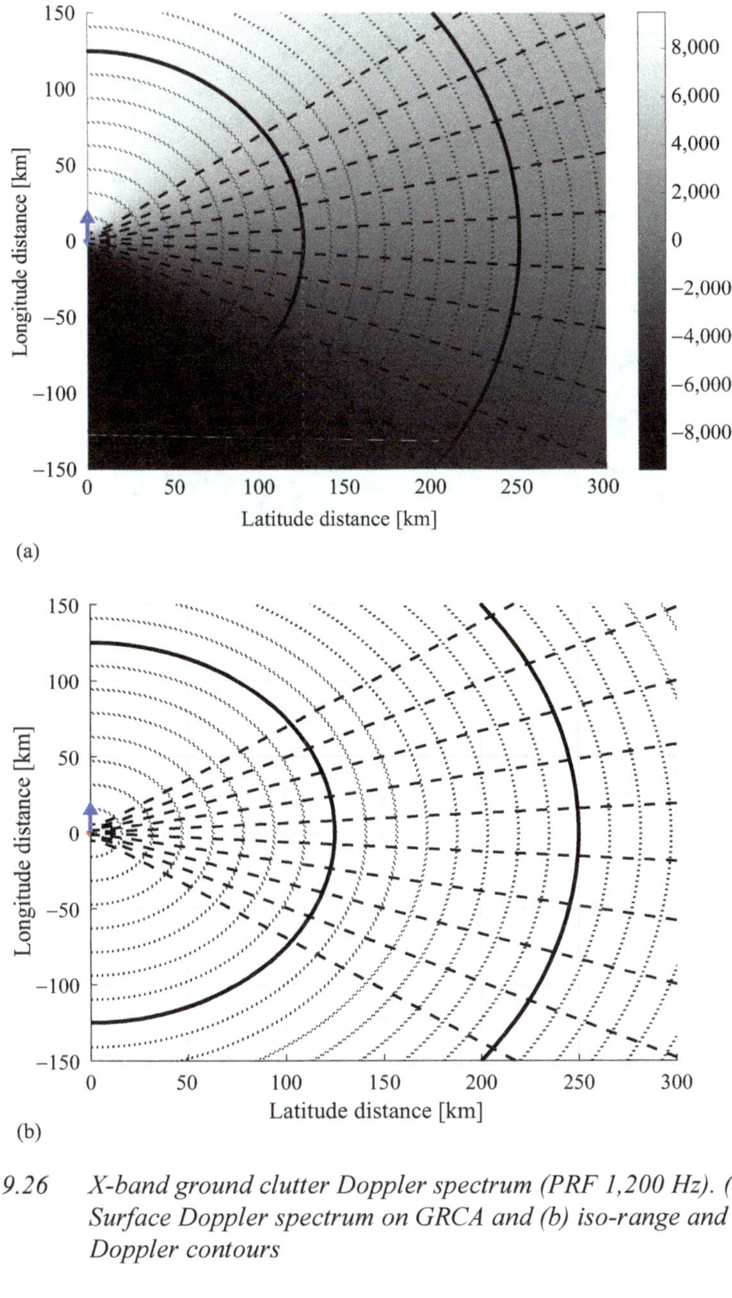

Figure 9.26 X-band ground clutter Doppler spectrum (PRF 1,200 Hz). (a) Surface Doppler spectrum on GRCA and (b) iso-range and iso-Doppler contours

SAR modes. However, the range ambiguities are brought into 125-km slant range at 1,200-Hz PRF and 75 km at 2,000-Hz PRF. Furthermore, the range eclipse regions are closer together. Despite the narrower range extent of the eclipsed areas, it is

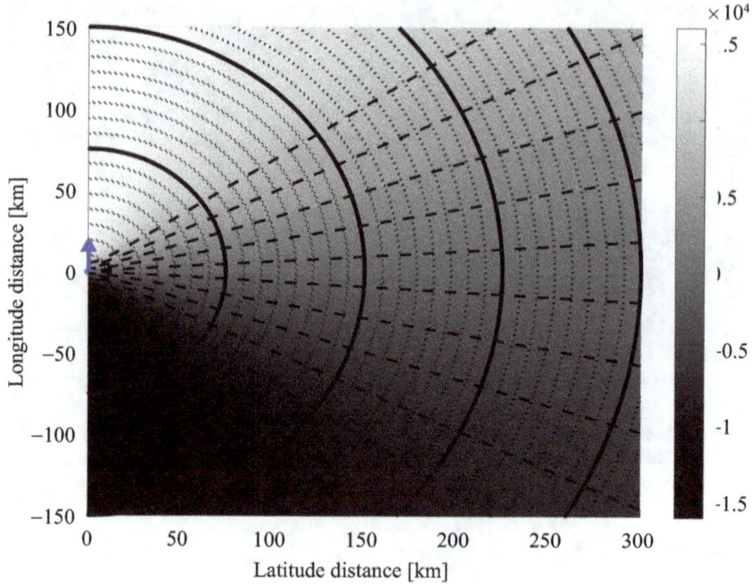

(a)

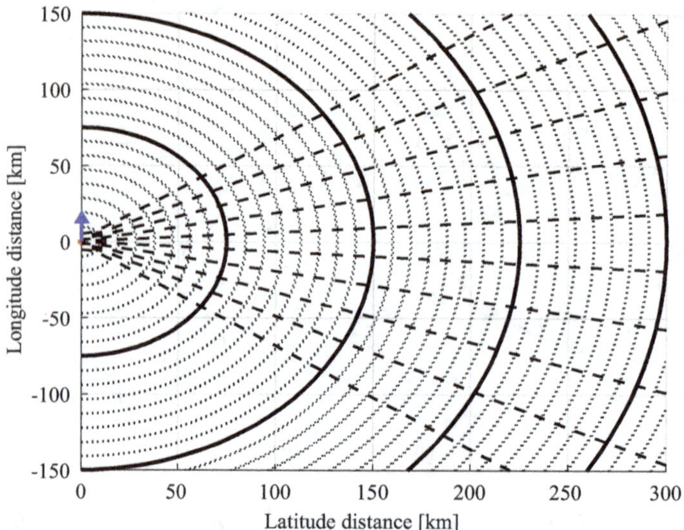

(b)

Figure 9.27 Ku-band ground clutter Doppler spectrum (PRF 2,000 Hz). (a) Surface Doppler spectrum on GRCA and (b) iso-range and iso-Doppler contours more important to understand the impact on both the SAR and GMTI operation with these reduced range regions.

The separation between the Doppler ambiguities is an important design boundary for SAR and GMTI considerations. It is important that the angular regions within one Doppler ambiguity can be isolated within a beamwidth, and the spatial filtering of the antenna provides sufficient cancellation of the targets and clutter to maintain the waveform multiplicative noise ratio (MNR).

Table 9.2 provides the angle spread in degrees from (9.16) between the Doppler contours for UHF, X-band and Ku-band, where the PRF is different for each case to enable separation of the iso-Doppler contours. The integration angle for SAR image formation is estimated from (3.16) and shown in Table 9.3 for five different cross-range resolutions. These angles are significantly lower than the Doppler ambiguity angular separation for the microwave SAR systems. However, the UHF SAR integration angles are borderline for achieving low MNR, if cross-range resolutions below 1 m are desired.

The positive conclusion from this simplistic analysis is that the integration angles are smaller than the Doppler ambiguity angles. So, it is possible that the system can be built at the lower PRF. However, the PRF needed to be increased for X-band and Ku-band for GMTI constraints, and for added ambiguity suppression for an operational MNR. And the angles are based on uniform azimuth weighting. The angles will increase as Taylor weights are applied for reducing the Doppler sidelobes. As the result, the design of the multibeam antenna needs to be tailored for meeting the antenna azimuth sidelobes and Doppler filter sidelobes in both SAR and GMTI modes.

Table 9.2 Doppler ambiguity parameters

	Doppler ambiguity angle (deg)		
V_p (m/s)	UHF	X-band	Ku-band
50	128	22.2	22
100	99	10.9	10.8
150	59	7.3	7.2
200	40	5.4	5.4

Table 9.3 SAR integration angles

	SAR integration angles (°)		
δ_{az} (m)	UHF	X-band	Ku-band
0.5	61.9	2.2	1.3
1	29.8	1.1	0.64
2	14.8	0.54	0.32
3	9.8	0.36	0.21
4	7.4	0.27	0.16

9.2.3 Multibeam ACR

A major factor in the evaluation of the common GMTI/SAR waveform is the ACR of the sensor, i.e., how fast can a surveillance volume of angular extent be covered during the CPI. The SAR operation requires a beamwidth sufficient to illuminate the surface for the angle required for cross-range resolution, as illustrated in Figure 9.28a. This pattern provides the wide illumination area but requires a very long integration time [12].

The primary measure of effectiveness for surveillance radar is ACR, typically defined in square kilometers per second. Multimode radar processing facilitates

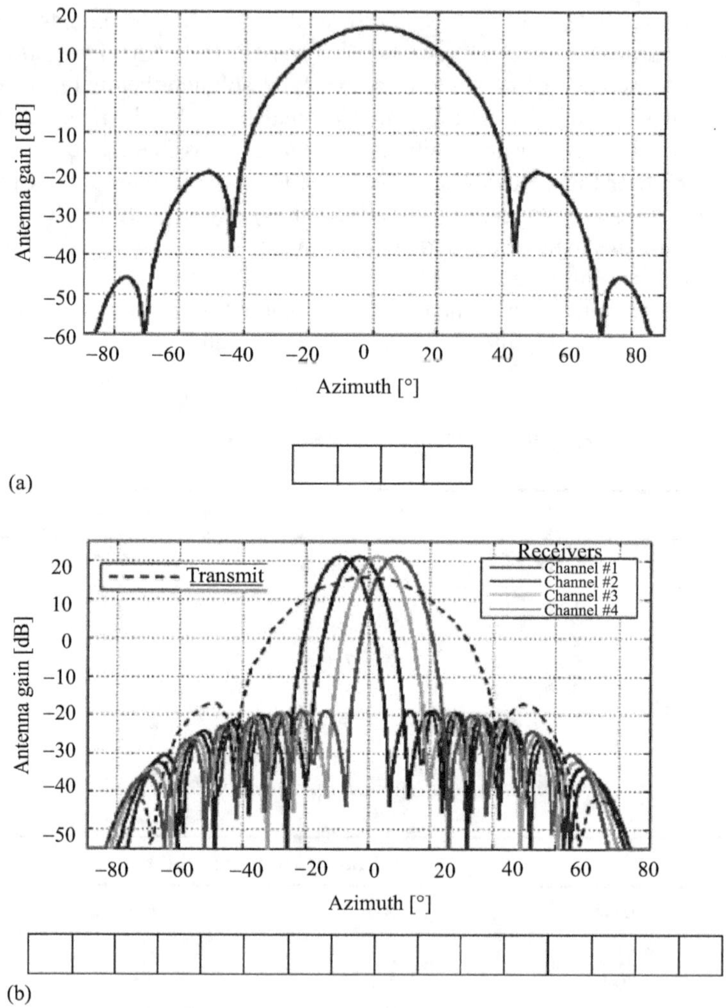

Figure 9.28 Antenna architecture for simultaneous radar modes [12]. (a) SAR Tx/Rcv pattern and (b) GMTI Tx/Rcv patterns

the simultaneous operation of multiple radar modes such as GMTI, SAR and ATI. Section 2.4 presented the ACR for a single dwell for range and angle search for a single CPI. This analysis will be expanded to the practical operation of the three surface surveillance modes, given the practical limitations of range sector surveillance.

The theoretical limit on GMTI ACR is a function of the received antenna pattern. However, in practice, true GMTI ACR is limited by the system's CPI, pulse width, PRI and the receive antenna pattern's projection on the ground. System parameters such as CPI, pulse width and PRF are chosen based on detection performance requirements:

$$\mathrm{ACR_{GMTI}} = \frac{\left(R_{\max}^2 - R_{\min}^2\right)}{2\ \mathrm{CPI}} \varphi_{3\ db_{\mathrm{B}}} \tag{9.17}$$

where R_{min} and R_{max} are the minimum and maximum ground ranges of the system in meters, respectively, $\varphi_{3\ db_B}$ is the azimuth 3-dB beamwidth in radians and CPI is the time of one CPI in seconds.

SAR images are normally generated for small regions via either spotlight or strip map processing. For strip map operation, the theoretical limit on ACR is a function of the received antenna pattern. However, this upper limit does not consider the required integration time for adequate image resolution. For strip map SAR, integration time is a function of cross-range resolution as shown in [13]:

$$t_{\mathrm{SAR}} = \frac{k\ R_{\mathrm{mid}}\lambda}{2V_{\mathrm{plat}}\delta_{\mathrm{CR}}} \tag{9.18}$$

where R_{mid} is range in meters to the mid-point of the image, k the beam broadening factor, λ system wavelength in meters, V_{plat} the speed of the platform in meters per seconds and δ_{CR} the desired cross-range resolution in meters of the SAR image. SAR ACR can be approximated as

$$\mathrm{ACR_{SAR}} = \frac{\left(R_{\max}^2 - R_{\min}^2\right)}{2\ t_{\mathrm{SAR}}} \varphi_{3\ db_{\mathrm{B}}} \tag{9.19}$$

where R_{min} and R_{max} are the minimum and maximum ground ranges of the system in meters, $\varphi_{3\ db_B}$ the azimuth 3-dB beamwidth in radians and t_{SAR} the SAR integration time in seconds.

Slow-moving target detection in ATI processing is a function of both sufficient SAR resolution and sufficient signal-to-clutter for correlation detection and velocity estimation. The theoretical ACR for ATI detection processing is given by [14]

$$\mathrm{ACR_{ATI}} \leq \frac{\sigma_{\mathrm{tgt}}\ B}{\sigma_{\mathrm{o}}\ \mathrm{SCR}} \tag{9.20}$$

where σ_{tgt} is the radar cross-section of the unresolved target, σ_0 the radar cross-section of the background, B the waveform bandwidth and SCR the signal-to-clutter ratio.

In practice, an SCR that provides suitable correlation for detection and phase for velocity estimates is achieved by adjusting the resolution of the SAR image. In the unresolved target case, adequate cell size will maximize SCR and therefore ATI detection probability. Cross-range resolution as a function of SCR is [15]

$$\delta_{\text{CR}} = \frac{\sigma_{\text{tgt}}}{\sigma_o\, \delta_{\text{R}} \text{SCR}} \tag{9.21}$$

where σ_{tgt} is the radar cross-section of the unresolved target, σ_0 the radar cross-section of the background, δ_R the range resolution of the system and SCR the signal-to-clutter ratio.

With adequate cross-range resolution for ATI detection, ACR can be estimated similar to standalone SAR image generation, but with lower cross-range resolution, and proportionately lower integration times.

$$\text{ACR}_{\text{ATI}} = \frac{\left(R_{\max}^2 - R_{\min}^2\right)}{2\, t_{\text{ATI}}} \varphi \tag{9.22}$$

where R_{min} and R_{max} are the minimum and maximum ground ranges of the system in meters, φ the azimuth 3-dB beamwidth in radians and t_{ATI} the SAR integration time for ATI processing in seconds.

Based on the parameters given in Table 9.3, a comparison of the ACR for the three independent modes can be constructed, as shown in Table 9.4. The receive architecture is a modest four-beam receive pattern. The ACR comparison of these mode parameters is shown in Figure 9.29. In this example, the ATI processing assumed a cross-range resolution four times the SAR, a sub-aperture of the full SAR integration time. The GMTI is seen to have at least two orders of magnitude ACR higher than the SAR, due to the larger search volume and much shorter CPI.

For a single-mode operation of the AESA antenna, the angular coverage and CPI time need to be matched to the waveform. If the GMTI search mode is interrupted for either SAR or ATI detection of targets, the ACR of moving targets will be reduced. Figure 9.30 shows the decrease in GMTI ACR when the search duty factor is reduced by the SAR imaging time. The curves are for 20, 40 and 80% allocation of the radar resources for either SAR or tracking waveforms.

When the combined SAR/GMTI mode is examined, GMTI SAR and ATI processing can be processed within the transmit beamwidth. A single beam can process the GMTI at multiple frequencies. However, the combination of beams can provide the angular coverage required for SAR and ATI. Digital beam forming will provide the ability to match the received angles with the spoiled transmit beam width.

Table 9.4 Common mode UHF evaluation parameters

Center frequency	400 MHz	Platform altitude	6.1 km
Bandwidth	$M \times 15$ MHz	Platform speed	60 m/s
Sub-pulses (M)	8	Slant range	15 km
PRF	300 Hz	Antenna width	6 m
Sub-pulses width (T_{pM})	41.667 µs	Number azimuth channels (N_c)	16
SAR resolution (range, cross-range) (weighted)	2.3, 6.7 m	GMTI range resolution (weighted)	18 m

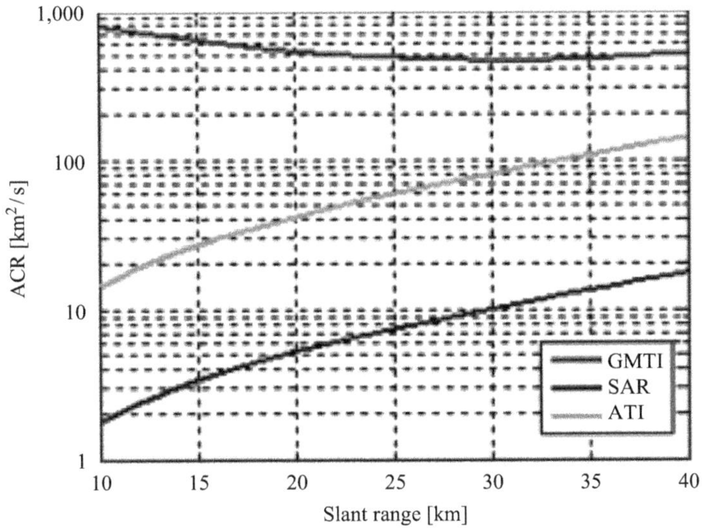

Figure 9.29 *ACR for three simulated cases [10]*

Figure 9.31 illustrates the combined ACR for the beams, along with the multiple beam GMTI ACR for 4 and 16 beams.

A direct advantage of the FJB waveform is frequency diversity. Depending on the frequency separation between the individual bursts, the target will transition from a Swerling-2 target to a nonfluctuating Swerling-0 return. This in effect has two benefits:

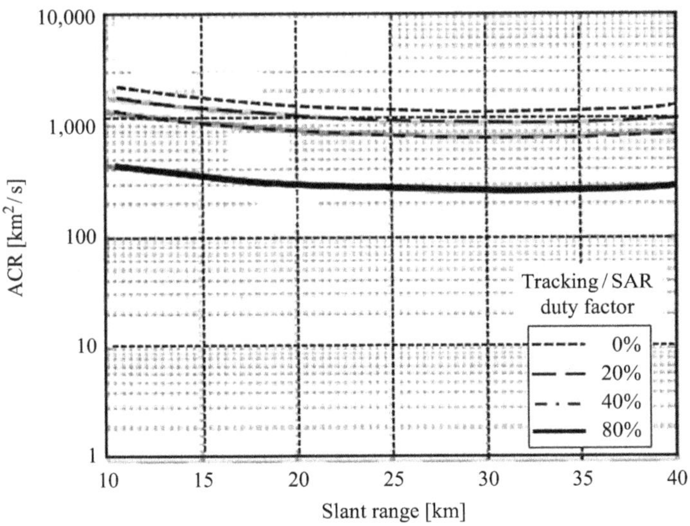

Figure 9.30 *Wide-area surveillance GMTI mode versus SAR/ATI interrupted scheduling [16]*

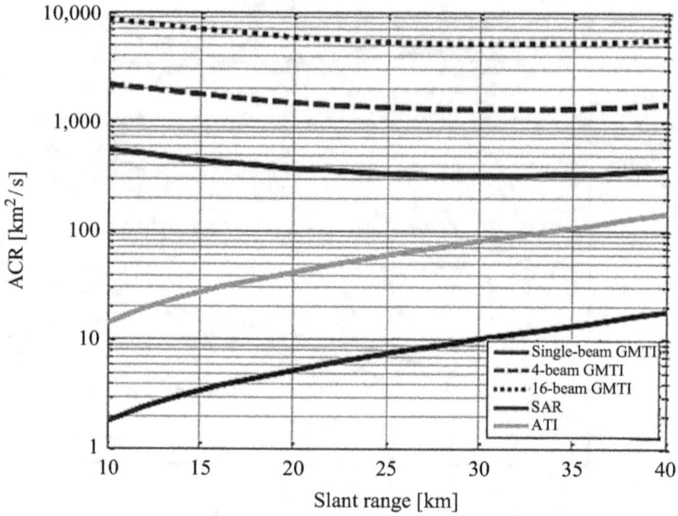

Figure 9.31 Simultaneous SAR/GMTI/ATI ACR [10]

first, the required SNR for detection will be decreased by approximately 4 dB, and second, there is a cumulative probability of detection by the multiple looks.

In addition, the frameworks for evaluating the ACR benefit of common radar modes and the multiple beam processing have been clearly demonstrated. As individual CPIs are scheduled for either SAR or high-resolution tracking, the effective ACR is greatly diminished. Figure 9.31 extends this analysis to show the effective ACR of GMTI, SAR and ATI with the simultaneous SAR/GMTI waveform. In addition, it shows the increase in ACR for the GMTI mode by using multiple beam operation.

9.2.4 Common mode evaluation

This section explores the potential multimode surveillance radar when a common waveform is designed. Given the advances in coherent waveform generation with long-term phase control between pulse and antenna phase centers, a multimode, multibeam radar is postulated. This is based on the ubiquitous radar suggested by Skolnik almost 20 years ago [1]. A combination of a frequency diverse waveform and UWB waveform and signal processing are assumed. The objective of this analysis was to determine the potential of rapid growth in component and waveform processing capabilities [6, 17].

GMTI clutter suppression requires an antenna architecture with sufficient along-track dimension to provide a moderate MDV. The illumination pattern is the same as for SAR, shown in Figure 9.28a, with a wide angle for covering both the SAR and GMTI modes. By using a multichannel design, multiple receive beams can be digitally formed, as shown in Figure 9.28b. However, simultaneous collection of the received energy by multiple receive beams is required to improve SNR, MDV and ACR.

The system challenge is placing requirements on low-sidelobe receive pattern implementation of digital beamforming and suppression of strong clutter returns in adjacent beams. Furthermore, the real-time processing of multiple beamformers requires an order of teraflops per second throughput. These digital processing capabilities are rapidly becoming a reality.

In order to evaluate the techniques for a common SAR/GMTI mode, a simulation scenario was used to determine the separability of point scatters in the scene, as well as the detection of moving targets. The simulation used a tool RLSTAP developed for realistic assessment of radars over distributed clutter, using DTED for terrain height variation and land-use land clutter models for the types of clutter in an area. The statistics of these clutter patches are based on well-documented clutter types at varying frequency and incident angles, as covered in Section 2.3.1 Because the modeling tool represents the position and velocity at each pulse, effects of navigation errors, range and Doppler walk, and waveform and beam shape contributions can be evaluated [16].

Figure 9.32 shows the scenario used in this analysis, with the basic simulation parameters are given in Table 9.5. A 1-km square region composed of forest and farmland has been chosen for this study. The simulation is provided by RLSTAP, similar to examples given in Chapter 7, but with a formulation of the FJB waveform. The analysis of the common SAR and GMTI waveform had multiple targets distributed, with velocities between +16 and – 16 m/s. The variation of target velocity provided Doppler spacing of the targets to analyze the endo- and exo-clutter limitations. Two arrays of four stationary targets were added to assess the SAR image resolution.

SAR performance

The monostatic SAR was formed by simulating a data cube from the 20-km slant range platform and "flying" the platform North at 75 m/s. The image shown in Figure 9.33 was constructed using a back-projection algorithm to the pixels in the scene. The image in Figure 9.33a shows the full 1-km^2 scene. The moving target groups in the open field are easily detected albeit with a cross-range displacement due to their velocities. For targets in the tree, line there was a signal attenuation due to the foliage, and the reduced SNR and CNR will make it difficult to detect.

The scenario was modeled using RLSTAP, a detailed pulse-by-pulse simulation of the transmitter and receiver characteristics, clutter and propagation characteristics and target model. Because the modeling tool represents the position and velocity at each pulse, effects of navigation errors, range and Doppler walk, and waveform and beam shape contributions can be evaluated. Table 9.6 shows the comparison of the resolution of the fixed target reflectors shown in Figure 9.34b. There was good correspondence of the resolution of the theoretical waveform and the back-projection images of these targets.

Significant challenges in combining the eight data cubes were required, due to the overlap between returns from individual sub-pulses. These changes were important to provide a realistic set of targets, noise and clutter for common mode

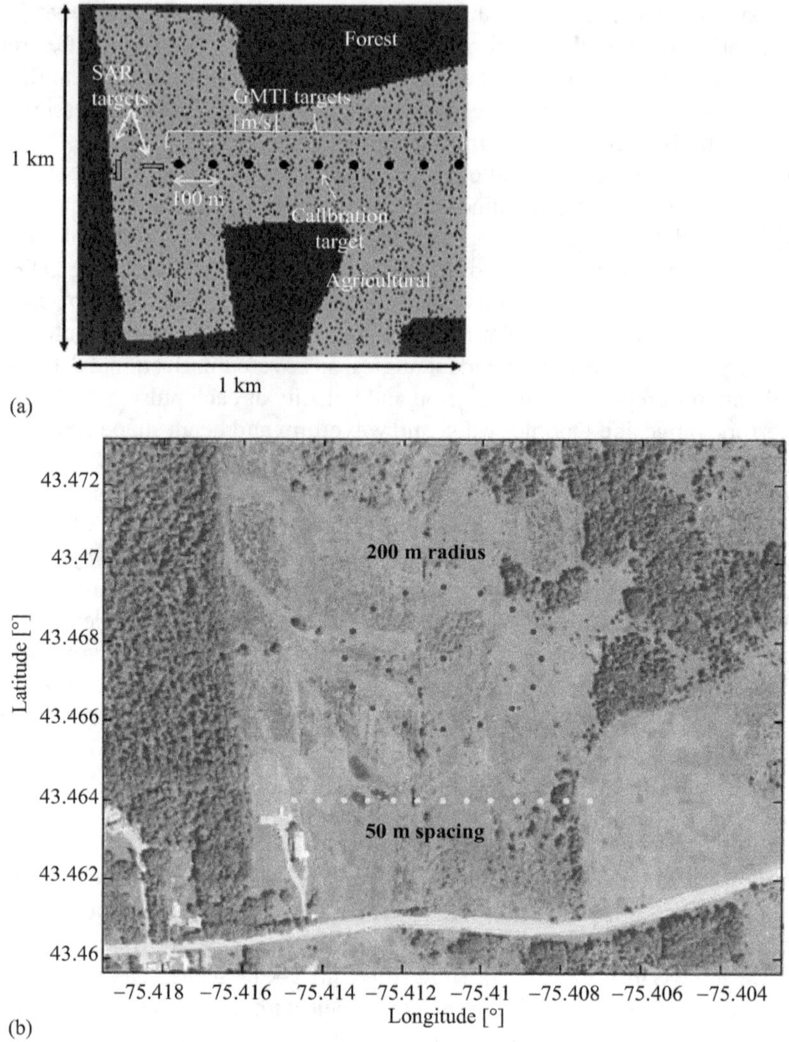

Figure 9.32 RLSTAP simulation of simultaneous SAR and GMTI modes. (a) Land-use land cover map and (b) Google Earth map and targets

processing. The important elements of this new approach was explored in processing the three simultaneous modes: GMTI, SAR and ATI. The first element is that target and noise data cubes were produced with the full range extent with a PRF of 2,400 Hz. Figure 9.34 depicts the eight frequencies generated by the target and noise simulation lineup.

At the signal bandwidth of 30 MHz and 2:1 oversampling, each data file was 3.2 GB, for 16 channels, 25,018 range bins and 64 pulses. Each sub-pulse must be delayed in the synthesis by 52 ms, for each FJB component. The full PRI of 416.67

Table 9.5 Multimode simulation parameters

Transmit		**Receive**		**Platform**	
PRF	300 Hz	Elements	16	Altitude	6.1 km
Pulse width	41.6 us	Sub-arrays	16	Velocity	60 m/s
Subpulse	8	Bandwidth	15 MHz	Slant range	14 km

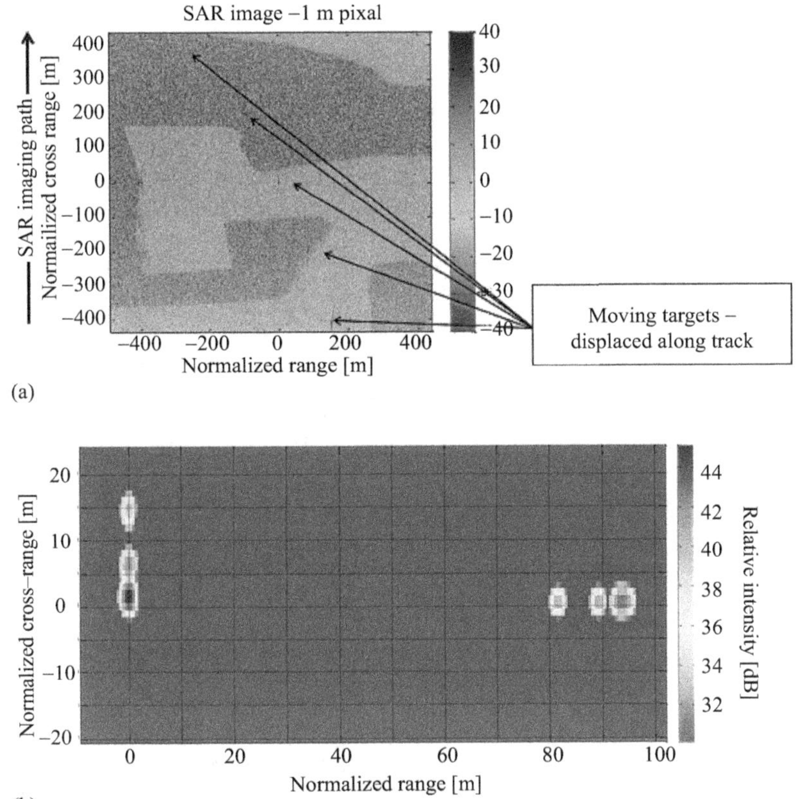

Figure 9.33 SAR image phenomena for fixed and moving targets. (a) Moving target SAR and (b) SAR stationary targets

ms in fast time is required to determine the combination of target, noise and clutter for each sub-pulse. Finally, the noise in the receiver is white bandwidth and contained in the PRI.

The FJB pulses were combined for 8K pulses to achieve the 6.7-m cross-range resolution. The SAR image of the scenario in Figure 9.32 is shown in Figure 9.33a. The distributed ground clutter and fixed targets are shown properly focused and geo referenced. The moving targets are displaced in cross-range proportional to the

Table 9.6 Comparison of SAR simulation resolutions

Resolution	Theoretical [m]	Simulated [m]
Range	1.3	2
Cross-range	6.7	7.1

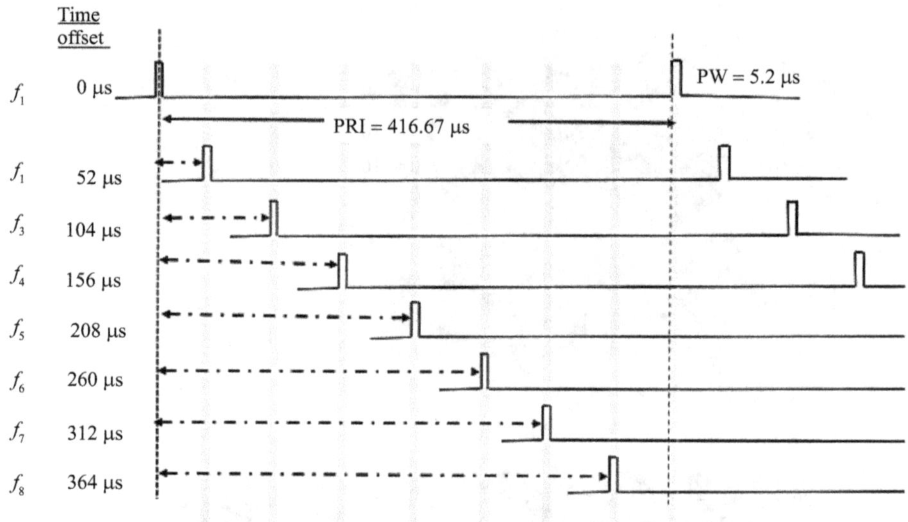

Figure 9.34 Common mode waveform scheduling from RLSTAP data cubes

radial velocity. Figure 9.34b shows the range and cross-range resolution of array of fixed targets, confirming the SAR resolution of the new waveform.

GMTI performance

The multiple receiver phase centers were used to form a narrow receive beam. Each receiver was sampling the returns at 120 MHz, so that the returns from each of eight frequencies are received. A complex zero-phase low-pass FIR filter is used on each channel to remove the other frequency components. The filter is an 80-dB stop band equal ripple, as shown in Figure 9.35. The return after FIR filtering of the multiple simultaneous frequency returns is shown in Figure 9.36, where the multiple frequency returns are significantly reduced. After filtering, a CPI of 64 pulses was used for the STAP processing with each of the eight sub-pulses.

A range-Doppler plot of the return after joint domain localized STAP processing is shown in Figure 9.37. Results from two of the eight frequencies are shown: frequency 2 at 348 MHz and frequency 6 at 438 MHz. The returns for the targets at lowest Doppler are attenuated, with the zero Doppler target not being detected. The SINR loss for the same two moving targets is shown in Figure 9.38. An MDV

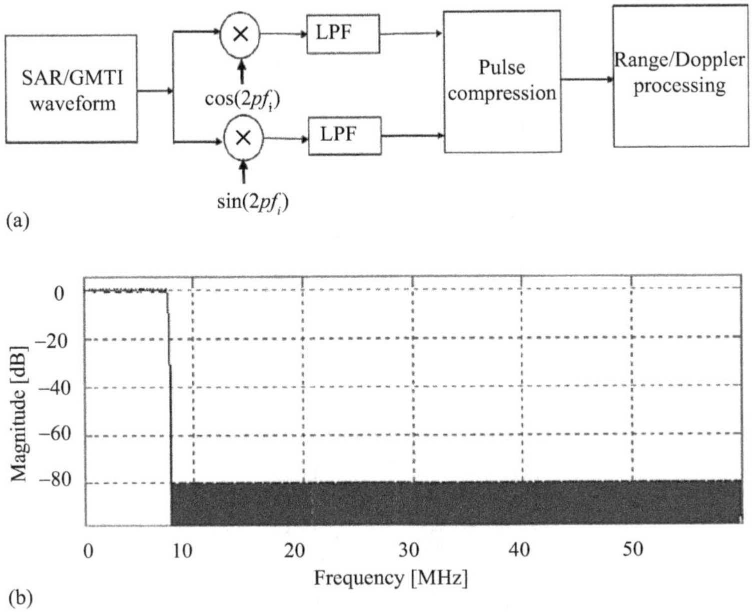

Figure 9.35 *Channel filtering for multiple frequency return removal [9]. (a) GMTI waveform baseband conversion and (b) low-pass filter for GMTI processing*

between 4 and 6 m/s was obtained after STAP, depending on the center frequency of the eight sub-pulses. Clearly the individual targets are attenuated by the STAP processing SINR loss. This technique is only viable for targets that are inside the edge of mainbeam clutter with sufficient radial velocity that SINR loss <10 dB.

These results are very encouraging for operation in a very wideband radar using the FJB with narrow-band sub-pulses. The SAR images shown above used the same transmit waveform, but with alignment for obtaining the wider required bandwidth. Most importantly, for these moving targets, the cross-track displacement has been overcome by using short CPI and STAP processing. Only the very slow targets are lost in the process.

Along-track interferometry

ATI has been shown to be a valuable technique for detection and repositioning slow-moving targets in high-resolution SAR images. ATI processing was accomplished utilizing the forward and aft sub-apertures of the multibeam antenna. With this configuration, a baseline separation of 4.5 m is achieved realizing an unambiguous radial velocity of 10 m/s. Figure 9.39 illustrates the MDV performance in an endo-clutter environment, with detection between 0.5 and 4 m/s as a function of signal-to-clutter.

For the system under evaluation, a mean SCR of 14 dB was observed. The resultant theoretical ATI MDV is 0.24 m/s. Given the MDV and phase ambiguities,

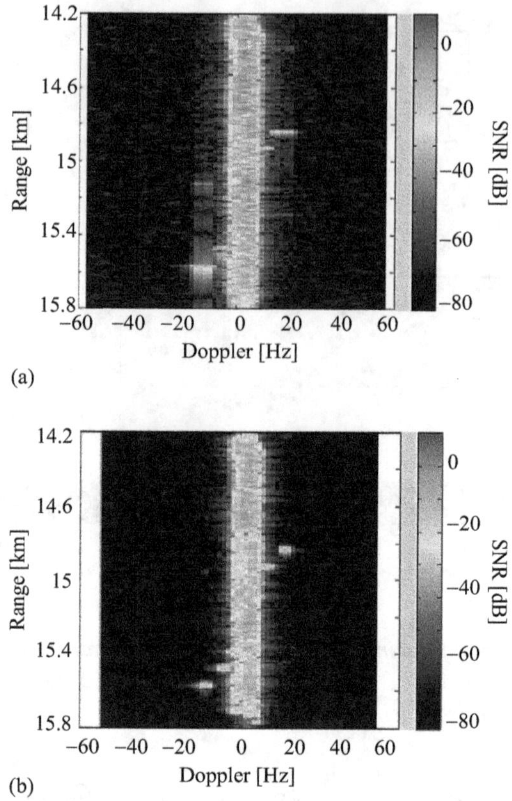

Figure 9.36 Narrow-band GMTI return with filter of wideband receiver returns [9]. (a) Full BW receiver return and (b) after low-pass filter

the system under evaluation should be sensitive to targets moving between 0.24 and 2.4 m/s radially. Given the geometry of the evaluation scenario, ATI should provide suitable detection and location performance for targets moving less than 2.7 m/s in the ground plane.

A cell averaging constant false alarm rate processing was used to indicate targets based on ATI correlation and phase. The velocity of these indications is estimated based on the detected phase as described in Section 7.2.2. These results clearly illustrate the potential utility of ATI in very slow-moving target detection, but questions such as SAR resolution and phase coherency requirements need to be investigated further.

Several features can be observed, especially when the forest clutter returns are significantly higher than the open field as expected. Using the sub-pulse bandwidth, the image resolution is comparable for range and azimuth, approximately 5 m. Only four of the individual scatterers are discernable at the 5-m resolution. The three closer space scatterers combine in an enhanced cross-section.

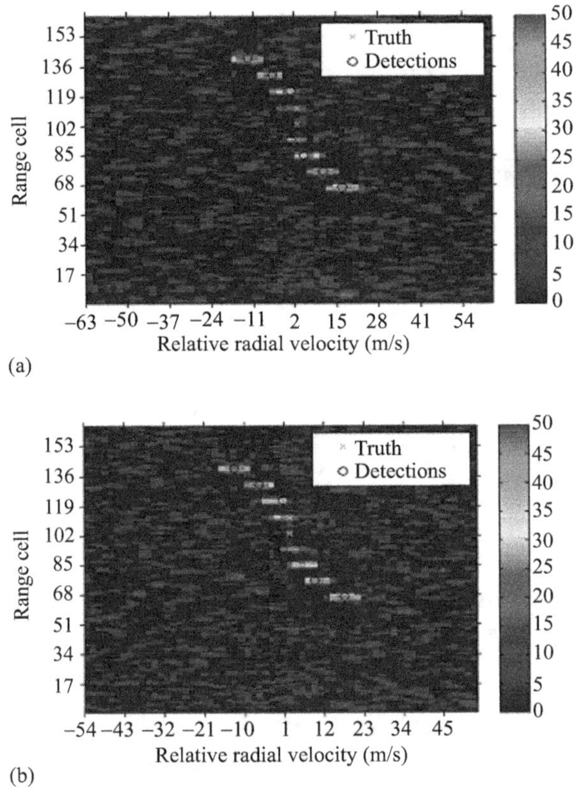

Figure 9.37 *GMTI detection after STAP processing [9]. (a) Frequency 2–348 MHz and (b) frequency 6–438 MHz*

The velocity of these interferometric phase measurements is estimated based on the detected phase as described. These results clearly illustrate the potential utility of ATI in very slow-moving target detection, but questions such as SAR resolution and phase coherency requirements need to be investigated further.

The complementary advantage of combining GMTI and ATI is shown in Figure 9.40. The top eight lines depict the eight sub-pulse frequencies, their associated MDV, along with the targets that were detected within the clutter Doppler bandwidth. It is clear that the STAP loss reduces the ability to detect endo-clutter targets. And the target Swerling-2 scintillation provides low detection probability. However, by combining an *m:n* logic, exo-clutter and faster endo-clutter target detection is improved. The second from bottom row of the figure shows the ATI detections. There were 100% detections out to 2.5 m/s. At that point, the targets are walking through range and Doppler lines during a dwell. And the bottom row shows the composite detection for the combined STAP and ATI on the targets spanning the clutter Doppler spectrum as shown in Figure 9.17. The next step in the evolution

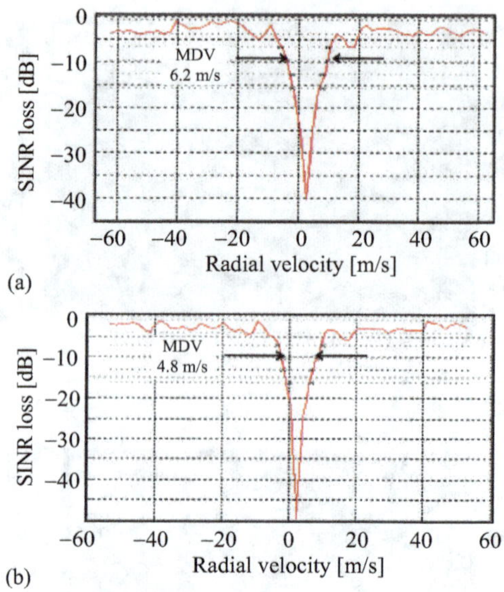

Figure 9.38 *SINR loss from STAP processing. (a) Frequency 2–348 MHz and (b) frequency 6–438 MHz [9]*

of this processing is to apply target coherence processing to improve the signal-to-noise level on these targets.

Future ISR systems will require a significant improvement in persistent detection and tracking of tactical targets. The density of targets and environments with more concealment and deception make it significantly more difficult to maintain long-term track on high-value targets. This initial investigation into simultaneous fixed and moving target detection has shown the feasibility for developing a new radar architecture.

9.2.5 Multiple beam multimode signal processing

In order to achieve multiple simultaneous receive beams and multiple spatial DOFs, digital beamforming is expected for any practical system. The onboard processing for the multibeam receiver must synchronize the multiple sub-pulses to the transmit waveform, provide waveform-matched filtering and provide the STAP to detect the target in the competing clutter. A first-order estimate of the processing real operations for each signal processing requirement can be made with the following estimates [16].

The beam forming for N_{Rbin} is estimated by the digital beamforming using the FFT as shown in

$$\text{Beamform} = N_{\text{Rbin}}\,[20N_{\text{Beam}}\log_2(N_{\text{Beam}}) + N_{\text{Beam}}] \tag{9.23}$$

The processing for Doppler of the number of pulses in a CPI N_{dop} with the number of range bins N_{Rbin} being calculated, and the number of beams is given by

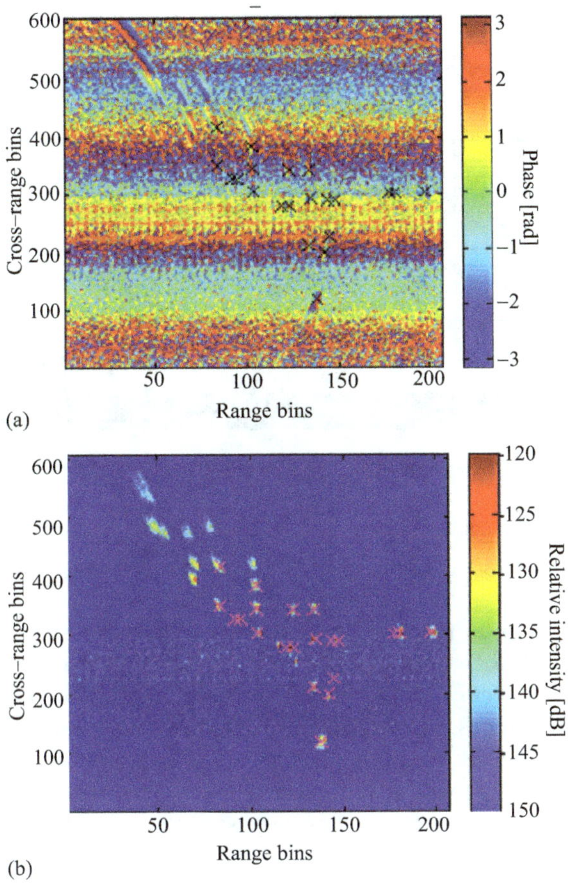

Figure 9.39 ATI results for moving target indication. (a) ATI phase and (b) ATI magnitude

$$\text{DopProc} = N_{\text{Beam}} N_{\text{Rbin}}[20 N_{\text{dop}} \log_2(N_{\text{dop}})] \tag{9.24}$$

For each range extent N_{Rbin}, the processing for a waveform pulse compression size of BT is given for the number of range bins and the number of digital beams as N_{Rbin}:

$$\text{PulseCompr} = N_{Beam}\, N_{Rbin}[20\, BT \log_2(BT)] \tag{9.25}$$

For the beams that are used for STAP of the moving targets, the processing requirements for the number of spatial B_{beams} (channels) and temporal (Doppler filters) degrees of freedom (DOFs), with the number of samples M_s, an estimate to process the covariance matrix is

$$\text{STAP} = 20 N_{Beam}[4\, M_S\, \text{DOF}^2 - 4/3\, \text{DOF}^3 + 2\, M_S\, \text{DOF} - 4/3\, \text{DOF}] \tag{9.26}$$

Freq	MDV	Target velocity [m/s]												
[MHz]	[m/s]	−16	−8	−4	−2	−1	−0.5	0	0.5	1	2	4	8	16
347.5	6.2													
362.5	5.9													
377.5	5.6													
392.5	5.3													
407.5	4.9													
422.5	4.6													
437.5	4.3													
4525	4.0													
ATI	0.2													
m:n detection														

Figure 9.40 Composite moving target detection GMTI and ATI [16]

The number of samples used to estimate the covariance matrix is classically taken as four DOFs.

This analysis can be projected into a multimode processing requirement for simultaneous GMTI, SAR and ATI. For GMTI, the number of range bins is proportional to the slant range in the surveillance region and the system bandwidth. The GMTI mode is processed using an FJB sub-pulsewith the number of sub-pulses given by N_{pulse}. If the multiple sub-pulses are used, as shown in Section 7.4, there is a requirement for a temporal factor in the number of simultaneous beams. For an SAR or ATI image, the range extent is necessarily smaller than the GMTI range swath. In this example, it is assumed that the range extent is divided by the number of sub-pulses, while the number of Doppler filters are increased by the same factor.

Since the number of range bins are proportional to the bandwidth of the signal, and the total time to accomplish this processing is the CPI, the total processing can be estimated as a function of the number of beams and the bandwidth of the fundamental waveform. Figure 9.41 provides a representation of the processing for the bandwidths and number of simultaneous beams indicated for the range of parameters [17].

It should be noted that this analysis is for STAP processing for the basic sub-pulse in an FJB waveform. If the number of sub-pulses is factored into the processing throughput, then the processing will be increased by N_{pulse}. An alternative approach would be to count the number of beams N_B as a factor of the spatial beam N_{beam} and the temporal beam N_{pulse}:

$$N_B = N_{beam} N_{pulse} \tag{9.27}$$

Given this factorization, the line for 64 FJB beams N_B in Figure 9.41 could be for $N_{beam} = 8$ and $N_{pulse} = 8$. A complete analysis of the processing throughput of an UWB, multibeam radar will increase the real-time processing requirements.

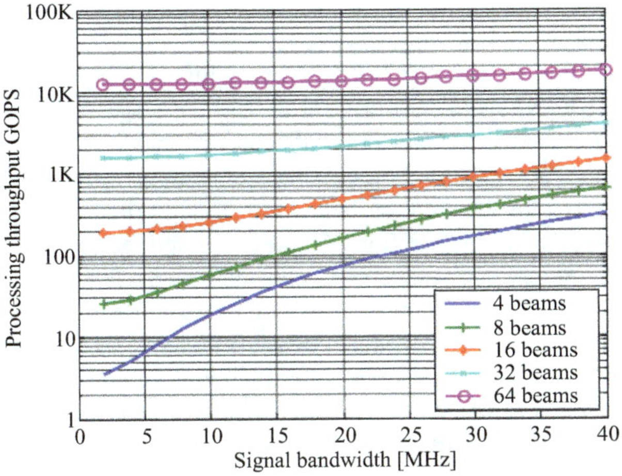

Figure 9.41 *Real-time processing estimates for multiple beam multimode processor*

For the simple example given in Figure 9.41, a processor of over 10 teraflops will be required. With the continued improvements for Moore's Law, this processing need is not that far in the future. An exact estimation of the processor will require a specification of how the multiple spatial beams and number of sub-pulses for bandwidth extension are factored for SAR and ATI images.

9.2.6 Simultaneous SAR/GMTI mode future research

The common or simultaneous SAR/GMTI waveform is a first step to the design of new radar capability. It also approaches the vision by Skolnik for an ubiquitous radar, enabled by digital beamforming and advanced high-speed signal processing. The waveform has been partially analyzed and simulated with a high-fidelity simulation to capture the effects of platform and target motion in strong distributed clutter. Three different radar frequencies have been considered: UHF, X-band and Ku-band. Each has peculiar range and Doppler characteristics, including ambiguities, that affect the efficient simultaneous operation in SAR and GMTI modes. The focus in this chapter has been on synthesizing the architectures that include spatial, spectral and temporal variations on the transmit and adaptive receiver processing.

To illustrate the complexity of the waveform and simulate real-world distributed clutter environments, the RLSTAP radar simulation tool has been used. A multiple channel antenna architecture was simulated to enable STAP, and ATI was analyzed with an FJB waveform to provide the UWB capability. The individual FJB pulses that will provide moderate-bandwidth spectral resolution for moving targets and wide-bandwidth resolution of fixed and slow-moving targets. Effects of receiver blanking on the wideband FJB operation for long-range surveillance need to be explored in terms of the dynamic range and signal processing throughput. Finally, the variation of PRF for Doppler ambiguity resolution should be explored in terms of effects on the SAR image formation when the PRF is changed during the collection. Because of the large data cube requirements, the simulation has been divided into multiple sub-pulse simulations and a high-fidelity combination of data cubes into a realistic received environment for further analysis.

The future research on UWB simultaneous SAR and GMTI architecture will require more sophisticated signal processing architectures. The data cubes need to be band-pass filtered for the medium-bandwidth STAP processing. The spatial pre-processing for wide azimuth detection of targets in a spoiled illumination beam will provide insights into the dynamic range and clutter cancellation techniques. The temporal processing of medium sub-pulses will provide the input for SAR processing at high-range resolution. The impact of sub-PRI range extent will be evaluated on limitation in the SAR swath width. Multiple azimuth phase center SAR image formation for ATI algorithms will determine the ability to detect a wide range of target velocities, by choosing multiple phase centers. A close accounting of the memory and operations count for each sub-mode will provide requirements for future high-performance computing developments.

9.3 Summary

A multibeam surveillance radar has many advantages in covering a wide area and providing an improvement in ACR. During the continued advances in digital signal processing and wide dynamic range ADCs, digital beamforming has become more practical. The improvements in surveillance radar since the multibeam MSR prototype radar have been significant. The ability to image and characterize both fixed and moving targets have given users a more useful tool for providing a persistent monitoring of both military and geoscience applications.

In addition to the digital processing advances, the microwave components have become significantly smaller in both size and cost. Single-chip radars have been advertised for short-range monitoring of scenes. Combining these monolithic microwave integrated circuits with high-speed signal processing has provided low-cost distributed surveillance sensors. When these small RF sensors have been integrated into higher power aperture radars, the active ESA arrays have been applied to volume- and power-constrained platforms. As a result, it should be practical to provide real-time processing in a multibeam radar on small airborne platforms.

In parallel with the development of a multibeam antenna, the ability to synthesize UWB waveforms has been widely demonstrated. These waveforms can change and adapt the characteristics to discriminate fine target and ground features. But more importantly, the rapid change in the waveform characteristics has the potential to adapt to the RF environmental spectrum constraints. More research on the use of arbitrary waveforms that are insensitive to platform Doppler effects is needed. If the radar community can emulate the communications waveforms and significantly reduce the cochannel interference, UWB surveillance systems will be more accepted in the regulated RF spectrum.

References

[1] Skolnik M.I. 'Attributes of the ubiquitous phased array radar'. *Proceedings IEEE International Symposium on Phased Array Systems and Technology*. Waltham, MA; 2003.

[2] Banister W., Grove C.E., Song J. '"Multibeam Antenna Development and Test Program", Final Report AFWAL-TR-3-1049'. *Contract F33615-79-C*. 1983;**1907**:AD078907.

[3] Mancini H.J. 'A multimode battlefield surveillance radar'. *Proceedings 1984 SAE Avionics Guidance and Control Conference*. Long Beach, CA; 1984.

[4] Hofmeister E.F., Szczenspanski W.E., Oot R.F., Dalpe D.C., Davis M.E. 'Modular surveillance radar for battlefield surveillance applications'. *Proceedings RADAR-82*. London England; 1982.

[5] Barker J., Davis M.E. 'Ku-band linear phased array'. *Microwave Journal*. 1977;**20**:44–57.

[6] Pietrzyk G., Samczynski P., Gorzelanczyk A., Kulpa K. 'Real time Implementation of doppler beam sharpening technique with simple motion estimation'. *Proceedings European Radar Conference*. Amsterdam, NL; 2004. pp. 185–8.

[7] Showman G.A. *Principles of Modern Radar – Advanced Techniques, Chapter 7 Stripmap SAR*. Edison, NJ: Scitech Publishing; 2013. pp. 269–85.

[8] Fowler C.A. 'The standoff observation of enemy ground forces from project peek to JointSTARS-A prolusion'. *IEEE Aerospace and Electronic Systems Magazine*. 1997;**12**(6):3–17.

[9] Davis M.E., Kapfer R.M., Bozek R.E. 'A common waveform for simultaneous SAR and GMTI'. *Proceedings 2011 IEEE Radar Conference*. Kansas City, MO; 2011.

[10] Kapfer R.M., Davis M.E. 'Ultra-wideband multi-mode radar processing'. *Proceedings 2012 IEEE Radar Conference*. Atlanta, GA; 2012.

[11] Davis M.E. 'Design alternatives for foliage penetration SAR ultra-wideband waveforms'. *Proceedings 2010 IEEE International RADAR Conference*. Washington, DC; 2010.

[12] Davis M.E. *"Quest for a simultaneous SAR/GMTI waveform", 2013 International Conference on Radar, Adelaide S.* Australia; 2013. pp. 134–9.

[13] Skolnik M.I. *Introduction to Radar Systems, Edition 1, McGraw Hill Book Co. 302*. New York, NY; 1962.

[14] Chen C.W. 'Performance assessment of along-track interferometry for detecting ground moving targets'. *Proc. 2004 IEEE Radar Conference*. Philadelphia, PA; 2004. pp. 99–104.

[15] Carrara W.G., Goodman R.S., Majewski R.M. *Spotlight Synthetic Aperture Radar: Signal Processing Algorithms*. Norwood, MA: Artech House Inc; 1995.

[16] Kapfer R.M., Davis M.E. '*Along track interferometry for foliage penetration moving target indication*'. *Proc. of the 2008 IEEE RADAR Conference*. Rome, Italy; 2008.
[17] Murthy V., Pillai S.U., Davis M.E. 'Waveforms for simultaneous SAR and GMTI'. *IEEE 2012 Radar Conference (RADAR)*. Atlanta, GA; April 2012. pp. 0051–6.

Index

www.ingramcontent.com/pod-product-compliance
Lightning Source LLC
LaVergne TN
LVHW011833200926
848484LV00002B/349

9781839530777